THE FUTURE O. ˙˙˙˙˙˙

THE FUTURE OF ENERGY

BRIAN F. TOWLER
University of Queensland
Department of Chemical Engineering
Brisbane, Qld, Australia

formerly with

University of Wyoming
Department of Chemical and Petroleum Engineering
Laramie, WY

AMSTERDAM • BOSTON • HEIDELBERG • LONDON
NEW YORK • OXFORD • PARIS • SAN DIEGO
SAN FRANCISCO • SYDNEY • TOKYO
Academic Press is an imprint of Elsevier

Academic Press is an imprint of Elsevier
32 Jamestown Road, London NW1 7BY, UK
225 Wyman Street, Waltham, MA 02451, USA
525 B Street, Suite 1800, San Diego, CA 92101-4495, USA

Notice
No responsibility is assumed by the publisher for any injury and/or damage to persons or
property as a matter of products liability, negligence or otherwise, or from any use or
operation of any methods, products, instructions or ideas contained in the material herein.
Because of rapid advances in the medical sciences, in particular, independent verification of
diagnoses and drug dosages should be made.

British Library Cataloguing-in-Publication Data
A catalogue record for this book is available from the British Library

Library of Congress Cataloging-in-Publication Data
Towler, Brian F., author.
   The future of energy / Brian F. Towler. – 1st edition.
      pages cm
1. Power resources.   I. Title.
HD9502.A2T694 2014
333.79–dc23
                                                    2014009788

ISBN: 978-0-12-801027-3

For information on all Academic Press publications
visit our website at elsevierdirect.com

Working together
to grow libraries in
developing countries

www.elsevier.com • www.bookaid.org

Contents

Preface

I was motivated to write this book because I have read many books, articles and news stories that suggest that we are running out of cheap energy and the world will soon be plunged into a never-ending cycle of wars designed to secure the few remaining energy resources. The second perception about energy is that the current energy sources are destroying the environment and we have to find safer, cleaner energy sources that will allow us a better, simpler lifestyle. This is the story of energy and what I found.

Energy fuels economies and lifestyles, and if we don't have enough, then the future could be bleak. So, is the world running out of energy, and do we have to fight wars to secure enough cheap energy supplies to maintain our current lifestyle or should we be lowering our standard of living so that the world has enough energy supplies? What are the alternatives? What is the future of energy? I will show in this book that we have plenty of energy supplies; we do not have to fight wars to maintain our lifestyle. There are some misguided politicians who think otherwise, and I hope this book will demonstrate to them the errors of their thinking.

The second and related question is: should we be developing cleaner forms of energy so that our energy sources are not destroying the environment? In seeking to answer this question, I came to a surprising conclusion: all forms of energy have an impact on the environment. That is because energy is part of the environment. If you manipulate energy sources, you will have an impact on the environment. The more successful an energy source is, the more impact it has on the environment. I formulated that idea into an important principle in Chapter 1 called *The Towler Principle*. We *do* have the ability to mitigate and minimize the environmental impacts, to lessen their impacts, and that should be our goal.

In examining these questions, I have tried to tell the story in simple and easily understood terms. In Chapter 2, I introduce the units used for energy, of which there are many. But they can be converted from one unit to another. In Chapter 3, I introduce the laws of thermodynamics, which are fundamental to understanding energy processes. These are not difficult principles to understand, but you cannot hope to understand energy if you do not have a rudimentary knowledge of these laws.

In each chapter on specific energy sources, I not only describe how the source is used to generate energy, but also show a lot of graphs and statistics to prove and illustrate my points. If this becomes too much information to absorb at once, the reader can skip to the end of the chapter where a summary of the conclusions will be found. Chapter 16 ties all these conclusions together and answers the questions raised in the book. It reaches optimistic conclusions. We have plenty of energy: we don't have to fight wars; we don't have to destroy the environment. The future is bright because we have enough energy, and we know how to access it and use it.

Brian Towler,
Laramie Wyoming, 2014

Acknowledgments

I would like to acknowledge the editorial assistance of Wynn-Anne Mourre and Shelley J. Leonard in preparing this manuscript, and I thank them for their help. My daughter, Renee Clayton, assisted with the cover design.

Dedication: I dedicate this book to my grandchildren, Alexander, Kristiana, Anne and Lucy Clayton. I wrote this book for them. They are the future, and I want them to know there will be plenty of energy for them to use and that the environmental impacts can be mitigated.

The History and Culture of Energy

Energy is the lifeblood of any civilization. Throughout history, the most successful civilizations have been those that have maximized their energy throughput and made good use of the resources at their disposal. Accessing this energy, however, is not always easy; in order to have access, it has to be cheap enough for the average person to afford and it has to be readily available. Without this access, people are condemned to live in poverty, without technology, and with a menial standard of living.

One of the major concerns of society today is the shortage of energy. People feel that the world is running out of cheap energy (particularly oil) and that there is a potential for energy wars in the future. These wars will be waged as civilization struggles to get access to the remaining sources of energy needed to fuel its economies and lifestyles. Moreover, as shortages develop, people feel that western civilization is being held hostage to hostile regimes (mostly in the Middle East) that control the energy supplies. Nothing could be further from the truth. There is plenty of cheap energy still available; however, circumstances are demanding that as cheap conventional oil runs out, there will be a need to switch to alternative energy sources.

Another major social concern is the rising carbon dioxide levels in the atmosphere which could potentially lead to runaway global warming. In order to solve this problem, some believe there is a need to switch to clean renewable sources of energy—particularly wind power, solar power, ethanol, and biomass. The truth is that all sources of energy, no matter how "green," impact the environment. This leads to a principle that I propose and will return to repeatedly in this book: Energy is the essence of the universe, and it is not possible to extract energy from the environment for use without having an impact on the environment. This is a fundamental principle that I call the *Towler Principle*.

The Future of Energy
http://dx.doi.org/10.1016/B978-0-12-801027-3.00001-4

THE TOWLER PRINCIPLE

It is not possible to extract energy from the environment without having an impact on the environment.

There are many forms of energy that can be converted into other forms of energy according to our needs and requirements. By examining the various forms and sources of energy, you will see that each of them has an impact on the environment as we extract them and use them; however the undesirable effects can be changed or minimized according to our needs. For example, the carbon dioxide problem may be reduced or eliminated by switching to wind power, solar power, ethanol, and biomass. Unfortunately, the end result is that these actions may have other intolerable effects that will be discussed later in this book.

What can we do to ensure minimal effect to the environment regardless of the energy sources chosen? How do we protect our environment and sustain it for future generations? An alternative solution to the carbon dioxide problem could be carbon capture and sequestration. The technology for this already exists and can be further improved. This technology only applies to point sources of carbon such as coal and gas-fired power plants. There is also a need to develop greener and more efficient building technologies and add cleaner renewable energy sources into the mix. In order to reduce the carbon effect of automobiles, there may be a greater need to move toward electric vehicles so that any carbon dioxide generated in the process can be captured at the point where the electricity is generated. The future is bright; there is plenty of energy available, but much change is afoot. This is the preeminent technical challenge of our time, one that will require significant political will and technical drive.

THE HISTORY OF ENERGY USE

According to physicists, the universe began with a "big bang." About 14 billion years ago, all the mass and energy of the universe was concentrated at a single point. This mass exploded, commenced expanding from that point and is still expanding today. Physicists have identified four fundamental forces of nature: gravity, electromagnetic force, strong nuclear force, and weak nuclear force. They also identified the physical principles that govern all the processes in the universe. Two of these physical principles are that mass and energy are always conserved. According to Albert Einstein, there is an equivalence between mass and energy and, under certain processes and circumstances, these two entities can be converted into one another. When he discovered this equivalence and published his famous equation ($E = mc^2$), Einstein was demonstrating that the very fabric of the universe is energy. If we extract energy from a source and use it, something will be left behind as a result of that action.

When homo sapiens first evolved in Africa, their great advantages were their large brains and their ability to outwit the fierce creatures that considered them prey. This large brain required increased energy resources to fuel itself, energy resources that came from an omnivorous diet of meat, grain, fruits, vegetables, and nuts. As civilizations developed about 15,000 years ago, in Mesopotamia and the Indus Valley, the successful civilizations were the ones that learned how to perform their tasks efficiently. To do this, they had to maximize the energy throughput of their society.

In his book *The Evolution of Culture (1959)*, anthropologist Leslie White acknowledges the key role played by the harnessing of energy in the development of civilizations. Initially, energy fueled the human bodies that provided the labor to hunt and gather and then to work the fields as mankind turned to agriculture. The next step in energy utilization was the domestication of animals such as horses and oxen that could be trained to do most of the heavy lifting and pulling that was required. This relieved some of the labor required of humans and also increased the energy throughput in the society. The animals required their own energy input in order to get them to do useful work; however, these animals generally ate foods that were inedible for humans, thus increasing rather than decreasing the efficiency of society.

Energy can be used for many different purposes and comes from many different sources. The principle uses of energy are for: doing work, heating buildings, cooking (a special use of heat), transportation, and communications. For early civilization, work and heat were separate functions of energy. The work was provided by human and animal labor and heat was provided by burning biomass. This heat was used to warm the living space and to cook food. It wasn't until the nineteenth century that mankind started to understand that heat and work were both forms of energy and that engines could be developed that converted heat into work.

As civilization further developed, some societies captured slaves to provide some of the labor. These slaves also required energy in order to do work, but they were able to perform tasks that animals could not. It took a long time before the immorality of slavery was recognized and bans on slavery were instituted; however, slavery is a significant part of human history that played a role in the rise of civilization.

The strength of a society is very dependent on its energy sources. The more human and animal labor that was conducted by a society, and the more fuel that could be gathered to provide heat, the more successful the society became. These societies knew they had to provide a source of energy to make the civilization work. Usually, this consisted of human labor (both free and slave), animal labor (usually horses and oxen) and biomass (usually wood). The human and animal labor had to be fed and efficient agriculture lay at the heart of its success. This was the fuel for the primary energy sources. The wood had to be cut and gathered and the energy for this process came from human and animal labor.

Some great societies collapsed when their energy sources failed. The Romans, Mayans, and Mesopotamians are examples of civilizations that failed in this manner. The Mayans were an extremely successful society that ruled around Central America and the Yucatan Peninsula from about 100 to 900 AD. They built many large cities with elaborate temples, surrounded by a network of buildings for living, agriculture, art, and manufacturing. They had the only well-developed written language of pre-Columbian America. They had a highly advanced mathematical system and very advanced astronomical knowledge. Their society was fueled primarily by the lush forests surrounding their cities and by their fields devoted to agriculture. Around 900 AD, the forests became severely depleted. Deprived of a fuel source, the Mayan civilization collapsed. The people dispersed in all directions seeking alternative energy sources. Their elaborate cities were abandoned, never again to be occupied.

Throughout history, some societies also found that coal, oil, and other hydrocarbons could be burned to provide heat. When heat engines were developed—particularly by Savery, Newcomen, and Watt in the eighteenth century—the hydrocarbon age was born. In China, for example, oil had been harvested from seeps and other locations for thousands of years and used for heating and lighting. Significant quantities for use in heat engines were not harvested, however, until "Colonel" Edwin Drake drilled a well on an island in Oil Creek near Titusville, Pennsylvania in 1860. This event ushered in the oil era, even though oil did not become the dominant energy source in the world until much later. In 1910, coal replaced wood as the dominant energy source. In 1962, coal was overtaken by oil which continues to maintain its worldwide dominance.

In *The Evolution of Culture (1959)* and a 1943 paper in the *American Anthropologist*, Leslie White argues that cultures evolve as the amount of energy harvested per capita per year is increased. If the parameter remains constant, the society stagnates; if it decreases, the society either moves backward or collapses. White puts this in a context of the efficiency of the technology used. Provided that societal technology is efficient in harvesting and deploying energy throughput, increased energy usage per capita per year is what will be necessary to move society forward. Alternatively, if the amount of energy used per capita remains constant, the culture can still move forward if the efficiency of energy use is increased.

White identifies five factors in the development of culture: the human organism, the habitat, the amount of energy controlled and expended by man, the ways and means in which energy is expended, and the human-need-serving product which accrues from the expenditure of energy.

Analyzing these factors led White to identify two laws of cultural development. First, other things being equal, the degree of cultural development varies directly as the amount of energy per capita per year is harnessed and put to work. Second, if the amount of energy expended per capita per unit of time remains constant, the degree of cultural development varies directly with the efficiency of the technological means with which the harnessed energy is put to work. He uses the following example to illustrate the two points of his thesis:

> A man cuts wood with an axe. Assuming the quality of the wood and the skill of the workman to be constant, the amount of wood cut in a given period of time, an hour say, depends, on the one hand, upon the amount of energy the man expends during this time: the more energy expended, the more wood cut. On the other hand, the amount of wood cut in an hour depends upon the kind of axe used. Other things being equal, the amount of wood cut varies with the quality of the axe: the better the axe the more wood cut. Our workman can cut more wood with an iron or steel axe than with a stone axe.

In this context, increased energy usage is a good thing. The United States is frequently criticized for using too much energy. According to White, if the United States is to advance and other nations are to aspire to a similar standard of living, energy use should be encouraged as long as it is not wasted. To ensure this high standard of living is maintained or increased, however, sufficient energy sources have to be identified. Where will the energy come from? Who or what will provide us with the energy required? Who are our biggest energy users? Who are our biggest energy wasters?

Some people believe that technology has a detrimental effect on the well-being of society. In the early part of the nineteenth century, the Luddites were a group of people who went around smashing up textile machinery because they saw it as a detriment to their skills and way of living. In today's modern society, those who resist technological advances are sometimes referred to as Luddites.

More recently, a former math professor turned sociopath by the name of Ted Kaczynski became known as the "Unabomber" because he sent out homemade bombs that killed three people and wounded 23 others. He did this because he became disillusioned with the technology of modern society. Kaczynski wanted to get attention and support for his view that technology was at the heart of society's problems and that we needed to eschew all technology. He wrote a manifesto that he demanded be published by leading newspapers, setting out these views. Many newspapers published his manifesto, which ultimately led to his capture. He was found living in a small shed in the backwoods of Montana without heat, electricity or running water and completely devoid of any useful technology. He was sentenced to life in prison without parole.

There are many others with less extreme and less violent views who believe that we need to simplify our society and voluntarily use less energy. There is nothing wrong with choosing a simpler lifestyle, but White's analysis of historical civilization suggests that to sustain and maintain our way of life and to move our society forward, society must use and develop the technology to provide sufficient energy to sustain and improve societal functions.

THE MOVEMENT TO OIL

As noted earlier, throughout history the most successful civilizations were the ones that maximized their energy throughput. The source of energy used has changed over time from human labor to animal labor to biomass to coal to oil and gas to nuclear energy. In today's civilization, hydrocarbon energy is the ruler. This type of energy is so cheap that it has profoundly raised the standard of living for the vast majority of people and has had a profound impact on the world.

Before oil was discovered and produced by Drake, the vast majority of people did not have access to cheap energy and consequently could not afford to move around or communicate effectively with one another. They had less food and less variety of food to eat. Rich people could afford to move around principally using horses for transportation, but if poor people wanted to move from one place to another, they usually had to walk. Consequently, people moved very little.

In the eighteenth and nineteenth centuries, artificial lighting was principally provided by whale oil. Whales were almost hunted to extinction in the nineteenth century because the whale blubber could be rendered into a fuel that could be burned in lamps and used as a light source. When Drake discovered oil, its first use was to make kerosene to replace the whale oil in lamps. This single event saved the whales from extinction. Kerosene was cheaper and better than whale oil for this purpose.

While the kerosene fraction was being widely used for lighting, there were no uses for the rest of the crude oil fractions. As automobiles were developed, gasoline and diesel fuel distilled from crude oil were made the principal transportation fuels. There were two principal internal combustion engines developed in the latter half of the nineteenth century. The first was the Otto cycle engine, developed by Nicholas Otto in 1862, which incorporated a spark plug to ignite the fuel mixture. Otto initially intended his engine to run on coal gas (or syngas), but his engine worked well on the gasoline (or petrol) fraction of crude oil. The second engine was a compression ignition engine developed by Rudolf Diesel in 1893. Diesel's initial intention was that his engine would run on vegetable oil, and he used peanut oil when he first demonstrated the engine to the public. When it was discovered that it would operate effectively on the diesel fraction in crude oil, this became the fuel of choice for Diesel's engines.

In the twentieth century, when jet engines were developed to drive airplanes and electric lights proved to be more effective than kerosene lamps, kerosene was found to be a useful fuel for jet engines. Oil has fueled a revolution in lighting, transportation, and communications that has profoundly raised the standard of living for the vast majority of people throughout the world.

Figure 1.1 is a graph that shows the primary sources of energy for the past 210 years. In 1800, the primary source of energy was biomass—mostly trees that were cut down and burned to provide heat for homes and factory furnaces. Some coal was also being burned in steam engines and in the few existing factory furnaces. This was a small and insignificant energy source compared to burning wood. During this time, and in the centuries prior, many of the forests of Europe were cleared and burned for fuel. The ash from the burning of trees was also an important resource for the manufacture of glass for windows and the making of soap for washing. It should be noted that human and animal labor does not appear on this graph, even though these energy sources were probably higher than all the others in 1800. The units on the Y-axis of this chart are exajoules, abbreviated EJ. This unit will be explained more fully in Chapter 2, where you will see that $1\ EJ = 10^{18}$ joules (J).

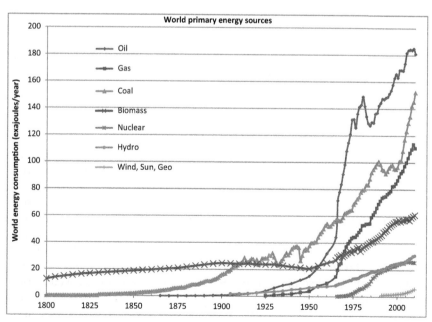

FIGURE 1.1 Primary sources of energy in the world from 1800 to 2010.

This figure also shows that around 1910 coal overtook biomass as the primary source of energy in the world. By this time coal was being widely used to generate electricity, to make steel and as the primary transportation fuel to drive locomotive engines. Biomass was still being used to heat homes. While biomass use has been relatively flat, it has not diminished. In fact, in the past 50 years, its use has increased again. In many developing countries in Asia and Africa, wood is the primary energy source.

In the latter half of the nineteenth century, coal use began to take off. It was finally being crowned "King Coal" around 1910. Oil had been discovered in the United States in 1860 and initially was used only to make kerosene for kerosene lamps. With the invention of the automobile, its use increased, both as gasoline in the spark ignition engines based on the Otto cycle and as heavier diesel fuel in the compression ignition engines developed by Diesel. It was not until the 1960s that oil overtook "Old King Coal" as the primary energy source in the world. It has not relinquished that title.

Natural gas was always seen as a by-product of oil production and often it was simply burned at the field unless a market for it could be found. Commencing around 1940, it became marketed as a fuel in industry and to generate electricity in gas turbines. Its use has expanded rapidly since then, and today it is the third leading energy source in the world, just behind coal. It is often seen as a more environmentally friendly alternative to coal because it produces less carbon dioxide and other pollutants than coal per unit of energy produced.

Hydroelectric power plants generate electricity from the release of water trapped in dams. This form of energy was first developed to drive grinding mills to grind grain. With the development of large hydroelectric generating plants around 1900, its use expanded greatly and it has now become the fifth leading source of energy. While hydroelectric power is clearly the cleanest and cheapest source of electricity, it can only be developed where dams are possible, limiting its usage. It should be noted, though, that there is still a lot of further potential for hydroelectricity.

Nuclear power, which results from the splitting of a particular isotope of uranium, was made possible by the theory of relativity developed by Einstein. Nuclear power plants that generate electricity were developed in the 1950s and, since then, nuclear power has increased rapidly to become the sixth largest source of energy behind oil, coal, natural gas, biomass, and hydroelectricity.

Wind, solar, and other sources of energy barely register on the graph because of their cost and unreliability. There is a great deal of interest in further developing these sources of energy because they are clean and renewable, but so far their use has been limited by their cost and reliability issues.

Table 1.1 compares the different energy sources that are used and shows the annual energy use per person per year in the United States.

TABLE 1.1   Annual Energy Use Per Capita for the U.S.

Total energy	306 GJ
Oil	22.3 barrels
Natural gas	77,045 ft^3
Coal	3.465 tons
Uranium	0.160 lbs
Electricity	13,811 kWh

Note the variety of units that are used there, which are the customary units for each energy source. To compare the sources properly, they would need to be converted to the same energy unit, the joule, which is the most appropriate universal energy unit. The total energy figure is given in gigajoules (GJ) or 10^9 joules. If and when you are able to convert these different energy sources to equivalent units, you will find that the five listed in the table do not add up to the total. There are two reasons for this: partly because the renewable energy sources are not listed and partly because coal, natural gas, and uranium fuels are used to make the majority of the electricity shown.

How does the U.S. per capita energy use compare with other countries? Table 1.2 shows this comparison for the top 66 countries in the world. Very few African countries appear on this list because their energy use per capita is so low and also more difficult to measure. They do not even appear in the statistical data prepared for the *BP Statistical Review*, the main data source used here.

It is interesting to note that the United States is not the top energy user in the world and comes in ninth. Four of the top eight are Middle Eastern countries, and one might argue that they use a lot of energy because they have a lot of oil and gas and can afford to use a lot. They are also hot countries that generate a lot of electricity to run air conditioners. Qatar's number might be affected by a low and inaccurate population count due to a lot of expatriate workers in the country that are not part of the official population statistics. Three of the top 10 countries on the list (Norway, Canada, and the United States) are cold countries that spend a lot of energy on heating in the winter time. All of the top 10 are highly industrialized countries that use a lot of energy through the manufacturing of energy intensive products. The energy use per capita for each country is a combination of these factors; it represents their economic activity, their standard of living and their efficiency of energy utilization.

China is only number 46 on the list at only 76 GJ per person per year (GJ/p/year), well below most of the western countries. This is a large increase from relatively recent times. In 1968, China was using

TABLE 1.2 Annual Energy Use Per Capita for the Top 66 Countries

Rank	Country	GJ/p/year
1	Qatar	1271.45
2	Trinidad and Tobago	749.34
3	United Arab Emirates	706.31
4	Singapore	616.89
5	Kuwait	493.54
6	Canada	389.93
7	Norway	373.18
8	Saudi Arabia	322.31
9	United States	305.74
10	Netherlands	249.05
11	Sweden	233.55
12	Finland	231.98
13	Australia	227.61
14	Korea, South	219.12
15	Turkmenistan	217.97
16	Russia	208.67
17	Taiwan	200.64
18	Kazakhstan	196.44
19	New Zealand	184.54
20	Austria	170.05
21	Czech Republic	169.67
22	Japan	165.93
23	Germany	164.29
24	France	161.92
25	Switzerland	158.85
26	Hong Kong	152.27
27	Denmark	147.89
28	United Kingdom	139.72
29	Spain	134.18
30	Israel	132.91

TABLE 1.2 Annual Energy Use Per Capita for the Top 66 Countries—cont'd

Rank	Country	GJ/p/year
31	Ireland	131.04
32	Greece	126.59
33	Slovakia	123.58
34	Venezuela	121.69
35	Italy	118.15
36	Iran	114.33
37	Ukraine	109.56
38	Belarus	106.74
39	Bulgaria	106.42
40	Portugal	105.39
41	Poland	104.37
42	South Africa	103.38
43	Hungary	98.30
44	Malaysia	91.81
45	Argentina	77.34
46	China	76.24
47	Uzbekistan	74.21
48	Lithuania	71.74
49	Chile	70.43
50	Thailand	67.78
51	Romania	65.98
52	Mexico	62.32
53	Turkey	58.97
54	Brazil	52.30
55	Azerbaijan	49.97
56	Algeria	49.25
57	Egypt	41.37
58	Ecuador	36.24
59	Colombia	30.18
60	Peru	26.24

Continued

TABLE 1.2   Annual Energy Use Per Capita for the Top 66 Countries—cont'd

Rank	Country	GJ/p/year
61	Indonesia	23.88
62	Vietnam	20.34
63	India	18.47
64	Pakistan	15.12
65	Philippines	11.37
66	Bangladesh	6.24

<6 GJ/p/year and, as recently as the year 2000, China's energy use was still below 30 GJ/p/year. It has since risen rapidly as its society and technology have modernized and expanded. China is now an economic superpower with massive and unprecedented growth rates. In 2010, it surpassed the United States as the largest energy consumer in the world and, as a consequence of that, it also emits the most carbon dioxide in the world. On a per capita basis its energy use is still below countries such as Portugal, Greece, and Argentina, even though it is now well ahead of its Asian neighbors such as India, Pakistan, and Thailand. Not only is China's economic activity increasing rapidly but its citizens aspire to a standard of living that is vastly higher than they have been used to, commensurate with other successful world economies. This means that its per capita energy use, as well as its total energy use, will continue to rise rapidly.

Figure 1.2 shows the per capita gross domestic product (GDP) of the countries in Table 1.2 as a function of the annual energy use per capita. A general trend appears: the higher the energy use, the higher the GDP. The data labels shown on the plot are the per capita energy use numbers shown in Table 1.2, rounded to zero decimal places. This helps link each country to its data point. The countries lying above the trend line are the more efficient ones, producing a higher than average GDP for a particular energy use. Conversely, the countries below the trend line are the less-efficient countries. The United States, at 306 GJ/p/year, is slightly above the trend line. The Chinese territory of Hong Kong is the highest above the trend line, producing a GDP of $45,900 per person while expending a per capita energy of 152 GJ/p/year. This may be because Hong Kong is a relatively focused industrial and commercial center with a large productive population living in a small area. It has a warm climate that has virtually no need for winter heat and, even though they have hot summers, they do not rely on much air conditioning for summer cooling. Most of their energy goes into producing goods for sale. The countries that are the lowest below the trend line are Turkmenistan and the United Arab Emirates.

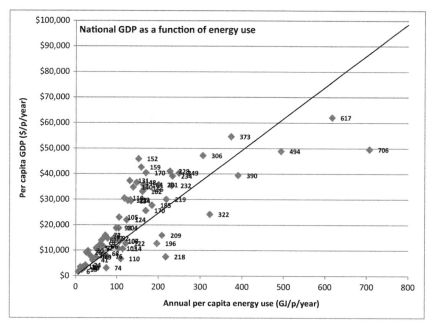

FIGURE 1.2   Per capita GDP as a function of energy use for countries listed in Table 1.2.

Two of the countries in Table 1.2 have been omitted from the plot in Figure 1.2 because they were outliers. These are Qatar and Trinidad and Tobago. Qatar's data point does fall close to the trend line but is a long way removed from the other data points possibly because their population numbers are artificially low. Trinidad and Tobago fall substantially below the trend line because of high energy use in its industries coupled with low wages and low productivity.

Figure 1.3 shows the total energy reserves for the top 20 countries in the world. Only the big four nonrenewable fuels (oil, coal, natural gas, and uranium) are included in this bar chart. By examining this chart, you can see that there are very few energy reserves in the form of uranium. These reserves are primarily located in Australia, Canada, and Kazakhstan. The uranium reserves shown in the figure only refer to the currently deployed nuclear power technology, which totally relies on the easily fissile U-235 isotope and represents only 0.7% of the uranium reserves in the world. The far more abundant U-238 and Th-232 isotopes can be used in fast breeder reactors, which have been tested but have not been built on a commercial scale yet. When breeder reactors are perfected and deployed, a much larger amount of nuclear fuel will appear in these charts. Despite this, there is still over 100 years of supply in the world, based on current usage, and this supply can be increased at a reasonable cost.

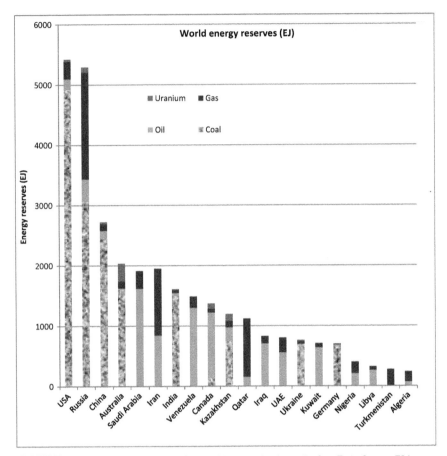

FIGURE 1.3   Energy reserves of the top 20 countries in exajoules. *Data Source: EIA.*

Natural gas is also underrepresented in Figure 1.3, particularly for the United States. In the past 5 years, the United States has identified and started to produce a large amount of shale gas. The U.S. gas reserves have started to increase again because of this gas, and when it is fully proved it will represent a large addition to the energy reserves of the United States. Shale gas also exists in many other countries around the world. When the fracture stimulation technology used to unlock these reserves is exported to other countries, they too will see a similar increase in their energy reserves.

The largest energy reserves appearing in Figure 1.3 are in the form of coal. Coal reserves at the current usage rate represent more than 120 years of supply, and this does not include the unproved coal reserves which are far more extensive. Looking at this chart, it is easy to see that coal has a very bright future. Some people may be horrified at that thought because coal is perceived as a dirty fuel. When it is burned without any environmental controls, it emits large amounts of carbon dioxide. In addition, it produces oxides of sulfur and nitrogen, mercury, coal ash, and a range of lesser known pollutants. In the past it has been responsible for acid rain, coal tars, and coal soot. It is also one of the carbon-based fuels primarily responsible for the rising CO_2 levels in the atmosphere.

Coal, however, can be burned without emitting any pollutants at all. The technology currently exists to do this, and it is known collectively as clean coal technology. While it is highly likely that all of the coal appearing in Figure 1.3 will get burned, it will have a minimal impact on the environment as long as clean coal technologies are used.

This figure shows that the United States has the largest coal reserves and the second largest coal production in the world. Most of this production is used domestically to generate electricity, and at least 45% of the U.S. electricity supply is generated from coal. China and Russia, the first- and third-largest coal producers, also use all of their production domestically. Australia is the largest exporter of coal even though they have a large domestic market for their coal in terms of both electricity generation and steel making. In terms of energy reserves, Australia has the fourth-largest reserves in the world, but the majority of this is tied up in coal. Australia also has the largest per capita energy reserves in the world.

SUMMARY

Energy use is a good thing as long as the usage is efficient. The higher the energy use, the higher the standard of living and the more the society progresses culturally. In order to produce enough food, water, clothing, and shelter for the world's current and future population, a large per capita energy use is required. The United States does not have the highest per capita energy use in the world, but a robust economy and a high standard of living demands adequate energy supplies at an economic price. There are adequate supplies of energy available, and the world does not need to be fighting wars over perceived energy shortages. Moreover, according to the Towler Principle espoused in this chapter, all energy sources have some effect on the environment. This cannot be avoided, but it can be mitigated. The current focus on renewable energy supplies eschews the real requirements needed in energy, which are reliability, cost and security of supply. While environmental effects are important considerations, the focus needs to be on how they can be mitigated.

How Is Energy Measured

The units for energy are surprisingly varied, but it is useful to understand the equivalencies between the different units and the reason for their existence. To derive the units for energy, it is necessary to review the equations from which energy is calculated. While it is not important to fully understand these equations, some equivalencies between the various units used for energy will be discussed later in this chapter. It is also useful to understand that there are various unit systems; however, in any one equation, it is necessary that a consistent unit system be used. The easiest and most widely used unit system is *Le Système Internationale d'Unités*, the SI unit system. Some people have also referred to this system as the metric system.

Table 2.1 shows the basic units for the common parameters in three different unit systems, the SI system, the British or American customary system, and the centimeter-gram-second (CGS) system.

The three basic units in any unit system are mass, distance, and time. In the SI system, the basic units are kilograms (kg), meters (m), and seconds (s). Velocity is the change in distance with time and is represented as meters per second (m/s). Acceleration is the rate of change of velocity and is represented as meters per second squared (m/s^2). According to Newton's second law of motion, force is equal to the mass times the acceleration. In the SI system, force is measured in newtons (N).

$$\text{Force} = \text{Mass} \times \text{Acceleration}$$

$$1\,\text{newton (N)} = 1\,\text{kg m/s}^2$$

Energy is derived from the equation for work (work and heat are forms of energy). Work is equal to force times the distance moved and, in the SI system, is measured in joules (J).

$$\text{Energy} = \text{Force} \times \text{Distance}$$

$$1\,\text{joule (J)} = 1\,\text{newton meter (Nm)}$$

The Future of Energy
http://dx.doi.org/10.1016/B978-0-12-801027-3.00002-6

TABLE 2.1   Units for Energy Parameters in Three Unit Systems

Parameter	SI	British	CGS
Mass	kilograms (kg)	slugs	grams (g)
Distance	meters (m)	feet (ft)	centimeters (cm)
Time	seconds (s)	seconds (s)	seconds (s)
Velocity	m/s	ft/s	cm/s
Acceleration	m/s^2	ft/s^2	cm/s^2
Force	newtons (N)	pounds (lbf)	dynes
Energy	joules (J)	ft-lbf	erg
Power	watt (W)	ft-lbf/s	erg/s

Power is the rate of doing work or the rate of energy conversion. In the SI system, it is measured in watts (W).

$$Power = Energy/time$$

$$1\,watt = 1\,joule/second\,(J/s)$$

When units are written out in full, capital letters are not used. This includes those units that are named after a person such as newtons, joules, and watts. That is the convention. When units named after a person are abbreviated, they are done so with capital letters. For example, newtons are abbreviated as N, joules are abbreviated as J, and watts are abbreviated as W.

When calculating the amount of energy used by a large country or by the whole world in a year, the numbers can be quite large. When you look at the energy required to accelerate a single subatomic particle in a nuclear reaction, the numbers are very small. That is why in the SI system a variety of prefixes are used to denote a unit being multiplied by a power of 10; for example, the prefix kilo means 1000, 10^3, or in scientific notation 1E+3. The prefixes for the powers of 10 are shown in Table 2.2. The large number prefixes are usually abbreviated with a capital letter and the small power prefixes are abbreviated with a small letter; for example, 1E+15 joules $= 10^{15}$ joules $= 1$ petajoule $= 1$ PJ. Similarly, 1E -15 joules $= 10^{-15}$ joules $= 1$ femtojoule $= 1$ fJ. In Figure 1.3, you can see the energy reserves for various countries in exajoules (EJ); 1 exajoule $= 10^{18}$ joules $= 1$ EJ. Later in the chapter you will see that 1 BTU is equal to 1055.056 J; 1 EJ is roughly equal to 1 quadrillion BTU, which is sometimes abbreviated as quads. The energy reserves for countries are often given in quads or exajoules, which are roughly equivalent to each other.

TABLE 2.2   Prefixes Used in SI Units

In Scientific Notation	In Full Numbers	In Words	SI Prefix	SI Symbol
1.0E+24	1,000,000,000,000,000,000,000,000	Septillion	Yotta-	Y
1.0E+21	1,000,000,000,000,000,000,000	Sextillion	Zetta-	Z
1.0E+18	1,000,000,000,000,000,000	Quintillion	Exa-	E
1.0E+15	1,000,000,000,000,000	Quadrillion	Peta-	P
1.0E+12	1,000,000,000,000	Trillion	Tera-	T
1.0E+9	1,000,000,000	Billion	Giga-	G
1.0E+6	1,000,000	Million	Mega-	M
1.0E+3	1000	Thousand	Kilo-	k
1.0E+2	100	Hundred	Hecto-	h
1.0E+1	10	Ten	Deca-	da
1.0E−1	0.1	Tenth	Deci-	d
1.0E−2	0.01	Hundredth	Centi-	c
1.0E−3	0.001	Thousandth	Milli-	m
1.0E−6	0.000,001	Millionth	Micro-	μ
1.0E−9	0.000,000,001	Billionth	Nano-	n
1.0E−12	0.000,000,000,001	Trillionth	Pico-	p
1.0E−15	0.000,000,000,000,001	Quadrillionth	Femto-	f
1.0E−18	0.000,000,000,000,000,001	Quintillionth	Atto-	a
1.0E−21	0.000,000,000,000,000,000,001	Sextillionth	Zepto-	z
1.0E−24	0.000,000,000,000,000,000,000,001	Septillionth	Yocto-	y

When you receive your electricity bill, the units for the energy used are not joules but kilowatt-hours (kWh). Given that a watt is a joule per second and there are 3600 s in an hour and there are 1000 watts in a kilowatt, 1 kWh=3,600,000 joules. A typical monthly electric bill might show that you have used 500 kWh charged out at about 10c/kWh. If they had shown the energy usage in joules, it would have been 1.8E+9 joules, or 1.8 gigajoules (GJ). The electric company uses kWh instead of GJ because people understand kilowatt-hours better than they understand gigajoules. As the electricity bill always uses kWh, electricity consumption and generation tends to be reported in various multiples of watt-hours instead of joules or multiples of joules.

A typical large coal-fired or nuclear-powered electric plant can produce about 1 billion joules of electricity per second, or 1 GJ/s, or 1 gigawatt (GW). Sometimes this is shown as 1000 megawatts (MW) but it really is just a different way of saying the same thing. In 1 day, the 1 GW plant can nominally produce 24 GWh of electricity. In 1 year, the same plant operating 24 h/day and 365 days/year can nominally produce 8760 GWh of electricity. You could also write this as 8.76 terawatt-hours (TWh), which is equivalent to $8.76E + 12 \times 3600$ joules of energy, or 31.536 PJ.

There are various other units for energy that are also used for various historical reasons. The British Thermal Unit (BTU or Btu) is a unit of energy that, as originally defined in Great Britain, is the amount of energy needed to heat one pound of water by one degree Fahrenheit (°F). Its value can vary depending on the temperature so the International Standards Organization (ISO) has recently set the value to be fixed at 1055.056 joules. A more widely used value, based on the International Steam Table (IT) calorie and defined by the *Fifth International Conference on the Properties of Steam* (London, July 1956), sets the value of the calorie at exactly 4.1868 joules. This makes the BTU equal to 1055.05585 joules; however, the value set by ISO (1 BTU = 1055.056 joules) is not uncommon.

The calorie is a unit of energy of French origin that, as originally defined, was the amount of energy needed to heat 1 g of water by 1 °C. Its value can also vary depending on the temperature. ISO has recently set the value to be 4.184 joules; however, the value based on the IT calorie was defined by the *Fifth International Conference on the Properties of Steam* (London, July 1956) to be exactly 4.1868 joules. This conversion factor is also called thermal calories, abbreviated as cal_{th}.

The calorie is most widely used today to specify the energy gained by human beings when food is digested. Interestingly, one calorie is such a small unit of energy that the food industry uses the kilocalorie to determine the quantity of energy in food; however, they call a kilocalorie a Calorie (with a capital C) and use Calories and kilocalories interchangeably. Understandably, this can create some confusion. I once knew an engineer who looked up the latent heat of melting for ice cream and found it to be 350 J/g. He then looked up the nutritional data on ice cream and found that it contained three Calories per gram of ice cream. He converted the 350 J/g to 84 Calories/g and came to the incorrect conclusion that it took more energy for the body to melt the ice cream than it gained from digesting it. He reasoned that he could eat as much ice cream as he wanted without gaining weight. After eating several gallons of ice cream and gaining about 15 pounds, he realized that something was wrong. He then discovered the confusion between Calories (which are really kilocalories) and Calories. While it does take 84 Calories/g to melt the ice cream, the energy gained from digesting the ice cream is

3 kilocalories/g, or 3000 Calories/g. I am sure that others have been similarly confused by this difference.

An electron-volt (eV) is another specialized unit of energy, particularly common in nuclear reactions. It is defined as the amount of kinetic energy gained by a single unbound electron when it accelerates through an electric potential difference of one volt. Its use in nuclear reactions will be seen later in Chapter 7.

$$1\,eV = 1.602176487\,E - 19\,\text{joules}$$

The erg is the unit of energy in the CGS unit system. While this unit system is not very commonly used today and was replaced by the SI unit system, you will occasionally see the erg unit used. It is easy to show that one erg is exactly equal to 10^{-7} J, or 100 nanojoules (nJ). These other (non-SI) units do not use the same prefixes as the SI units, but instead use scientific notation or words for powers of 10, shown in Table 2.2; for example, 7×10^{15} BTU can be written as 7E+15 BTU or written out as 7 quadrillion BTU, which can be abbreviated as 7 quads.

Table 2.3 shows the conversions between the important energy unit values. The numbers in the table show how many of the units at the top of the column are equal to one of the units on the left-hand side; for example, 1 Calorie = 4.184 joules.

Sometimes when energy values of different sources are compared, people convert the energy from different fuels into tons of oil equivalent or tons of coal equivalent. To do this, they have to define the energy content of one ton of standard oil. The energy content of crude oil varies from oil to oil but, for the purpose of this conversion, the equivalent oil has been defined as 10 Gcal$_{th}$/ton, or 41.868 GJ = 1 ton of oil equivalent (toe). Similarly, one ton of coal equivalent (tce) is defined as 7 Gcal$_{th}$/ton of coal; therefore, 29.3076 GJ = 1 tce.

The average energy content of various fuels in BTUs is shown in Table 2.4.

Many homes in the United States and throughout the world are heated with natural gas. One cubic foot of natural gas (measured at standard atmospheric conditions) contains, in approximate round numbers, 1000 BTUs or 1 million joules (1 MJ) of energy when burned. When natural gas is bought and sold, it is usually measured in volumes of one thousand cubic feet (1 MCF for short). The M comes from the Latin word mille, which means thousand, so 1 MCF of natural gas contains ~1 million BTUs, or 1 billion joules (1 GJ) of energy. In the United States, invoices sent to homeowners by natural gas utilities showed the gas usage in CCF, which stood for one hundred cubic feet. The Latin word for hundred is centum; hence, one hundred cubic feet was abbreviated CCF. It makes more sense to charge the user for the energy content of the gas used rather than the volume. By consensus, the utility companies switched to charging for

TABLE 2.3   Energy Conversions

	BTU	ft-lbf	Joules	Calories	Electron-volts	kWh
1 British Thermal Unit (BTU)	1	778.169371	1055.056	252.1644359	5.9158089E+15	2.9307111E−04
1 Foot-pound (ft-lbf)	1.2850673E−03	1	1.355817948	0.32404827	4.6035013E+18	3.7661610E−07
1 Joule (J)	9.4781699E−04	0.737562149	1	0.239005736	6.2415096E+18	2.7777778E−07
1 Calorie (cal)	3.9656663E−03	3.085960033	4.184	1	1.4917566E+18	1.1622222E−06
1 Electron-volt	1.6903859E−16	2.1722596E−19	1.6021765E−19	6.7035064E−19	1	4.4504902E−26
1 Kilowatt-hour (kWh)	3.4121412E+03	2.6552237E+06	3.60E+06	8.5984523E+05	2.2469435E+25	1

TABLE 2.4   Average Energy Content of Various Fuels

1 Kilowatt-hour of electricity ≈ 3412 Btu

1 Cubic foot of natural gas ≈ 1000 Btu

1 Therm of natural gas = 100,000 Btu

1 Gallon of crude oil ≈ 138,095 Btu

1 Barrel of crude oil ≈ 5,800,000 Btu

1 Gallon of fuel oil ≈ 149,690 Btu

1 Gallon of gasoline ≈ 125,000 Btu

1 Gallon of ethanol ≈ 84,400 Btu

1 Gallon methanol ≈ 62,800 Btu

1 Gallon of gasohol (10% ethanol, 90% gasoline) ≈ 120,900 Btu

1 Gallon of e-85 (85% ethanol, 15% gasoline) ≈ 90,500 Btu

1 Gallon of kerosene ≈ 135,000 Btu

1 Gallon of diesel fuel ≈ 138,690 Btu

1 Gallon of (lpg) ≈ 95,475 Btu

1 Pound of coal ≈ 8100-13,000 Btu

1 Ton coal ≈ 16,000,000-26,000,000 Btu

1 Ton wood (dry) ≈ 16,000,000-18,000,000 Btu

1 Standard cord of wood ≈ 17,000,000-30,000,000 Btu

1 Face cord of wood ≈ 6,000,000-8,000,000 Btu

the BTUs used. They adopted the therm as the basic unit of energy, and this is what is now displayed in customer invoices. As shown in Table 2.4, 1 therm = 100,000 BTUs. One CCF of gas contains about 100,000 BTU. So, the CCF value previously used and the therm unit currently used are approximately the same.

HYDROCARBON FUELS

Table 2.5 shows the energy content of the widely used hydrocarbon fuels. The gallon used here is the U.S. gallon, which is equal to 0.96894 imperial gallons. One barrel, a unit widely used in the oil business, is equal to 42 U.S. gallons, which is 5.61458 cubic feet and 158.98 liters.

TABLE 2.5 Average Energy Content of Hydrocarbon Fuels

Fuel	BTU/Barrel	BTU/Gallon
Crude oil	5,855,795	139,424
Motor gasoline	5,250,000	125,000
Aviation gasoline	5,005,224	119,172
Jet fuel	5,434,926	129,403
L.P.G.	4,054,470	96,535
Propane	3,836,000	91,333
Ethane	3,082,000	73,381
Butane	4,326,000	103,000
Kerosene	5,670,000	135,000
#1 Distillate	5,706,000	135,857
#2 Distillate	5,825,000	138,690
#4 Distillate	6,062,000	144,333
Residual oil	6,287,000	149,690

WOOD FUELS

Table 2.6 shows the energy content of various types of woods when burned. Soft woods such as pine, fir, willow, and aspen have less energy per unit volume. In terms of weight, most woods have similar energy content per unit weight: 8250 BTU/pound for dried nonresinous woods (such as aspen, oak, and maple) and 9150 BTU/pound for dried resinous woods (such as pine and western red cedar). The cord is the common unit for wood and one cord is defined as a volume of 128 cubic feet ($4' \times 4' \times 8'$ is a typical arrangement to give a volume of one cord). In Table 2.6, the lower value of the range assumes 70 cubic feet of wood per cord (i.e., 58 ft^3 of the cord is air). The higher value of the range assumes 90 cubic feet of wood per cord (i.e., 38 ft^3 of the cord is air). The green weight assumes the wood has not been dried and the moisture content is ~50%. The dry weight assumes the wood has been dried to a moisture content of 12%. Green wood is harder to burn because of the high moisture content.

UNITS OF POWER

Power is simply the rate of doing work, or the rate of energy conversion. In the SI units system, this is measured in watts (which is equal to joules per second). In the British units system (also known as American

TABLE 2.6 Wood Energy Content and Density Values

Species	Million BTU/Cord	Wood Density (Pounds/Cord) (Dry)	Wood Density (Pounds/Cord) (Green)
Alder, Red	18.4-19.5	2000-2600	3200-4100
Ash	24.5-26.0	2680-3450	4630-5460
Aspen	17.0-18.0	1860-2400	3020-3880
Beech	28.6-30.4	3100-4000	4890-6290
Birch	25.9-27.5	2840-3650	4630-5960
Cedar, Incense	17.8-20.1	1800-2350	3020-3880
Cedar, Port Orford	20.7-23.4	2100-2700	3400-4370
Cherry	22.3-23.7	2450-3150	4100-5275
Chinquapin	23.2-24.7	2580-3450	3670-4720
Cottonwood	15.8-16.8	1730-2225	2700-3475
Dogwood	28.6-30.4	3130-4025	5070-6520
Douglas-Fir	23.5-26.5	2400-3075	3930-5050
Elm	22.3-23.7	2450-3150	4070-5170
Eucalyptus	32.5-34.5	3550-4560	6470-7320
Fir, Grand	17.8-20.1	1800-2330	3020-3880
Fir, Red	18.3-20.6	1860-2400	3140-4040
Fir, White	18.8-21.1	1900-2450	3190-4100
Hemlock, Western	21.6-24.4	2200-2830	4460-5730
Juniper, Western	23.4-26.4	2400-3050	4225-5410
Laurel, California	24.6-26.1	2690-3450	4460-5730
Locust, Black	29.5-31.4	3230-4150	6030-7750
Madrone	29.1-30.9	3180-4086	5070-6520
Magnolia	22.3-23.7	2440-3140	4020-5170
Maple, Big Leaf	21.4-22.7	2350-3000	3840-4940
Oak, Black	25.8-27.4	2821-3625	4450-5725
Oak, Live	34.4-36.6	3766-4840	6120-7870
Oak, White	26.4-28.0	2880-3710	4890-6290
Pine, Jeffery	19.3-21.7	1960-2520	3320-4270
Pine, Lodgepole	19.7-22.3	2000-2580	3320-4270
Pine, Ponderosa	19.3-21.7	1960-2520	3370-4270

Continued

TABLE 2.6   Wood Energy Content and Density Values—cont'd

Species	Million BTU/Cord	Wood Density (Pounds/Cord) (Dry)	Wood Density (Pounds/Cord) (Green)
Pine, Sugar	17.3-19.6	1960-2270	2970-3820
Redwood, Coast	17.8-20.1	1810-2330	3140-4040
Spruce, Sitka	19.3-21.7	1960-2520	3190-4100
Sweetgum	20.6-21.9	2255-2900	4545-5840
Sycamore	21.9-23.3	2390-3080	4020-5170
Tanoak	25.9-27.5	2845-3650	4770-6070
Walnut, Black	24.5-26.0	2680-3450	4450-5725
Western Red Cedar	15.4-17.4	1570-2000	2700-3475
Willow, Black	17.5-18.6	1910-2450	3140-4040

customary units) it is given in ft-lbf/s. Another common power unit is horsepower (hp), which is defined as 550 ft-lbf/s. As its name implies, this unit was meant to estimate the approximate peak rate at which a horse can work. Other than horsepower, the other power conversion factors are the same as energy conversions if time is measured in the same units (seconds, hours, days, or years).

$$1 \text{ watt} = 1 \text{ J/s} = 0.73756 \text{ ft} - \text{lbf/s} = 0.001341 \text{ hp}$$

A little calculation exercise is useful here. A human being can work at the maximum rate of about ¼ hp, so in 1 h a human can do:

$$\frac{1}{4} \times 550 \times 3600/778 \text{ BTU of work} = 636 \text{ BTU/h of work}$$

If that person is paid $10/h for their labor, you will be getting 63.6 BTU of work for each dollar spent. In Table 2.5, you will see that one gallon of gasoline contains about 125,000 BTU. At $3.50/gallon, energy in the form of gasoline can be purchased for about 35,700 BTU for each dollar spent. At this rate, gasoline is 560 times cheaper than human labor. Put another way, compared to human labor at $10/h, gasoline is worth $1965/gallon. This illustrates the tremendous value that the discovery and refining of crude oil has had on the well-being of humans. Energy was made much cheaper and society has advanced because of this discovery.

In his 1943 paper in the *American Anthropologist* discussed in Chapter 1, Leslie White makes this point even more dramatically. He says, "In the US the energy expended is now about 13.5 horsepower hours per day per capita, which is the equivalent of 100 human slaves for each person."

Energy Science and Thermodynamics

Energy utilization is greatly affected by the laws of thermodynamics. The first law states that energy cannot be created or destroyed, merely transformed from one type to another. If energy cannot be destroyed, why do we need new energy supplies continuously? What do we mean when we say we use energy? The answer lies in the second law of thermodynamics, which can be stated in many forms. One form of this law says that it is impossible to convert heat into work without rejecting some of that energy to a heat sink. Moreover, the heat that is transferred to and from a heat engine depends on the temperature of the heat source and the heat sink. This means that there is a thermodynamic limit on the amount of energy that can be captured from energy sources. Eventually, these energy sources are degraded into less usable forms of energy—particularly heat that is dissipated into the atmosphere and eventually out to the rest of the universe.

THE FIRST LAW OF THERMODYNAMICS

The first law of thermodynamics states that energy cannot be created or destroyed; it can only be transformed from one type of energy to another. This seems like an obvious statement, one with which most students of science and engineering are eminently familiar. It leads to a relatively simple equation: during any process, the energy in equals the energy out. This law was not always obvious to scientists. In fact, up until the nineteenth century, scientists were under the mistaken belief that heat was not a form of energy at all, but that heat had mass. This was part of the caloric theory of heat, which had its origins in the medieval notion that there were only four elements that made up the universe: earth, air, fire and water. There was considerable evidence to support this theory. For example, if you burned a piece of wood you would see smoke and other vapors emanating from it that would disappear into and became part of the air. During the burning, you would see heat and flames, which clearly indicated that fire was one of the elements making up the wood. The end result would be ash which would become part of the earth. This ash could not be burned and weighed less than the original piece of wood because of the air and fire that it had lost. There are also many substances out of which you can squeeze water, including fruit, vegetables and the earth itself. These observations led to the mistaken belief that heat or fire was an element and as such had mass. There were many observations that disputed this belief, but scientists had explained away these observations.

The big breakthrough in the development of the first law of thermodynamics was made by an American engineer called Benjamin Thompson. Thompson was born in rural Woburn, Massachusetts, on March 26, 1753; his birthplace is preserved there as a museum. He was educated mainly at the Woburn schools, although he sometimes managed to get to Cambridge to attend lectures by Professor John Winthrop of Harvard College. It was there he learned about the caloric theory of heat. In 1769 when Thompson was 16, he began to conduct his own experiments to try to understand the nature of heat, and he corresponded with friends about these experiments (these records of his thoughts are still in existence today). In 1785 at the age of 32, he moved to Bavaria in Germany where he spent about 12 years. Some of his duties there were to reorganize the army and to establish workhouses for the poor. He made studies on the fuels used for lighting, heating and cooking, including their relative costs and efficiencies. He also constructed the English Garden in Munich in 1789; it remains there today and is known as one of the largest urban public parks in the world. Thompson was greatly admired for this work, and in 1791 he was named a Count of the Holy Roman Empire and granted the formal title of Count Rumford.

In 1797 while Thompson was working in Munich (the capital city of Bavaria), he was made responsible for the boring of cannon barrels using sharp cutting tools in a lathelike machine. He observed that when the blade became blunt, the lathe had to work harder using more energy. At the same time, the barrel became hotter. This suggested to him that there might be a link between work and heat and that heat could be a form of energy, not mass. He made a series of measurements on the cannons, linking the amount of work to the rise in temperature of the cannon. He began to realize that the link between work and heat was that heat was a form of energy, not the caloric matter of the current scientific thinking. In a paper published in 1798 entitled, *An Experimental Enquiry Concerning the Source of the Heat Which Is Excited by Friction*, Thompson argued that heat was not caloric matter; instead, heat, work and motion are all forms of energy. Thompson (now Count Rumford) described how he immersed a cannon barrel in water and arranged for a blunted cutting tool. He showed that while he was cutting the barrel the water could be boiled and that the supply of frictional heat continued as long as the machine kept boring. He pointed out that the weight of the cannon and the material removed remained the same no matter how much work was done or how hot the barrel became. He also noted that the only thing communicated to the barrel was the energy of motion. This was a fundamental breakthrough that rejected the caloric theory of heat, the accepted theory of the day. While his work did cause some hostility, it was subsequently important in establishing the law of conservation of energy, or the first law of thermodynamics.

While the term "energy" had not been coined yet, it appears that it was first used by Thomas Young in 1807—a few years after Benjamin Thompson's observations on the cannon. A further step in the development of the first law of thermodynamics on the conservation of energy was formulated in 1837 by Karl Mohr, who stated:

> ...besides the 54 known chemical elements there is in the physical world one agent only, and this is called energy. It may appear, according to circumstances, as motion, chemical affinity, cohesion, electricity, light and magnetism; and from any one of these forms it can be transformed into any of the others.

Mohr had not included heat in his conservation principle because, despite Thompson's breakthrough, many scientists still clung to the caloric theory that maintained that heat could neither be created nor destroyed. Thompson had shown that mechanical energy can be converted into heat. Savery, Newcomen, Watt, and other purveyors of the steam engine had shown that heat can be converted into mechanical energy, which contradicts and disproves the caloric theory. Thus, the principle of conservation of energy assumes that heat is a form of energy and that heat and mechanical work can be made to be interchangeable.

The first explicit statement of the first law of thermodynamics was given by Rudolf Clausius in 1850, 53 years after Thompson's breakthrough. Clausius stated:

> There is a state function E, called "energy," whose differential equals the work exchanged with the surroundings during an adiabatic process.

This is a rather precise but tricky statement, requiring some explanation. The word "adiabatic" means that there is no heat transferred in the process. Clausius recognized that heat is a form of energy that could be transferred into other forms of energy. If the process is adiabatic, any change in the energy of the system is equal to the work done by the system. If the process is not adiabatic, the heat transfer must also enter the equation.

A more modern explanation of the first law of thermodynamics is that there are many forms of energy—kinetic, mechanical, potential, chemical, electrical, internal, nuclear, heat, and work. All of these can be converted from one form into another, but the total amount of energy is always conserved. The implications of this are very important. When fuel is burned, the chemical energy of the chemical bonds is released as heat. If the main purpose is to heat up the living space, energy must be directed into heating up air or water. The heated medium is pumped around the living space to distribute the energy to where it is wanted and in the correct form. If the goal is to create mechanical energy or work, heat has to be converted into work. This can be easily done in heat engines such as the internal combustion engine or the steam engine, which is an external combustion engine. If the goal is to create electrical energy, heat must first be transformed into mechanical energy and then into electrical energy.

The sun generates energy through nuclear reactions. In this process, hydrogen atoms are fused together under intense heat and pressure to form helium atoms. The helium atoms contain slightly less mass than the hydrogen atoms used to create them. This extra mass is converted into energy according to Einstein's famous equation, $E = mc^2$. This energy is then stored in the sun as internal energy and nuclear energy and can be transferred to the earth as heat and radiation.

One of the most efficient and clean forms of energy on the earth is hydroelectric power. In this process, the sun vaporizes water from the ocean, and it falls on high mountains as rain or snow and flows into rivers. At high altitudes, the water has potential energy that originally was driven by the nuclear energy of the sun. The potential energy of this water is captured by building a dam and then letting the water flow through a mechanical turbine, converting the potential energy into mechanical energy and ultimately into electrical energy. At each step of this process, energy is being conserved. It should be noted that most of the energy utilized on the earth originated on the sun. The main exception to this rule is the geothermal energy deep in the earth, which is discussed in Chapter 11. This energy, which is due to nuclear processes under the earth, is captured via heat transfer and utilized.

Wind energy can also be converted into electrical energy. Wind energy is actually kinetic energy due to the movement of the air molecules, which are moved by the sun's energy. This energy is used to turn the blades of wind turbines, effectively converting the kinetic energy of the air molecules into the kinetic energy of the blades, which in turn is converted into the electrical energy used to power our homes and industries.

None of these conversions are 100% efficient. Some energy is always lost due to friction and other such inefficiencies; however, that energy is not destroyed. It usually ends up as low-grade heat energy that serves to heat up the world or the universe slightly. There are other inefficiencies in the energy conversion process, and these are dealt with in the second law of thermodynamics.

If energy cannot be created or destroyed, merely transformed from one type to another, why do we need new energy supplies continuously and what do we mean when we say we use energy? This answer lies in the second law of thermodynamics.

THE SECOND LAW OF THERMODYNAMICS

The second law of thermodynamics puts a limit on how efficient the energy conversion processes can be. Even though energy transferred as heat is indeed energy transfer, there is something different about heat energy. The way scientists define heat is somewhat different than the way most people understand heat. If two bodies of different temperatures are placed in contact with one another, there will be a transfer of energy from the hot body to the cold body and that transfer will continue until the two bodies are the same temperature. This transfer of energy is called heat. The temperature of each body is due to its "internal energy." This energy is stored as the kinetic energy of vibrating and moving molecules and is a function of the temperature and pressure of the material. The hotter the material, the faster the molecules move. When energy is transferred as heat, the hotter body loses some of its internal energy and the colder body gains some internal energy. It is therefore incorrect to say a hot body contains heat; it contains internal energy and it can transfer some of that energy to a colder body as heat. The colder body then converts that energy to internal energy. The transfer of that internal energy is what is called heat. The reader can now go back to the previous section on the first law of thermodynamics and see all my deliberate errors written about the word heat. For instance, I said that in the sun "hydrogen atoms are fused together under intense heat and pressure to form helium atoms." This is not correct; I should have said that "hydrogen atoms are fused together at very high temperatures and pressures to form helium atoms."

There are three different heat transfer methods: conduction, convection and radiation.

How much heat can be transferred between two bodies depends on their temperatures and the method of heat transfer. Heat energy cannot be converted entirely into work or mechanical energy, electricity or any other form of energy. This has a big impact on the efficiency of heat engines. When fuel is burned in a heat engine, the chemical energy that is stored in the chemical bonds of the fuel molecules is released. This energy is transferred to the engine as heat; however, not all of that heat can be converted into mechanical energy. When people were developing and trying to perfect heat engines, such as the internal combustion engine, they discovered this limitation and had to understand it. The result was the formulation of the second law of thermodynamics.

The seminal work in this area was due to a French engineer called Sadi Carnot. In 1824, he published a paper entitled, *Reflections on the Motive Power of Fire and the Machines Needed to Develop This Power*. This paper presented the idea that the amount of work done by a heat engine is due to the flow of heat from a hot to a cold body. Carnot's understanding of heat was still mired in the incorrect caloric theory of heat, but his conclusions were still valid. His analysis determined that the theoretical heat that could be transferred to the heat engine was proportional to the temperature difference between the heat source (the hot body) and the heat sink (the cold body). This analysis allowed him to calculate the theoretical efficiency of a heat engine, which turned out to be much lower than the efficiency of other energy conversion processes.

Using Carnot's analysis, several people were able to deduce different statements of the second law of thermodynamics. Some of these are:

1. It is impossible to produce work in the surroundings using a cyclic process connected to a single heat reservoir (Thomson, 1851).
2. It is impossible to carry out a cyclic process using an engine connected to two heat reservoirs that will have as its only effect the transfer of a quantity of heat from the low-temperature reservoir to the high-temperature reservoir (Clausius, 1854).
3. In any process, the entropy of the universe increases, causing it to tend towards a maximum (Clausius, 1865).

This third statement introduces the concept of entropy and puts the law on a more mathematical basis. Since the amount of heat transferred in any process depends on the temperature of the body transferring the heat, entropy is defined as the heat transferred divided by the temperature, T, at which it is transferred. Giving entropy the symbol S and the heat transferred the symbol Q, by definition:

$$S = Q/T$$

A certain understanding of entropy is required to fully understand the limitations of energy usage. This is illustrated in Appendix A and discussed in the next section.

One of the consequences of the second law of thermodynamics is that, when you burn fuel to drive a heat engine, only some of the heat from the fuel can be converted to work in the engine. The rest must be rejected to a heat sink, which is usually the atmosphere surrounding the engine. Consequently, heat engines are inherently inefficient. Another consequence is that heating your house with an electric heating element is going to be much more expensive than using most other fuels. This is because the electricity has been created using an inefficient heat engine where some of the heat had to be rejected to the surroundings. If your goal is to increase the temperature of some space (such as your house), it is better to burn a fuel directly and capture as much of that heat in your house as possible. The efficiency of direct heating by burning a fuel is much higher than creating electricity where some of the heat must be lost. All of this was analyzed by Sadi Carnot using his Carnot cycle and published in 1824.

Sadi Carnot was the eldest son of a French Revolutionary named Lazare Carnot and was born on June 1, 1796, during the height of the French Revolution. Sadi studied at the École Polytechnique beginning in 1812. By the time Sadi graduated in 1814, Napoleon's empire was on the run and European armies were invading France. During Napoleon's return to power in 1815, Sadi's father, Lazare Carnot, was Minister of the Interior for a few months. Following Napoleon's final defeat later that year, Lazare fled to Germany, never to return to France.

Sadi Carnot was an army officer for most of his life, but in 1819 he semi-retired from the army and began to devote his attention to designing steam engines. These engines were the main workhorses of Europe, particularly Britain, and were used for pumping water from mines, dredging ports and rivers, grinding wheat, and spinning and weaving cloth; however, they were somewhat inefficient. The import of the more advanced British steam engines into France after the war showed Carnot how far the French had fallen behind in their technology. He was particularly dismayed that the British had progressed so far through the genius of a few engineers who lacked any real scientific education. British engineers had also accumulated and published reliable data about the efficiency of many types of engines under actual running conditions; they argued about the merits of low and high pressure engines and of single-cylinder and multi-cylinder engines.

Carnot understood implicitly that great civilizations need to harness energy to advance their technology. Convinced that France's inadequate utilization of steam was a factor in its downfall, he began to write a nontechnical work on the efficiency of steam engines. Other workers before him had examined the question of improving the efficiency of steam engines by comparing the expansion and compression of steam with the production of work and consumption of fuel. In his essay, *Réflexions sur la puissance motrice du feu et sur les machines propres à développer cette puissance (Reflections on the Motive Power of Fire and the Machines Needed to Develop This Power)*, published in 1824, Carnot gave a lot of attention to the theory of the process not concerning himself, as others had done, with its mechanical details.

Carnot stated that, in a steam engine, motive power is produced when heat "drops" from the higher temperature of the boiler to the lower temperature of the condenser, just as water, when falling, provides power in a waterwheel. He worked within the theoretical framework of the caloric theory of heat, assuming that heat was a gas that could be neither created nor destroyed. Though this assumption was incorrect and Carnot himself had doubts about it even while he was writing his essay, many of his results were nevertheless true. One of these was his prediction that the efficiency of an idealized engine depends only on the temperature of its hottest and coldest parts and not on the substance (steam or any other fluid) that drives the mechanism.

Carnot understood that every thermodynamic system exists in a particular thermodynamic state. When a system is taken through a series of different states and finally returned to its initial state, a thermodynamic cycle is said to have occurred. In the process of going through this cycle, the system may perform work on its surroundings, thereby acting as a heat engine. The cycle that he proposed and used in his analysis is now known as the Carnot cycle. A system undergoing a Carnot cycle is called a Carnot heat engine, although such a "perfect" engine is only theoretical and cannot be built in practice.

The mathematical details of the *Carnot cycle* are shown in Appendix A, but it is not necessary to fully understand those details to appreciate its usefulness. The Carnot cycle when acting as a heat engine, consists of the following four steps:

1. *Reversible and isothermal expansion of the working fluid at the "hot" temperature, T_H (isothermal heat addition)*. During this step, the fuel is burned creating the hot temperature and causing the working fluid or gas to expand. The expanding gas makes the engine's piston do work on the surroundings. As the piston is forced to move, it drives a shaft which converts the work into kinetic energy. The gas expansion is propelled by the absorption of heat from the high temperature reservoir created by the burning fuel.

2. *A reversible and adiabatic (isentropic) expansion of the working fluid (isentropic work output)*. Remember that adiabatic means that there is no heat transferred. Isentropic means that the entropy of the system remains constant. For this step, the piston and cylinder are assumed to be thermally insulated (adiabatic), thus they neither gain nor lose heat. The gas continues to expand, working on the surroundings. When gas expands it also cools, losing energy. Since the process is insulated, however, it cannot lose that energy as heat. This forces the gas to continue to do work by driving the piston. This expansion of the gas causes it to cool to the "cold" temperature, T_C.

3. *Reversible isothermal compression of the gas at the "cold" temperature, T_C (isothermal heat rejection)*. In this step, the surroundings do work on the gas, which causes a quantity of heat to flow out of the gas to the low temperature reservoir.

4. *Isentropic compression of the gas (isentropic work input)*. Once again, the piston and cylinder are assumed to be thermally insulated

(or adiabatic). During this step, the surroundings, through the piston, do work on the gas, compressing it and causing the temperature to rise to T_H. At this point, the gas is in the same state as at the start of step one.

The antithesis of a heat engine is a refrigerator. A heat engine burns fuel as part of a thermodynamic cycle to create heat that is converted into mechanical energy. A refrigerator sends the cycle in the opposite direction and uses electrical energy to create mechanical energy that then pumps heat from the cold body to the hotter body.

The efficiency of the heat engine, η, is defined as the work produced divided by the heat input from the hot reservoir. In Appendix A the efficiency is calculated as follows:

$$\eta = \frac{W}{Q_H} = 1 - \frac{T_C}{T_H} = \frac{T_H - T_C}{T_C} \qquad 3.1$$

Where,

W is the work done by the system (energy exiting the system as work).
Q_H is the heat put into the system (heat energy entering the system).
T_C is the absolute temperature of the cold reservoir.
T_H is the absolute temperature of the hot reservoir.

This efficiency describes the fraction of the heat energy extracted from the hot reservoir and converted to mechanical work. A Rankine cycle is usually the practical approximation of a Carnot cycle for a steam engine. It is shown, in Appendix A, that for any cycle operating between temperatures T_H and T_C, none can exceed the efficiency of a Carnot cycle.

Carnot's theorem is a formal statement of this fact: *No engine operating between two heat reservoirs can be more efficient than a Carnot engine operating between those same reservoirs.* Equation 3.1 gives the maximum efficiency possible for any engine using the corresponding temperatures. A corollary to Carnot's theorem states that: *All reversible engines operating between the same heat reservoirs are equally efficient.* The right-hand side of Equation 3.1 gives what may be a more easily understood form of the equation: the theoretical maximum efficiency of a heat engine equals the difference in temperature between the hot and cold reservoir divided by the absolute temperature of the hot reservoir. To find the absolute temperature in degrees Kelvin, add 273.15° to the Celsius temperature. To find the absolute temperature in degrees Rankine, add 459.6° to the Fahrenheit temperature. Looking at the formula in Equation 3.1, an interesting fact becomes apparent. Lowering the temperature of the cold reservoir will have more effect on the ceiling efficiency of a heat engine than raising the temperature of the hot reservoir by the same amount. In the real world, this may be difficult to achieve since the cold reservoir is often an existing ambient temperature, such as the atmosphere.

In other words, maximum efficiency is achieved if no new entropy is created in the cycle. In practice, the required dumping of heat into the environment to dispose of excess entropy leads to a reduction in efficiency. Equation 3.1 gives the efficiency of any theoretically reversible heat engine.

Carnot realized that in reality it is not possible to build a thermodynamically reversible engine. Real heat engines are less efficient than indicated by Equation 3.1. Nevertheless, Equation 3.1 is extremely useful for determining the maximum efficiency that could ever be expected for a given set of thermal reservoirs.

There are four practical heat engine cycles in wide use today, each trying to approximate the Carnot thermodynamic cycle. They are

1. The Otto cycle, which is the basis of the gasoline engine.
2. The Diesel cycle, commercialized in the Diesel engine.
3. The Rankine cycle, the basis for steam engines widely used today in power plants to generate electricity.
4. The Brayton cycle used in gas turbines that are used to generate electricity or provide thrust.

There is also the Stirling cycle that can be used to make a practical external combustion heat engine, but this engine has never been commercialized. Despite this there is a lot of interest in developing Stirling engines because a large variety of fuels can be used to drive such engines, including solar energy. The Stirling engine is an alternative to the Rankine cycle engine.

The entropy statement of the second law also allows scientists to analyze chemical reactions, the phase behavior of fluids, and many other seemingly unconnected processes. It also explains why people say they use energy when they are actually converting energy from one form into another. When fuel is burned to generate energy, chemical energy is converted into heat and then some of that heat energy is converted into electricity. Some of it is also rejected to the atmosphere where it is no longer usable. This electricity creates light in a lightbulb, which is also lost as heat to the atmosphere.

If the fuel is used to power an internal combustion engine to drive an automobile after some of the heat is rejected to the atmosphere, the rest of the fuel's energy creates useful and usable kinetic energy. All of that kinetic energy is eventually lost as frictional heat, which is also lost to the atmosphere. All the energy we "use" becomes lost as heat that has been mostly transferred to the atmosphere, some of which is then radiated through space to other parts of the universe.

Another consequence of the first and second laws of thermodynamics is that perpetual motion machines are not possible. The first law simply states that if you set a machine in motion by supplying it with energy it could keep running forever in a frictionless environment. You could not extract more energy back out of it than you put in because that would violate the first law. The second law says that you cannot even get as much out as you put in because some of the energy is lost as heat via friction. Perpetual motion machines fall into two categories: those that violate the first law of thermodynamics and those that violate the second law of thermodynamics.

The entropy parameter is also a measure of the randomness of the universe, and the second law states that the randomness of the universe is increasing. In other words, as processes unfold, the elements of the universe tend to a more disordered state.

The answer to the question "What do we mean when we say we use energy?" is that the available energy is used and then converted into unavailable energy. Electrical energy, potential energy, kinetic energy and chemical energy in fuels are all available forms of energy. Energy lost to the atmosphere as heat becomes mostly unavailable energy. It is hard to extract energy from the atmosphere because the temperature is not high enough. The cumulative effect of energy lost to the atmosphere or the ocean is that it is also continuously radiated to the rest of the universe where it becomes completely unavailable. The second law of thermodynamics governs this process.

The laws of thermodynamics have many more applications than have been shown here, but that is beyond the scope of this book. Here it is simply necessary to have a little understanding of the laws of thermodynamics so that the energy processes can be understood a little better.

Environmental Issues

Whenever energy is captured for use, there are always impacts on the environment. Hydrocarbon sources generate not only air pollution but carbon dioxide, one of the greenhouse gases that are most likely to lead to global warming. Nuclear energy generates radioactive waste. While that has been the subject of intense public debate, this type of energy was slowly coming back into favor due to the fact that it does not generate carbon dioxide. Many people are strongly in favor of renewable energy sources because they are perceived as being clean and as having less undesirable effects; however, they too have an impact on the environment. Large wind generation devices create noise and visual pollution, and there is even a perception (undeserved) that they kill a lot of birds. The noise, both audible and infrasound, has impact on the health of the people that live and work near the devices. Hydroelectric power is surprisingly clean and cheap, but some people object to this form of energy because it requires dams that alter the ecology of river basins and interrupt the migration paths of fish. The truth is, as discussed in Chapter 1, you cannot extract energy from the environment without having an impact on the environment. The reason for this is that energy is the essence of the fabric of the universe. As we manipulate the flow of energy, an impact is left behind. The goal should be to work on ways to minimize that impact. So what are the impacts and how do we go about mitigating their environmental influence?

THE CARBON DIOXIDE ISSUE

In the last 200 years, carbon dioxide (CO_2) levels in the Earth's atmosphere have risen from about 280 parts per million (ppm) to about 400 ppm. Figure 4.1 shows the CO_2 levels (in ppm by volume) for the past 800,000 years. Ice ages occurred about every 100,000 years. Each time the Earth got cold, the CO_2 level dropped to below 200 ppm, and when the Earth warmed up the CO_2 level built back up to near 300 ppm.

The Future of Energy
http://dx.doi.org/10.1016/B978-0-12-801027-3.00004-X

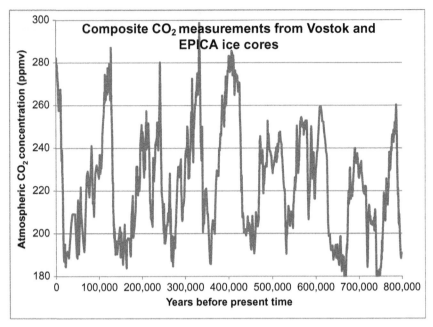

FIGURE 4.1 CO_2 levels in the atmosphere for the past 800,000 years. *Data source: NOAA compilation of data from Vostok and EPICA ice cores in Antarctica.*

Two hundred years ago, the CO_2 level was near 280 ppm. Fifty-five years ago in 1958 when accurate measurements commenced on the Mauna Loa Weather Station in Hawaii, the level had risen to 318 ppm (see Figure 4.2), a small but significant rise. By 2010, the CO_2 value had risen to 390 ppm. This rise in the CO_2 level occurred very rapidly (compared to changes in geological time). During that time period, large amounts of hydrocarbon-based fuels that emitted large amounts of CO_2 into the atmosphere had been burned, and there is a strong presumption that the two situations are cause and effect, or at least strongly linked. There is also a theory that this rise in CO_2 levels is causing or will lead to global warming, a general rise in the Earth's temperature and a disruption of the weather patterns around the Earth. Because of the general acceptance of these theories and facts, people are demanding that CO_2 emissions be restricted and that alternative energy sources be adopted.

While it is easy to conclude that CO_2 has a negative impact on the Earth and its use should be restricted, not all scientists agree with these conclusions, and it is necessary to examine both sides of the issues. It does seem like a logical conclusion (a no-brainer, if you will) that the rising CO_2 levels in the atmosphere have been caused by the vast amounts of hydrocarbons that mankind has been burning in the past 200 years. When hydrocarbons are burned, CO_2 is emitted; however, it is not as simple as

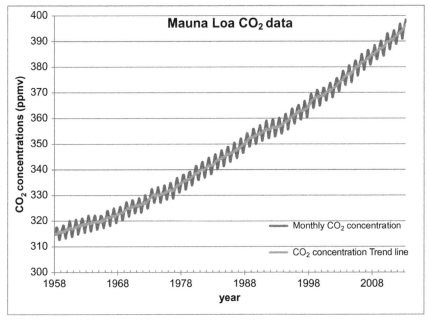

FIGURE 4.2   Atmospheric CO_2 concentrations (ppmv) measured on Mauna Loa Observatory, Hawaii, Since 1958. *Data source: NOAA-ESRL.*

that. The Earth's carbon cycle is actually very complicated. Carbon dioxide is not only generated by burning hydrocarbons, but also by transpiration by the animal life on the planet. We breathe in oxygen and breathe out carbon dioxide. It is also generated by the decay of animal and vegetable matter when they die. Carbon dioxide is also consumed by plant matter as they grow. The CO_2 in the atmosphere is photosynthesized by plant life during growth, and this removes CO_2 from the atmosphere. In fact, it has been shown that new-growth forests and crop growth remove significant quantities of CO_2 from the atmosphere each year. On the other hand, old-growth forests can be carbon neutral because the CO_2 that is being absorbed by the tree growth is balanced by the decay of the trees that die each year.

The oceans also contain vast stores of CO_2 dissolved in its waters. The amount of CO_2 dissolved in the ocean is inversely proportional to its temperature. As the water temperature rises, the oceans release CO_2 into the atmosphere; as the water temperature decreases, it absorbs CO_2 from the atmosphere.

Keeping track of all these effects can be very complicated. Those that argue against a hydrocarbon source for the rising CO_2 levels believe that the oceans are a much larger cause of this increase. They argue that the Earth's temperature is dependent on the amount of heat insolation that is radiated to the Earth by the Sun. As the solar activity rises and falls (solar activity does follow cycles), the Earth's average temperature rises and falls. As the temperature rises, the CO_2 level in the atmosphere rises because of the CO_2 that is being released by the sun-warmed oceans. They point to the fact that throughout the geological past, the CO_2 levels have been much higher than they are now. In the Cambrian period, 500 million years ago, the CO_2 levels in the atmosphere were more than 7000 ppm. Prior to that, it has been reported that the CO_2 levels could have been even higher. They were only reduced by the rapid plant growth of the Carboniferous period 300 million years ago. This was also a time when the Earth was significantly warmer than it is now. During the Jurassic period, the CO_2 levels rose to more than 6000 ppm. As recently as the Cretaceous period 65 million years ago, the golden age of the dinosaurs, the CO_2 levels were over 2000 ppm. The cause of these high CO_2 levels was clearly not the burning of fossil fuels; one of the causes was CO_2 emitted by volcanoes, but it was also affected by changing solar insolation which caused the Earth's temperature to rise, which caused the CO_2 levels to rise. Higher temperatures and higher CO_2 levels led to increased growth of plant life on Earth during the Carboniferous period. Plants thrive in such an environment. This is illustrated in Figure 4.3, which shows the estimates of CO_2 levels for the past 550 million years.

In the movie, *An Inconvenient Truth,* former Vice President Al Gore makes the argument that throughout geological history every time the CO_2 levels increased, so did the temperatures. What he did not explain, or did not understand, was that in almost every case the temperatures rose

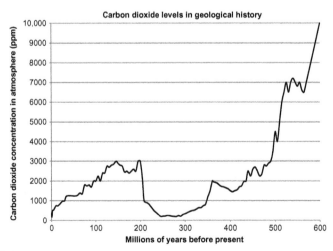

FIGURE 4.3   Estimates of CO_2 levels throughout geological history. *Source: Robert A. Rohde, 2009.*

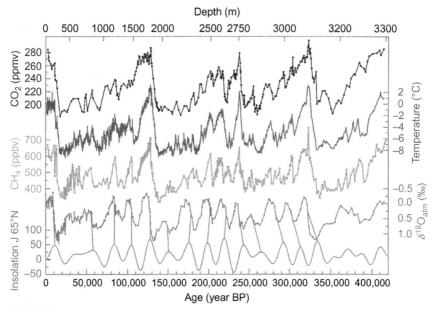

FIGURE 4.4 CO_2, CH_4, solar insolation, and temperatures on Earth for the past 400,000 years. *Source: U.S. Government: http://www.usgcrp.gov/usgcrp/images/Vostok.jpg.*

before the CO_2 levels rose, begging the question as what is cause and what is effect. Recent geological history certainly supports the argument that a rise in global temperatures leads to a rise in CO_2 levels, rather than the other way around. Figure 4.4 illustrates this point. It shows a similar plot as that used by Gore. The average global temperature increases before the CO_2 level increases and the global temperature also decreases before the CO_2 level decreases.

Militating against this argument and in favor of a hydrocarbon source for the rising CO_2 levels is the isotopic analysis of the atmospheric CO_2. This shows that most of the additional CO_2 in the atmosphere came from hydrocarbon sources. Whether this is leading to or will lead to global warming is a separate issue. The argument is that CO_2 is a greenhouse gas and that its presence in the atmosphere has led to rising temperatures. The evidence for this is less convincing because there are many factors that have a direct impact on the global temperatures. The Earth's temperature rises and falls in cycles all the time. Five hundred years ago the Earth endured a mini-ice age, when much of the Northern Hemisphere endured bitterly cold winters that had not previously existed. Since then there has been a gradual warming that is continuing and was probably not caused by CO_2 levels. During this warming period, there have been cycles of warming and cooling. In 30 of the past 40 years, there was a period of rising global temperatures that have been blamed on the CO_2 levels; however, in the past 15 years the global temperatures have stopped rising and have commenced to decrease again, suggesting that global warming has not yet taken hold or other factors are more important. Comparing Figures 4.1 and 4.2 you can see that there is no doubt the current CO_2 levels are now higher than they have been for the past 800,000 years and indeed one has to go back about 3 million years before we find atmospheric CO_2 levels this high.

It is generally accepted that CO_2 is a greenhouse gas, which means that CO_2 helps trap heat from the Sun and keep it from being reradiated from the Earth back into the surrounding universe. It is not the strongest greenhouse gas; methane, nitrous oxide, chlorofluorocarbons (CFCs), and water vapor have more intense greenhouse effects. Figure 4.5 shows how during the last 33 years not only have CO_2 levels been continually rising but so have methane (CH_4), nitrous oxide (N_2O), and CFCs. All of these are greenhouse gases. CFCs had been widely used as a propellant in aerosol spray cans and as a refrigerant for many years. Because it has both a greenhouse effect and depletes the ozone layer, it was banned from use in 1990. The graph shows its levels have stabilized and are now decreasing. CH_4 and N_2O have a much stronger greenhouse effect than CO_2; however, because the CO_2 concentration is much higher than methane and nitrous oxide, it is having or going to have the largest greenhouse effect. As the CO_2 level increases, its greenhouse effect will also increase.

There is also a concern that there is an unstable feedback loop in the global warming effect that could lead to runaway global temperatures. The reasoning is that, as the CO_2 levels rise, the Earth starts warming up due to the greenhouse effect. As the seas warm, more CO_2 and water vapor will be released from the oceans. This increases the CO_2 and water vapor concentrations in the atmosphere that will lead to more global warming. Even if this were true, the CO_2 coming from the oceans would eventually deplete, leading to a plateau in the CO_2 concentration. By that time, all of the ice in Greenland and Antarctica would have melted and the ocean levels would have risen by about 50-100 feet. This is speculation, however, and scientists do not fully understand this process yet.

Figure 4.6 shows an estimate of the global carbon emissions for the past 100 years. The majority of the carbon emissions have come from coal, but in the past 50 years oil had become both the primary energy source and the primary CO_2 source. But since 2005 the emissions from coal have again moved above those from oil, even though they are both about equal. The CO_2 emissions from coal can be reduced more easily than the emissions from oil. This will be discussed more fully in the next two sections of this chapter.

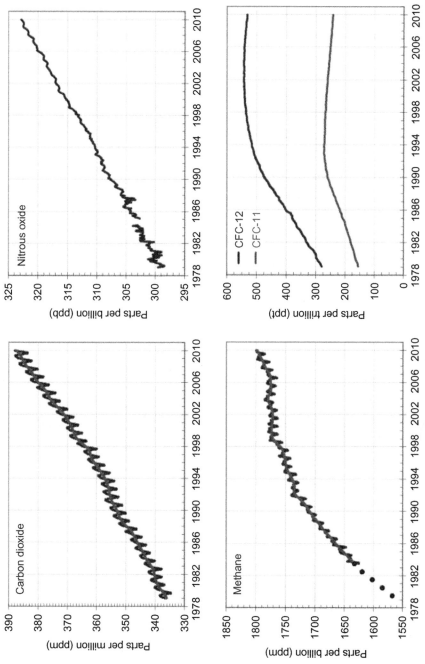

FIGURE 4.5 CO₂, CH₄, N₂O, and CFC levels in the atmosphere since 1978. *Source: NOAA.*

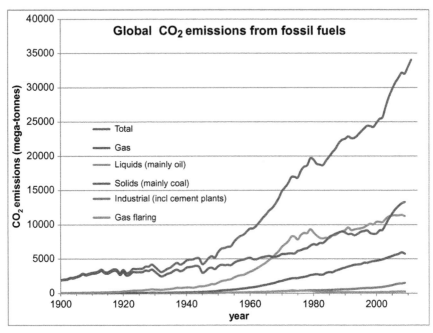

FIGURE 4.6   Global carbon emissions for the past 112 years. *Data source: Boden, Marland, and Andres, Oak Ridge National Laboratory.*

The preponderance of the evidence supports the theory that the rising CO_2 levels in the atmosphere is being caused by the burning of fossil fuels, oil, coal, and natural gas. The principal data used to support this postulate has been an examination of the ^{13}C isotope in the carbon molecules of coral growth. This evidence shows that the level of the ^{13}C isotope in the coral has been decreasing over the past 100 years. Coal in particular has lower levels of the ^{13}C isotope in its carbon molecules, leading one to conclude that the burning of coal is one of the causes of the rise in the CO_2 levels. However, I have developed some more direct evidence linking the burning of fossil fuels to the atmospheric CO_2 issue. I have taken the carbon dioxide emissions data of Boden, Marland, and Andres, shown plotted in Figure 4.6, and calculated the cumulative carbon dioxide emitted globally since 1750 and plotted it on the same plot as the Mauna Loa CO_2 measurements, shown plotted in Figure 4.2. This combined plot is shown in Figure 4.7. The two plots line up very closely. This is very strong evidence that the carbon emissions from the fossil fuels is the main cause the for the rising CO_2 levels in the atmosphere.

There is less evidence that this has caused global warming. Global warming is the concept that the rising CO_2 level in the atmosphere is

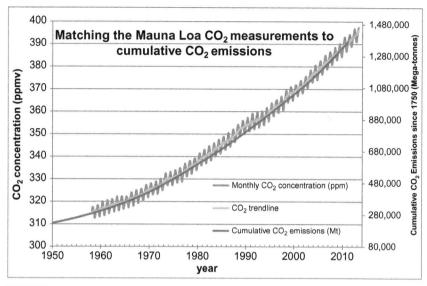

FIGURE 4.7 Matching the cumulative global carbon emissions for the past 262 Years with measured atmospheric CO_2 levels from Mauna Loa. *Data sources: Boden, Marland, and Andres, Oak Ridge National Laboratory and NOAA-ESRL.*

causing the current average temperature around the Earth to rise. This is based on rather complicated research done by Arrhenius and others in the nineteenth century which showed that CO_2 is a greenhouse gas. Therefore higher levels in the atmosphere will cause warming. If we take the CO_2 data shown in Figure 4.1 and plot it with the temperatures from the same period, we see a very close correlation. This is illustrated in Figure 4.8. Every time the temperature goes up, the CO_2 levels rise in concert with it, and vice versa. Over the past 800,000 years these two parameters have been very strongly linked however the question remains as to which is cause and which is effect. We certainly have not been burning fossil fuels over that period. We have only been doing that in appreciable amounts for the past 100 years. If the data in Figure 4.8 is examined very closely, it can be seen that the temperature rises or falls first and the CO_2 levels follow the temperatures up and down. So it is the global temperature which is the primary cause of the change in the CO_2 level. The reason for this is: as the global temperature rises, CO_2 dissolved in the ocean is vaporized from the ocean and increases the CO_2 level in the atmosphere. The reverse happens when the Earth cools: CO_2 from the atmosphere is dissolved in the ocean, decreasing the level in the atmosphere. These effects have caused the CO_2 levels over the past 800,000 years to cycle between 180 and 280 ppm. Since the end of the last ice age (10,000 years ago) the CO_2 level has remained relatively constant between 260 and 280 ppm as shown in Figure 4.9.

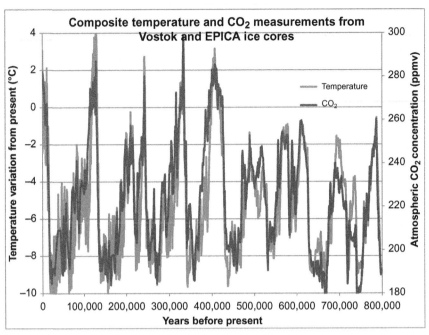

FIGURE 4.8 Global temperatures and CO_2 levels in the atmosphere for the past 800,000 years. *Data source: NOAA compilation of data from Vostok and EPICA ice cores in Antarctica.*

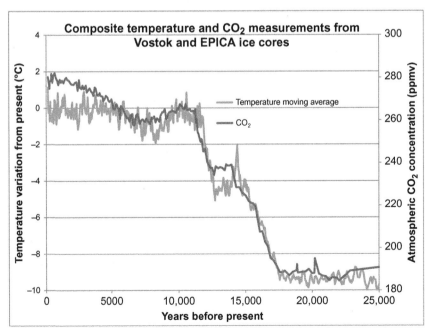

FIGURE 4.9 Global temperatures and CO_2 levels in the atmosphere for the past 25,000 years. *Data source: NOAA compilation of data from Vostok and EPICA ice cores in Antarctica.*

It can also be seen in Figure 4.9 that the global temperature varies more widely than the CO_2 levels, but on average the CO_2 levels tracked the temperatures as they moved upward. Compared to the last 1 million years we are currently living in a relatively warm period. However, compared to the last 500 million years we are currently living in a relatively cold period. Over the past 500 million years most of the time there were no polar ice caps and the sea levels were much higher than they are today. With those caveats in mind, Figure 4.10 shows the average global temperatures over the past 132 years using the temperature dataset measured and compiled by the NASA-Goddard Institute for Space Studies (GISS). Clearly the temperatures have been moving upward during that time, but is that rise being caused by the rising atmospheric CO_2 levels, or vice versa? We only have reliable CO_2 data from 1958 until the present. But Figure 4.7 demonstrates that there is a strong relationship between cumulative CO_2 emissions and atmospheric CO_2 concentrations. We have a reasonably reliable estimate of cumulative CO_2 emissions since 1750 because we have kept track of how much fossil fuel was burned during that time. So the question now is, "Is there a strong and predictable relationship between cumulative CO_2 emissions and average global temperatures?" This is answered in Figure 4.11 where the two datasets are overlain as before.

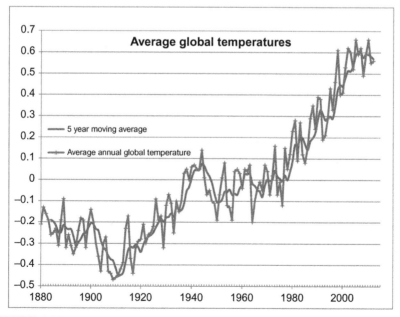

FIGURE 4.10 Measured average global temperatures for the past 132 years. *Data source: NASA-Goddard Institute for Space Studies (GISS) Temperature Dataset.*

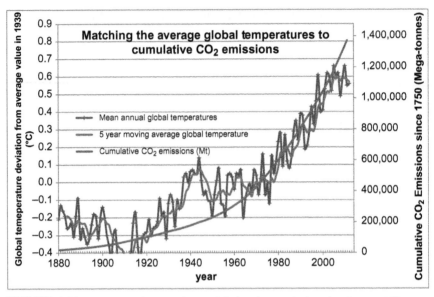

FIGURE 4.11   Matching the cumulative global carbon emissions for the past 132 years with measured global temperatures. *Data sources: Boden, Marland, and Andres, Oak Ridge National Laboratory and NASA-GISS Temperature Dataset.*

While there is a weak correlation between the two, there are clearly other factors involved. From 1880 to 1910 the global temperature was decreasing. From 1910 to 1945 the temperatures rose more rapidly than could be predicted by cumulative CO_2 emissions. Then, from 1945 to 1965 the temperatures actually fell, the Earth got colder, while the cumulative CO_2 emissions continued to rise rapidly. From 1965 to 1997 the average global temperatures rose in a similar manner to the cumulative CO_2 emissions, if we ignore the local temporal oscillations. But since 1997 the global temperatures have been relatively constant while the cumulative CO_2 emissions have continued to rise rapidly. So the relationship between the average global temperatures and cumulative CO_2 emissions is relatively weak. Discussion of the other factors that influence global temperatures is beyond the scope of this book, but solar activity and the resulting insolation are a major factor, as well as are particulates and sulfur dioxide emissions. But there are many other very influential factors, as well.

It is likely, however, that global warming will continue to be an issue in the future if the rising CO_2, CH_4, and N_2O concentrations in the atmosphere are not abated. The solution to this problem is not to stop using fossil fuels, but to stop putting the CO_2 and other greenhouse gases into the atmosphere. Instead, these gases should be captured at point sources and sequestered.

COAL

When coal is burned, it emits carbon dioxide, various oxides of sulfur dioxide (collectively known as SO_x), various oxides of nitrogen (collectively known as NO_x), as well as fine particulates of coal, ash, mercury, cadmium, and various other toxins. Because of these by-products, the emission limits for coal burning have been restricted for many years. This has forced the coal-powered industry to use capture technologies to prevent the SO_x, NO_x, ash, mercury, and cadmium from being released into the atmosphere. Once they are captured, they can be sold or sequestered in some manner. Prior to deploying these capture technologies, coal had a reputation as a dirty fuel. It was not only responsible for soot and ash raining on the countryside, but also for the acid rain that damaged rivers, soil, buildings and statues. Today most coal-powered plants are relatively clean and have little impact on the environment, except for the CO_2 emissions. At the moment, no limits on CO_2 emissions have been set; however, it is likely that limits will be eventually put in place at some time in the future.

Coal produces twice as much CO_2 when burned as does natural gas for the same amount of energy generated. This is because coal has a much lower hydrogen to carbon (H:C) ratio than does methane (CH_4). In methane, the H:C ratio is 4:1. For coal, the H:C ratio is less than 1:1.

There are various ways to capture CO_2 from the flue gas of a power plant. Figure 4.12 shows a process diagram for the most common method in use today. In this process, the CO_2 laden gas (often termed sour gas) is

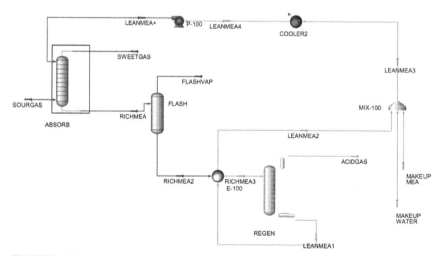

FIGURE 4.12   Process diagram for recovering CO_2 or H_2S from a gas stream. *Diagram courtesy of John E. Myers, created using HYSYS.*

input to a contactor column, where it is contacted with an absorber, which strips the CO_2 out of the gas. The sweetened gas then exits the top of the contactor column. The most common absorbing material used is monoethanolamine (MEA), although there are many different materials used and the process has many variations. The CO_2 laden MEA is then directed to a stripping column, where the MEA is heated and the CO_2 is driven off at the top of the stripping column. The regenerated MEA is then pumped across to the top of the contacting column ready for the next cycle. Using this technology, CO_2 can be stripped out of flue gas from a power plant for about $50 per ton of CO_2 recovered. It costs another $25 per ton to compress the CO_2 and sequester it underground. This increases the cost of coal-fired or gas-fired electricity by about 30%. Some data on the cost of CO_2 capture was published in 1999 by Professor Herzog from MIT. The table of data from his paper is reproduced here as Table 4.1.

Once it is captured, the CO_2 can be sequestered in porous formations deep within the Earth. The technology for doing this is reviewed and discussed in Chapter 14. It is quite feasible using current technology to build a coal-fired power plant that has zero emissions. If all the CO_2, NO_x, SO_x,

TABLE 4.1   Cost of CO_2 Capture and Separation

Cycle Technology		IGCC Today	IGCC 2012	PC Today	PC 2012	NGCC Today	NGCC 2012
Data Description	Units	Value	Value	Value	Value	Value	Value
Reference plant							
Coe: capital	mill/kWh	30	26	26	25	12	12
Coe: fuel	mill/kWh	10	9	10	10	18	17
Coe: O&M	mill/kWh	6	6	6	6	2	2
Capital cost	$/kW	1300	1145	1150	1095	525	525
Net power output	MW	500	500	500	500	500	500
CO_2 emitted	kg/kWh	0.74	0.65	0.77	0.73	0.37	0.33
Thermal efficiency (LHV)		42.0%	47.8%	40.3%	42.4%	54.1%	60.1%
Heat rate (LHV)	Btu/kWh	8124	7137	8462	8042	6308	5677
Cost of electricity	¢/kWh	4.6	4.1	4.3	4.1	3.3	3.1

Herzog (1999)

mercury, and cadmium are captured, the valuable products can be separated and sold. The rest can be sequestered underground along with the CO_2. The technologies for capturing the waste products and sequestering them are collectively known as clean coal technologies. Such a zero emissions plant would still have an impact on the environment, but the impact would be moved to deep underground formations that do not affect our quality of life. The impact would be reduced considerably. This is further discussed in Chapters 13 and 14.

OIL

Oil has many of the same environmental impacts as coal. For oil, the H:C ratio is about 2:1, which means that it emits about two-thirds (2/3) the CO_2 per unit of energy produced compared to coal. The emissions are 50% more than natural gas per unit of energy produced; however, because more oil is burned in the world than coal, oil often produces more CO_2 than coal. Moreover, much of the oil is burned as transportation fuels, which are distributed moving sources of CO_2. This means that it is very difficult to capture and sequester the CO_2 being produced. Coal does not have this problem. As coal is burned in power plants, which are fixed point sources, the emissions can be captured and sequestered at the source. One way to overcome this problem with transportation fuels is to build electric- or hydrogen-powered cars, trucks, and trains. Airplanes would also have to be fueled by hydrogen. (The topic of hydrogen will be discussed in more detail in Chapter 15, but there are major technical difficulties with the development of a hydrogen-fueled economy.) Hydrogen is expensive, difficult to store and transport and its energy density on a per volume basis is one-sixth that of gasoline. The attraction with hydrogen is that, when burned, it only produces water; however, this may create another problem. Water vapor is a stronger greenhouse gas than CO_2. If all transportation vehicles became hydrogen powered, the end result could be a problem that is as bad as or worse than the current CO_2 problem.

Some oils, like some coals, have significant quantities of sulfur associated with them, and when burned they can produce SO_x, as well as NO_x. These pollutants are easily removed at the refinery so that most transportation fuels made from refined oil are today relatively clean burning.

NATURAL GAS

Natural gas is actually more abundant than oil and today is widely used both for space heating as well as to generate electricity. It can also be processed into transportation fuels such as gasoline, diesel, and jet fuel.

For natural gas, the H:C ratio is about 4:1, which means that it emits about one-half (1/2) of the CO_2 per unit of energy produced, compared to anthracite-coal. The emissions are three-fourths (3/4) of that of oil, per unit of energy produced. Consequently, natural gas is promoted as a clean-burning fuel that is better for the environment. The CO_2 that is generated could be captured from the flue gas and sequestered in the same manner as from coal plants.

Some natural gas fields, like some oils and some coals, have significant quantities of sulfur associated with them, and if burned in their raw state they could produce SO_x and NO_x. These pollutants (such as hydrogen sulfide, carbon dioxide, and nitrogen) are required to be removed by processing at the production facilities in the field so that by the time they are compressed and put into a pipeline for transportation to the user, the gas will be free of all such pollutants. Consequently, when it is burned, the only pollutant it will produce is $CO_2.$

WIND

Wind power is very popular amongst environmental activists because it is perceived as a clean, renewable energy source; however, it too has environmental impacts that cause concern to some people. The main environmental issues are materials of construction, destruction of the viewshed, noise and health effects. It also has an issue of unreliability (because the wind does not blow all the time), but this is not an environmental issue and will be discussed in more detail in Chapter 9.

Modern wind generators consist of large turbines that make significant noise in the low frequency range. This frequency ranges from the audible to the infrasonic frequencies. The noise is not more than that caused by busy highways, but to people living and working near them, it can be very disconcerting. In addition, people also feel that the noise in the inaudible infrasonic range causes health effects that are significant and not fully understood. Consequently, people do not want to live and work close to them. These effects are still being studied. On the other hand, if they are placed in remote areas away from people, others complain that they destroy the viewshed in otherwise pristine environments. The mitigation and solution to these complaints are not immediately obvious.

Some have also claimed that wind generators kill birds. While this is true, the environmental impact is quite minimal. Wind generators do kill about 0.4 birds per gigawatt-hour of electricity generated. This turns out to be a small number compared to the number that die as a result of automobile traffic, hunting, electric power transmission lines, high-rise buildings, and cats. In the United States, turbines kill about 100,000 birds per year, compared to 80,000 killed by aircraft, 60 million killed by cars, 100 million killed by collisions with plate glass in buildings, and 200 million killed by cats. The impact of wind turbines on birds is really negligible compared to these other hazards. The Peñascal Wind Power Project in Texas and the Altamont Pass Wind Farm in California are two projects that have had particular issues with bird mortality, but they have taken steps to minimize the hazard to the local bird populations.

SOLAR

Solar Energy, like wind power, is also very popular among people in environmental circles because it is perceived as a clean, renewable energy source. Solar power is unlikely to be anything more than a small niche player, but if it ever became large and widespread, its environmental impacts would become more obvious. Solar energy is very diffuse and to concentrate it into a more usable form requires large land areas. Whether photovoltaics or large solar collectors are used, they have to be spread over a large land area to generate enough energy to supply a large number of people. This can destroy the viewshed as well as whole ecosystems in pristine environments such as deserts. All of the plants and most of the animals that live in the ecosystem that gets covered by the solar collectors would die. Whole ecosystems would be wiped out. The other significant impact is that when installed on roofs of houses or other small buildings, trees have to be removed, not planted or inhibited in some way lest they decrease the collection of the solar energy.

HYDROELECTRIC

Hydroelectric power is also a clean, renewable energy. Interestingly enough, when states such as California require that electric utilities provide 20-50% of their power from renewable sources, they will not allow hydroelectric power to be counted as renewable in the calculation. Ostensibly, this is because they recognize that it has some environmental impact. While this is true, as I have explained in this text, there are no environmentally benign energy sources.

To generate hydroelectric power, rivers have to be dammed and valleys have to be flooded. This has a clear environmental impact. The ecology of the entire river valley is altered forever. This particularly affects any anadromous fish (such as salmon, trout, sturgeon, and lamprey) that are born in the upper reaches of the river and move downstream to the ocean to mature and return to the river to spawn and die. The dam will disrupt their migration paths. The solution to this is to construct fish chutes beside the dam that allow the fish to still travel up and down the river, although many are not completely happy with this method. An example of these effects is illustrated by the Columbia River system in Northwest United States and discussed in Chapter 10.

Another well-known environmental impact is that when river valleys are flooded, the vegetation (particularly trees) that get flooded and die decay in an anaerobic environment, which produces methane. This methane finds its way into the atmosphere, where it acts as a powerful greenhouse gas. Some of the rise in atmospheric methane levels shown in Figure 4.5 is due to the construction of hydroelectric dams.

NUCLEAR

Nuclear power produces radioactive waste, which has to be sequestered. This is not an easy task, demonstrated by the controversy over the Yucca Mountain Nuclear Waste Repository. No one wants to live near a nuclear waste repository and no one wants nuclear waste being transported through their city to the repository. The nuclear industry, by necessity, has an excellent safety record, but three incidents in particular illustrate the public's concern. These were the accidents at Three Mile Island in Pennsylvania, at Chernobyl in the Ukraine and the recent accident at the Fukushima facility in Japan. Accidents happen, and continued use of nuclear power will likely lead to further releases of radiation. This is discussed in more detail in Chapter 7.

Crude Oil

Oil is the number one source of energy in the world and has held this place since the middle of the twentieth century. The world is very dependent on oil for transportation fuels, petrochemicals and asphalt. The ever increasing demand has caused the price to spike in recent years and only the world economic crisis has been able to temper demand and bring the price down to more reasonable levels. Once the economy recovers, the demand and price are likely to shoot up again. At the same time, the peak oil theory of King Hubbert predicts that world oil production is likely to peak soon, and this will place even more upward pressure on the price of oil. This raises two questions: when is oil production likely to peak and what source of energy will come to the forefront if and when oil is not able to keep up?

The theories of Hubbert have garnered a lot of credibility after his successful prediction of the rise and fall of oil production in the United States. His prediction, however, that world oil production would peak in 2000 and fall rapidly after that has not proven true. Eleven years later, the world oil production continues to rise according to the demands of the world economy. Despite evidence to the contrary, Hubbert's theories still have a lot of support and many people still expect that oil production is about to peak and will fall rapidly in the immediate future. They then expect that the world will be starved of energy and resource wars will follow that will plunge the world into a bitter struggle for control of the remaining resources that are rapidly depleting. Oil is a finite resource, and there is no denying the fact that at some point world oil production will peak and at some point in the distant future we will run out of it altogether. In this chapter, you will see that the world still has plenty of oil and that, if and when it does eventually peak, it will not decline as rapidly as Hubbert predicts.

Probably the leading exponent of Hubbert's theories is Ken Deffeyes, a geologist who worked at the Shell Research Labs in the early 1960s as a colleague and protégé of Hubbert. Deffeyes has published three books that espouse his theories: *Hubbert's Peak: The Impending World Oil Shortage*

(Deffeyes, 2001); *Beyond Oil: The View from Hubbert's Peak* (Deffeyes, 2005); and, *When Oil Peaked* (Deffeyes, 2010). After his early work with Hubbert, Deffeyes became a Professor of Geology, first at the University of Minnesota and then in 1967 at Princeton University. His first book (*Hubbert's Peak*) forecast in 2001 that the oil shortage was about to start and forecast dire consequences for the world economy. His actual projection for the peak at that time was August 2004. His second book (*Beyond Oil*) slightly revised the forecast for the world oil production peak to occur in late 2005. In the preface to the paperback edition (written in 2006), he nominated December 16, 2005 as the actual date that world oil production peaked. In his third book (*When Oil Peaked*), he ignored the fact that the world oil production was still rising and continued to insist that production peaked in 2005. He pointed to the rapid increase in the oil price over the last six years and, in the preface to the 2008 edition of *Hubbert's Peak*, he triumphantly says, "I told you so."

Another popular book that espoused the sentiment behind Hubbert's "peak oil theory" is Matthew Simmons' *Twilight in the Desert*, also published in 2005 (Simmons, 2005). In this book, Simmons argues that the oil production of Saudi Arabia—the world's largest oil exporter—has peaked and will soon begin to decline rapidly. He further predicted that this would precipitate a rapid decline in world oil supply, causing the price of oil to shoot up. In separate predictions, he made bets with people that the price of oil through the entire year of 2010 would average over $200/barrel. This bet was looking pretty safe when prices jumped up to $140/barrel in the middle of 2008, fueled by a robust world economy. Under the influence of the financial crisis of late 2008, however, prices plunged again and in 2010 the price averaged about $80/barrel. Simmons, an investment banker, made his predictions after being given access to the Society of Petroleum Engineers database of technical articles. He researched all of the articles that had come from Saudi Aramco (the national oil company of Saudi Arabia) and came to the conclusion that Saudi's main oil fields were depleting rapidly and would soon begin to decline in production. Saudi Aramco, of course, denied his claims and asserted that Saudi Arabia had plenty of oil and the capacity to meet world demand for the foreseeable future. In fact, in May 2011 an influential Saudi Prince—Al-Waleed bin Talal—said that he wanted oil prices to drop so that the United States and Europe did not accelerate efforts to wean themselves off his country's supply. If Saudi Arabia was in fact facing their "twilight in the desert," they would not be trying to keep the rest of the world addicted to oil because they would not be able to maintain the supply. Six years after the publication of Simmons' book, world oil production continues to increase. Incidentally, Simmons also made a number of public statements about the Deepwater Horizon disaster in the Gulf of Mexico in 2010 that also proved to be untrue. For instance, in late April 2010 he predicted that BP would not be able to cap the well and would have to declare bankruptcy by July 2010. Simmons died of accidental drowning in early August 2010.

Another purveyor of the doom and gloom scenarios is Michael Klare, who has published a series of books on the issues including: *Resource War: The New Landscape of Global Conflict* (2001), *Blood and Oil: The Dangers and Consequences of America's Growing Dependency on Imported Petroleum* (2005) and *Rising Powers, Shrinking Planet* (Klare, 2008). Klare—a journalist who writes for *The Nation, Harper's, Foreign Affairs* and *The Los Angeles Times*— forecasts a bleak future "... of surprising new alliances and explosive danger." His concern is not just for the shortage of oil and natural gas, but also for uranium, coal, copper and several other key resources.

There are many such books and articles, too numerous to review here, that follow a similar theme—the world is about to peak in oil production and, on the other side of the peak, as we slide down the back side of Hubbert's curve, we will experience severe economic dislocation and terrible conflicts. What does this Peak Oil Theory really refer to and does it have any basis in truth?

U.S. WORLD OIL PRODUCTION AND THE PEAK OIL THEORY

The idea of a "Peak Oil Theory" grew out of a 1956 paper published by Marion King Hubbert , who was a geologist/geophysicist who worked at the Shell Research Lab in Houston, Texas. Hubbert used his middle given name and thus was universally known amongst his colleagues as King Hubbert, which euphemistically reflected the high regard that his colleagues held for his intellectual abilities. He was born in San Saba, Texas, in 1903 and attended the University of Chicago, where he received his B.S. in 1926, his MS in 1928, and his PhD in 1937, studying geology and geophysics. He then taught geology at Columbia University in New York City for seven years. In 1943, he joined Shell Oil Company, working at their Shell Research Lab in Houston. Even before joining Shell, he had a consuming interest in the finite limits to national and world oil production. He examined the data for many years before publishing his landmark paper in 1956 in which he proposed that any finite resource—oil, gas, coal or uranium—follows a bell-shaped curve in its production history. At some point, it reaches a peak and begins to decline and the decline in production will mirror the rise in production on the way up. The peak production rate and the timing of the peak production depend on the total reserves that exist and are to be discovered in the future. Given that the bell-shaped curve is a mirror image of itself, when half of the world's reserves are produced, the production rate will begin to decline from that point. Hubbert turned to two of his colleagues, Wallace Pratt and Lewis

Weeks (eminent geologists of their time both of whom worked for subsidiaries of ExxonMobil), to give him estimates for the ultimate recovery of oil and gas in the world and in the United States in particular. He used these estimates to project the forecasts that he published. Determining the ultimate reserves that are to be discovered and produced in the future was somewhat of a guessing game that made Hubbert uncomfortable. This led him to realize that if he had the right model he did not need to guess; the production data itself could be fitted to the model to determine what the ultimate reserves would be. From this model, he worked out some very elegant mathematical methods for forecasting the ultimate reserves. These reserve methods depended on the model he assumed for production and the exact equations for the bell-shaped curve, but this technique removed some of the uncertainty about the ultimate reserves.

After retiring from Shell in 1964, he became a senior research scientist for the United States Geological Survey until his second retirement in 1976. He also held positions as an adjunct professor of geology and geophysics at Stanford University from 1963 to 1968 and at UC Berkeley from 1973 to 1976. While working for the United States Geological Survey, he produced an update to his forecast for the world oil production in 1969.

Three of Hubbert's bell-shaped curves that he presented are reproduced as Figures 5.1, 5.3 and 5.4. Figure 5.1 (which is figure 21 from Hubbert's 1956 paper) shows his forecast for oil production in the United States. He used two estimates of the ultimate oil reserves (150 billion and 200 billion barrels). One of these was from Wallace Pratt and the other was the estimate from Lewis Weeks. Using the higher estimate, he forecast that U.S. oil production would peak in 1970 at a production rate of 3 billion barrels per year (8.2 million b/d). Figure 5.2 shows the actual oil rate in

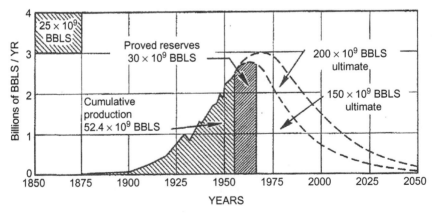

FIGURE 5.1 Hubbert's 1956 Forecast for U.S. Oil Production.

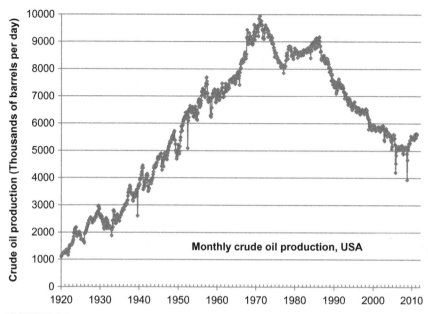

FIGURE 5.2   Actual Monthly U.S. Oil Production. *Data Source: EIA.*

the United States on a monthly basis. The U.S. oil rate did actually peak in November 1970 at 10 million b/d and in 1970 the United States actually produced 3.5 billion barrels. Hubbert's prediction was remarkably successful and this gave a lot of credibility to his methodology of what has now come to be known as Peak Oil Theory, or "Hubbert's Peak."

Two deviations on the monthly oil plot in Figure 5.2 deserve some explanation. After peaking in November 1970, U.S. oil production declined for six years in the manner predicted by Hubbert. From January 1977 to February 1986, the oil production rose again forming a secondary peak in that month at 9.14 million b/d. This deviation and secondary peak was due to the discovery and production of oil in the Prudhoe Bay and Kuparuk oil fields on the north slope of Alaska. Hubbert might argue that he did not take this area into consideration and that, excluding the Alaska oil production, the U.S. oil production from 1970 did decline in the manner that he predicted. The other deviation on the plot is the incline that has been occurring since late 2008 until the present time. This is due to oil production that is ramping up in the deep water of the Gulf of Mexico and the unconventional oil that is coming from the Bakken oil play in North Dakota. If allowed to continue, these oil sources are likely to cause U.S. oil production to keep increasing to form a tertiary peak sometime after 2020. Once again, Hubbert might argue that he was talking about conventional onshore oil and he did not take these unconventional plays into

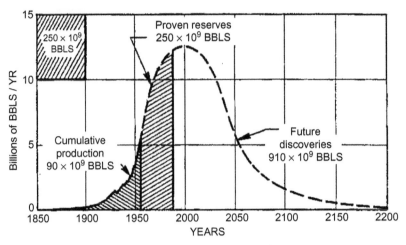

FIGURE 5.3 Hubbert's 1956 Forecast for World Oil Production.

consideration. Despite these two deviations, Hubbert's predictions for U.S. oil production proved to be remarkably accurate.

When Hubbert turned his attention to the forecast of world oil production, he produced the estimate shown in Figure 5.3 (his figure 20). This was based on ultimate oil reserves of 1.25 trillion barrels. His forecast showed a peak production of 12.5 billion barrels per year (34.25 million b/d), and it occurred in the year 2000. Later in his 1969 paper, he revised his numbers to show two estimates of ultimate recovery: 1.35 trillion barrels and 2.1 trillion barrels. The more optimistic 2.1 trillion barrel estimate again peaked in the year 2000 at 37.5 billion barrels per year (103 million b/d).

It is obvious that Hubbert's forecasts depend on knowing how much oil can be discovered and produced in the world. Critics of his method say that this number is only a guess. How can you know how much oil is going to be eventually discovered in the world? Supporters point out that the timing of the peak is not sensitive to the actual number assumed for the ultimate recovery. Indeed, if you examine Figures 5.3 and 5.4 where three numbers are assumed for the ultimate recovery ranging from 1.25 trillion to 2.1 trillion barrels, the peak still occurs around the year 2000. Hubbert himself tackled this uncertainty and, borrowing from the mathematics of biological population growth and decay, he devised a mathematical method for forecasting the number for ultimate oil recovery. This mathematical method is shown in more detail in Appendix B. There you will see that Deffeyes predicts the ultimate world oil recovery will be 2 trillion barrels and that the peak oil production had already occurred in 2005.

World oil production did not peak in 2000 or 2005. Figure 5.5 shows annual world oil production from 1965 to 2010, and at that time the

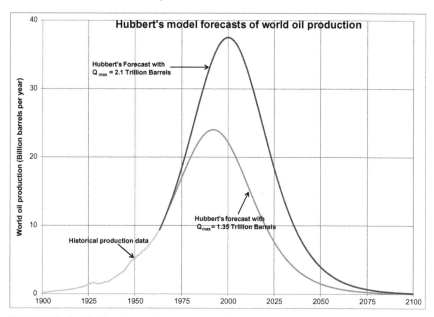

FIGURE 5.4   Hubbert's 1969 Forecast of World Oil Production. *Derived from Hubbert, 1969.*

production was still increasing. Figure 5.6 shows the world oil production on a monthly basis from the beginning of 1994 to early 2011, and though there are occasional declines, the overall trend on both plots is upwards and there is no evidence that a peak has occurred or will occur. The occasional dips in production that have occurred in the past 45 years are due to decreases in demand rather than decreases in supply. This does not stop the pundits from declaring that world oil production has peaked every time the production decreases from one year to the next or even from one month to the next. Figure 5.5 and Figure 5.6 use data from two different sources. The data for Figure 5.5 comes from the BP Annual Statistical Review, while the data for Figure 5.6 comes from the U.S. Energy Information Agency (EIA). Any discrepancies that are perceived between the two plots are due to the different data sources. What is important is that both plots show a consistent upward trend in world oil production. A secondary peak did occur in 1979 and, from 1979 to 1983, production did decrease. This was due to a steep increase in the world oil price in the 1970s, which resulted in decreased demand in the early 1980s as people reduced their use of transportation fuels. From 1983 to the present day, there has been a steady increase in oil demand which has been matched by supply. The world economic crisis that was precipitated in the middle of 2008 also tempered demand, causing a decrease in production in late

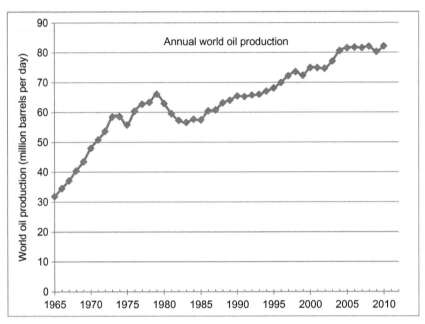

FIGURE 5.5   Annual World Oil Production Since 1965. *Data Source: BP Annual Review.*

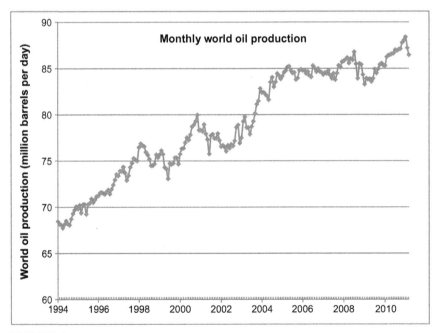

FIGURE 5.6   Monthly World Oil Production Since 1994. *Data Source: EIA.*

2008. From January 2009 to January 2011, there was a persistent increase in production. The uprising in Libya and other countries in the Middle East in early 2011, which is being called the "2011 Arab Spring," has once again caused a new round of price increases and a resulting decrease in oil demand; however, this too is expected to be temporary. Using Hubbert's model and the latest production data, you will see demonstrated in Appendix B that the peak in world oil production will not occur until at least 2018, and probably will be later than that.

So what is wrong with the Hubbert model and the analysis of his supporters that has caused their predictions about the timing of peak oil to be off? The Hubbert model is essentially correct under a constant price and constant technology scenario; however, when the oil price increases, the value of the ultimate recoverable oil (Q_{max} in Hubbert's model discussed in Appendix B) also increases as more oil becomes economic to produce. In accordance with the cultural theories of Lesley White, discussed in Chapter 1, technology also has an impact.

From time to time, technology breakthroughs bring more oil reserves into production even under a constant price scenario. Additionally, when the price of oil increases, there is an incentive to develop new technologies which also increases the oil supply. Price increases and technology breakthroughs seem to unlock large volumes of oil that were previously not economic to produce. In the past 20 years, large reserves in the Canadian oil sands (a.k.a. tar sands, bitumen, ultra-heavy oil) have come on stream. New fracture stimulation and horizontal drilling technologies have also unlocked large reserves of tight oil, tight gas and shale gas. The Canadian oil sands in Alberta have currently booked reserves of 170 billion barrels, and the possible reserves reach as high as 1 trillion barrels. Compare that to the 1.3 trillion barrels of oil that have been produced in the world up until the present time and the 1 trillion barrels of conventional oil that Deffeyes said were left to be produced in 2005. Hundreds of billions of barrels of oil in the Bakken formation in North Dakota, Montana, Manitoba and Saskatchewan are also being unlocked with horizontal wells and multi-zone fracture stimulations. The success in the Bakken is also opening up more formations to successful production such as: the Eagle Ford in Texas; the Granite Wash in Oklahoma and the Texas Panhandle; the Niobrara in Colorado and Wyoming; the Monterey in California; the Utica formation in Eastern Ohio; and, the Shublik formation on the North Slope of Alaska. Vast quantities of oil have also been discovered in the deep waters of the Gulf of Mexico and in the Santos offshore basin in Brazil. As these technologies are developed, they will be applied to many other areas around the world.

Figure 5.7 shows the average price of crude oil in the world since 1989 as determined by the EIA. The price increases of 2007-2008 and 2011 have accelerated the deployment of these new technologies as well as the

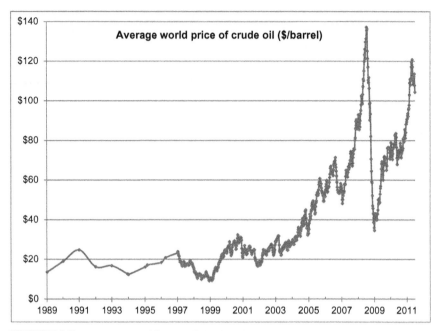

FIGURE 5.7   Average World Price of Crude Oil. *Data Source: EIA.*

economic development of other marginal reserves. From 1986 to 2005, the price of oil remained relatively constant at around $20/barrel, but the loud voices of the peak oil theorists in the twenty-first century probably caused the market to start anticipating supply shortages. This, and the political upheavals in the Middle East in 2011, has driven up the price of oil. At the same time, they have driven up the value of the world's ultimate oil reserves, Q_{max}, and pushed out the timing of the peak production, t_{peak}. These developments not only affect Q_{max} and t_{peak} but will also affect the decline in production on the back side of the Hubbert curve. Oil will not be sliding down the other side of the curve in the manner that the Hubbert model forecasts or that the theorists expect. To illustrate this point, it is necessary to examine the U.S. natural gas production curves, which will be touched on in the next section of this chapter and examined in more detail in the following chapter. The history of the U.S. natural gas production has considerable relevance to the trends in world oil supply.

At this point, you might also ask why the Hubbert model worked so well in the prediction of the U.S. oil supply but did not seem to be working, and will not continue to work, for predicting the peak and decline in world oil supply. The answer is simple: the oil market is a world market and the U.S. oil production peaked and declined amidst a relatively constant world oil price. Even though the U.S. supply was decreasing, the world supply of oil was stable and the price remained relatively constant from 1986 to 2005. The Hubbert model works as long as the oil price and technology of extraction remains stable. The world price is now ramping upwards and this will profoundly change the dynamics of the Hubbert model.

Ken Deffeyes is fond of saying that, "The economists all think that if you show up at the cashier's cage with enough currency, God will put more oil in the ground." God did not put any more oil in the ground, but between 2004 and 2010, the price of oil increased from $20/barrel to $100/barrel. This caused the tight oil in the Bakken and Eagle Ford formations, the heavy oil sands in Canada and the deepwater oil in the Gulf of Mexico and the Santos Basin to become economic to find and produce. In fact, another half trillion barrels of oil became economic to produce. In Appendix B, it is shown that if the monthly data published by the EIA and shown plotted in Figure 5.6 is used in the Hubbert model, an even more startling revelation emerges. When this data is extrapolated as shown in Appendix B, the predicted ultimate oil recovery, Q_{max}, is 3.055 trillion barrels. Of this 3 trillion barrels 1.3 trillion barrels has already been produced, leaving about 1.7 trillion barrels still to be found and produced. The current listed world-oil proved reserves according to the EIA is 1.3 trillion barrels, meaning that there are only 400 billion barrels still to be found. Remember, these numbers are a function of oil price and are likely to be conservative. Given that the data used for that extrapolation is probably the most reliable available data, this estimate is probably the most reliable estimate using current economics. It would seem that more oil is appearing all the time and this is due to economics. If and when the price of oil goes up again and stays up, these numbers will increase again.

The Hubbert model and the latest data give a peak oil date of October 23, 2018. Note that I am not saying that the world oil production will peak in October 2018. I am saying that using the Hubbert model and the latest and most accurate monthly world-oil production data the current projection for the peak oil rate is October 23, 2018. If the price of oil remained constant for the next seven years and there were no big technological breakthroughs during this time, this may prove to be an approximately accurate forecast. I don't expect that this will be the case and so this date will probably move again.

THE U.S. GAS SUPPLY ANALOGY TO WORLD OIL PRODUCTION

The U.S. gas supply trends have a lot of relevance to world oil supply behavior. In his 1956 paper, Hubbert made a forecast of U.S. natural gas production, which is reproduced in Figure 5.8 (his figure 22). This showed the nation's gas production also peaking in the early 1970s

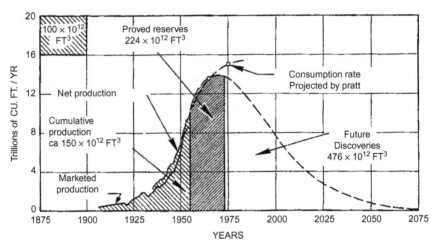

FIGURE 5.8   Hubbert's 1956 Forecast for U.S. Natural Gas Production.

(around1972) at about 14 trillion cubic feet per year (38 billion cubic feet per day), with an ultimate cumulative production (Q_{max}) of 850 trillion cubic feet. His figure also showed a line for the projected gas consumption rate which continues to increase after the nation's gas production peaks and starts to decline. This was a projection using data supplied by Wallace Pratt. Figure 5.9 shows the actual U.S. dry gas production. The units of production on this plot are a billion cubic feet of gas per day, abbreviated as BCF/day. The data shows that the U.S. peaked in 1973 (almost exactly when Hubbert forecast it) at just below 60 BCF/day. His forecast of the peak year was quite accurate even though his forecast of peak rate was low. The consumption of gas continued to increase as his plot showed and the shortfall that occurred after 1973 was met by imports by pipeline from Canada.

In Figure 5.9, the production rate is shown on an annual basis up until 1997 and on a monthly basis from 1997 until the present time (2011). The decline in production after the peak in 1973, however, did not happen according to Hubbert's forecast. The rate began to decline for two years then remained constant for five years. The decline then resumes, finally bottoming out at 44 BCF/day in 1986. Since then, there has been a steady increase in production. The production started to increase more sharply in 2006 and today we are making about 68 BCF/day of total gas including 63 BCF/day of dry gas. This is more gas than the United States was making at the peak in 1973. If the production had declined in a mirror image of the front side of the peak, in 2010 the gas production rate should have been the same as in 1936 when the production was only 6 BCF/day. Instead, more than ten times that amount was produced.

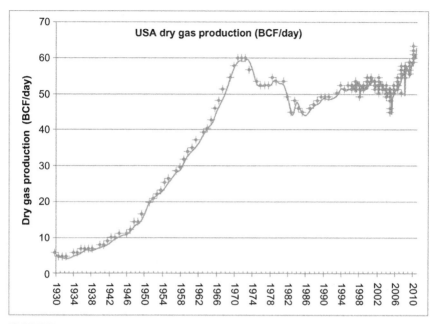

FIGURE 5.9   Actual U.S. Dry Gas Production. *Data Source: EIA.*

The Hubbert method has failed miserably to forecast what has happened on the back side of the curve. The reason for this is economics. Once the nation's gas supply peaked and a gas shortage developed, the price of gas began to increase. The increased price shook loose gas that was previously uneconomic to produce and people began to develop technologies to find and produce coal-bed methane (also known as coal seam gas), tight gas and shale gas. This type of gas is now known collectively as unconventional gas and it accounts for the majority of the current U.S. gas supply. The United States surpassed Hubbert's estimate of the total ultimate recovery of 850 trillion cubic feet of gas in 1997, and so far has produced more than 1,120 trillion cubic feet without reaching any new peak. The production is still increasing. Figure 5.10 shows the gas production repeated with the average annual gas price, which illustrates the effect that price has on the supply. The steep increase in the gas price after the peak has created new supplies of gas, causing the Hubbert model to fail on the back side.

The same thing did not happen to the U.S. oil supply curve because the oil market is a global market. The price for U.S. oil could not be pushed up in the same way that the gas price was pushed up because the United States has always imported a significant portion of its oil and so the U.S. oil producers were constrained to the world price of oil. The world price of oil remained relatively low until recently;

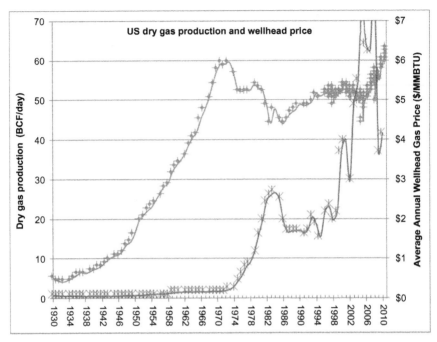

FIGURE 5.10   U.S. Dry Gas Production with Average Annual Wellhead Price. *Data Source: EIA.*

however, apart from Canadian supplies and a small and expensive Liquefied Natural Gas (LNG) market, the U.S. gas market is relatively insular.

This analogy illustrates that economics affects the supply of commodities such as oil and gas. In the case of the world oil supply, the price of oil has recently been pushed up before the peak of the oil production has been reached. This in turn has pushed out the timing of the peak and indeed the total cumulative production. When the world production of oil does eventually peak (as it must), the back side of the curve will not be a mirror image of the front side of the curve as Hubbert had forecast. It probably will not look like the back side of the U.S. gas supply curve either. It will be controlled by economics and technology, which at this stage are a little difficult to predict.

In his 2010 book *When Oil Peaked*, Ken Deffeyes presents several proofs and analogies to show that the production curves must be symmetrical, that the back side of the curve must be a mirror image of the front side. All of these proofs depend on a constant oil price, which clearly will not happen. If and when oil production does peak, the shortage of supply will push the price of oil up and the back side of the curve will not be a mirror image of the front side. The fact that the U.S. gas production curve is not symmetrical is an indication of the fallacy of Hubbert's assumption and the subsequent proofs by Deffeyes. Scenarios can be envisaged where the back side of the world oil production curve might be a mirror image of the front side. At $100/barrel, Gas to Liquids (GTL) technology (see Chapter 6) and Coal to Liquids (CTL) technology (see Chapter 13) is economic. If the world began to shift rapidly to these technologies and they were able to maintain their supply at a high enough rate to keep the price of oil constant, the back side of the oil production curve might then be a mirror image of the front side.

NEW SUPPLIES OF OIL IN THE UNITED STATES AND THE IMPLICATIONS FOR THE WORLD OIL MARKET

As mentioned earlier, an increase in the world oil price has begun to create new supplies of oil in the United States and this is likely to further increase the world supply. There are a number of areas in North America that are being actively explored, and some of these are likely to contribute significant production to the U.S. oil supply in the immediate future. They include: the Canadian oil sands in Alberta and Saskatchewan; the Bakken formation in North Dakota, Montana, Manitoba and Saskatchewan; the Eagle Ford formation in Texas; the Granite Wash in Oklahoma and the Texas Panhandle; the Niobrara in Colorado and Wyoming; the Monterey in California; the Utica in Ohio; and the Shublik formation on the North Slope of Alaska. The large quantities of oil that have been discovered in the deep waters of the Gulf of Mexico are expected to also play a large role.

Consultants Purvin and Gertz have examined the onshore unconventional oil production from the Bakken, Eagle Ford, and Niobrara plays, and they expect that these three alone will be producing 900,000 b/d by 2015 and exceed 1.3 million b/d by 2020. Currently, oil production from the Bakken and Eagle Ford alone is 350,000-400,000 b/d. These forecasts are pretty conservative. They also leave out the other formations listed above, which will also contribute additional quantities of crude oil and liquid production.

The Bakken formation extends from North Dakota and Montana in the United States to Manitoba and Saskatchewan in Canada (see Figure 5.11). The Bakken is sometimes called the Bakken shale, but the oil production primarily comes from the Middle Bakken zone, which is a low permeability, shale-bearing sandy dolomite. The North Dakota Geological Survey has estimated that there are 250 billion barrels of oil in place in the Bakken and the associated Middle Forks formation just in the North Dakotan part of the play alone. It is not unreasonable to believe that there is an estimated

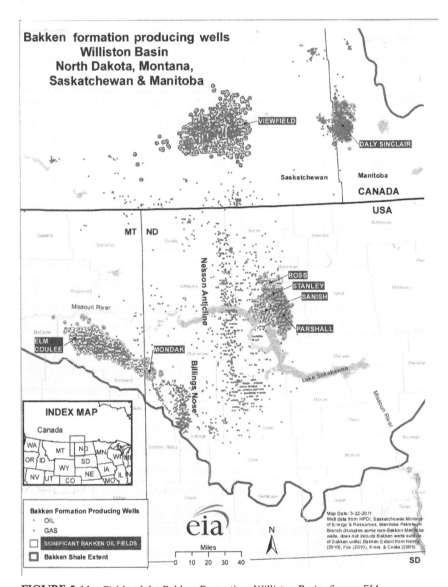

FIGURE 5.11   Fields of the Bakken Formation, Williston Basin. *Source: EIA.*

500 billion barrels of total oil in place in the entire four state/province regions and it is essentially one continuous field. The areas that have been developed there are subdivided into separate fields: the Elm-Coulee in Montana; the Mondak, Parshall, Sanish, Stanley and Ross fields in North Dakota; the Viewfield in Saskatchewan; and the Daly Sinclair in Manitoba. Owing to the low porosity and permeability, the primary recovery is

expected to be very low at about 7%. While this number may seem low, 7% of 250 billion barrels is a lot of oil (17.5 billion barrels). When enhanced oil recovery methods such as carbon dioxide miscible flooding are applied, more of the oil can be recovered. This is not proved oil; however, if 17.5 billion barrels were eventually recovered, it would make this the largest oil field ever discovered in North America and one of the largest oil fields in the world.

A recent EIA report, released in 2011, estimated that 3.5 billion barrels are recoverable from the U.S. part of the basin. Based on their methodology, this is a very conservative estimate mainly because they have only examined the currently leased acreage. This number changes by using the method of volumetrics to determine the total amount of oil originally in place in the middle Bakken formation. The method is known as volumetrics because you identify the oil in place by calculating the volume that it occupies. This method gives an estimate of original oil in place of 250 billion barrels. The following parameters were used to make this estimate: all of the Bakken area in Figure.5.11 covers more than 35 million acres; the thickness of the Middle Bakken zone is 22 feet; the porosity averages 8% and the oil saturation averages 68%; and the value for the formation volume factor (B_o) is 1.3. The result is 250 billion barrels and this is just the oil in place. To get the recoverable oil, you have to multiply by the recovery factor of 7%, which then gives 17.5 billion barrels of recoverable oil by primary recovery means. The Middle Fork formation, which is below the Bakken formation, will give additional oil, and enhanced oil recovery methods will also increase this number. Even this estimate of 17.5 billion barrels is still relatively low.

In May 2011, the largest operator in the Bakken play—Continental Resources—estimated the primary recoverable reserves in that area to be 24 billion barrels of oil. This was a large increase from previous estimates and was attributed to continuing improvements in drilling and production methods and other technological advances. According to an article in the Oil and Gas Journal (6/6/2011), most operators are drilling 18,000-foot to 21,000-foot wellbores that include 9,500-foot laterals and generally apply 18-30 frac stages per well. Continental based its 24 billion barrel estimate on the following assumptions about two areas of continuous oil reservoirs:

- A recoverable oil volume of 500,000 barrels/well
- Middle Bakken and Three Forks act as separate reservoirs (500,000 barrels/reservoir)
- Dual-zone development (both Middle Bakken and Three Forks reservoirs)
- 320-acre spacing/well (4 wells per zone with 8 wells per 1,280-acre spacing unit)

- Estimated Area 1 having 10,314 square miles (6.6 million acres) in the area where the source rocks are thermally mature, and
- Estimated Area 2 having 4,357 square miles (2.8 million acres) in the area where the source rocks are marginally mature but where mature oil has migrated.

In the article in the *Oil and Gas Journal* (6/6/2011) Harold Hamm—Continental chairman and chief executive officer—said the industry has completed 3,600 horizontal wells in the Bakken and plans to add 2,100 wells per year using 170-180 drilling rigs that are actively running. Continental restricts initial production rates on many of its North Dakota wells to minimize natural gas flaring and to maximize the volume of rich gas that can be marketed. "Some wells have initial flowing tubing pressures of more than 3,000 psi for several days, so they could easily have tested at double or more their announced rates, if we have opened them up," Hamm said. As of March 31, Continental had 878,900 net acres leased in the Bakken. As of early May, Continental operated 22 drilling rigs in North Dakota and two in Montana. According to Hamm, six of Continental's 8 best wells during the first quarter of 2011 were in the Williston prospect where Continental has leased 110,000 net acres.

Type curves for the production from typical Bakken wells are shown in Figure 5.12. The average Bakken well makes about 550,000 barrels ranging from 350,000-700,000 barrels. This number is consistent with Continental's assumption above. The huge potential of this play is obvious.

To produce oil from the Bakken, the operators have to drill horizontal wells through the formation and then conduct massive fracture stimulation jobs to get the oil to flow out. Oilmen have known about the oil in the Bakken formation since 1950, but they could not produce very much of it at economic rates until these new horizontal drilling and fracture stimulation methods were deployed in the past 5 years. Prior to that, many of the Bakken wells were poor wells, only marginally economic or subeconomic. The new high-tech Bakken wells cost between $5.5 million to $8.5 million to drill and the initial rates are often more than 2,000 barrels per day. Even though these rates decline quickly, at $100 per barrel for the oil, the wells are now very profitable and there is still a lot of oil to be developed there. Activity in the Bakken has pushed North Dakota's oil production from less than 80,000 barrels per day in 2004 to more than 350,000 barrels per day in 2011. This production will continue to increase. In mid-2011, there were 178 rigs drilling wells in the Bakken play.

The hottest oil play in the United States in 2011 is the Eagle Ford Shale Play in southwest Texas, which is shown in Figure 5.13. It actually has an oil leg on the northern side of the play, a gas leg on the southern side of the play, with a gas-condensate zone in-between those areas. Gas condensate means that it is a gas phase fluid but there are a lot of liquids volatile in the

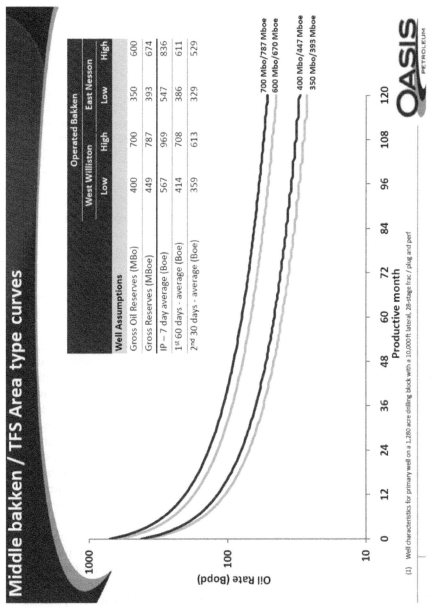

FIGURE 5.12 Type Curve for Typical Bakken Oil Wells. *Source: EIA.*

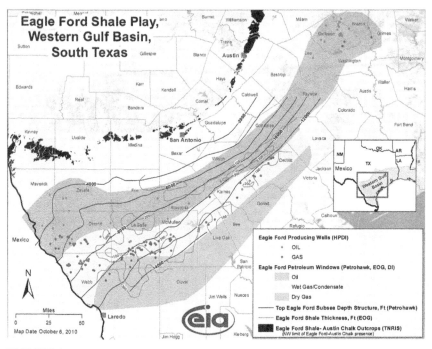

FIGURE 5.13   The Eagle Ford Shale Play in Texas. *Source: EIA.*

gas. When the gas is produced, light oil-like liquid called gas condensate drops out of the gas. The gas condensate can be sold as oil, and the gas is sold separately. Due to the fact that oil prices are very high and gas prices are moderately low, the producers derive much more revenue from the oil and condensate production. At the moment, the producers are concentrating on the oil and gas condensate areas of the Eagle Ford play. There is high hope that this area will eventually be producing 500,000 barrels per day of oil and condensate. In mid-2011, there were 195 rigs drilling wells in the Eagle Ford play making it the hottest exploration target in the nation. The EIA estimates of recoveries from the Eagle Ford are 3 billion barrels of oil and 21 trillion cubic feet of gas. Again, these estimates are conservative.

The Niobrara play is barely getting started, but there are high expectations for this play. The area of current interest is in the Denver-Julesberg basin, just north of Denver, Colorado, extending up into Wyoming and over into Nebraska. This is shown in Figure 5.14. The Niobrara formation actually extends beyond this area up into the Powder River Basin in the northeast of Wyoming, and into the Green River basin in the west of Wyoming. If this play proves successful, it is likely to be extended further afield.

FIGURE 5.14 The Niobrara Play. *Source: EIA.*

Another recent entry into the shale oil exploration is the Utica shale in eastern Ohio (see Figure 5.15). In late July 2011, Chesapeake Energy Company announced the discovery of a major new liquids-rich play in that area. No estimates of total oil and gas in place or recoverable reserves were made but, based on its geo-scientific, petro-physical and engineering research during the previous two years and the results of six horizontal and nine vertical wells it has drilled, Chesapeake announced that its 1.25 million net leasehold acres in the Utica play could be worth $20 billion in increased value to the company. The company's dataset on the Utica Shale includes approximately 2,000 well logs on approximately 200 wells, 3,200 feet of core samples from nine wells and production results from three wells. As a result of its analysis, the company believes the Utica Shale

FIGURE 5.15 The Utica Shale Play in Eastern Ohio. *Source: Chesapeake, modified from Rowan, Ohio Geological Survey.*

will be characterized by a western oil phase, a central wet gas phase and an eastern dry gas phase. It is likely most analogous to the Eagle Ford Shale in South Texas. Chesapeake went so far as to say that it believes that the Utica is economically superior to the Eagle Ford play, and if this proves true, it would make it a very hot exploration target. Chesapeake is currently drilling in the Utica Shale with five rigs to further evaluate and develop its position and anticipates increasing its rig count to eight by the end of 2011 and reaching at least a range of 16-20 rigs by late 2012 and 40 rigs by year-end 2014. Given the 195 rigs currently drilling the Eagle Ford play, this could prove a conservative estimate for the entire play. There will undoubtedly be other such plays in the future that have not been heard of yet.

Out in California, there is another shale formation called the Monterey Shale that contains a lot of recoverable oil, and it is also beginning to heat up. The active area for the Monterey shale play is approximately 1,752 square miles in both the San Joaquin and the Los Angeles Basins (see Figure 5.16). The depth of the shale ranges from 8,000 to 14,000 feet deep and is between 1,000 and 3,000 feet thick. The very thickness of the shale in this play allows for a great deal of oil in place. The Monterey is almost pure shale that is naturally fractured in many places. It has been produced through vertical wells for many years, but with the new horizontal drilling

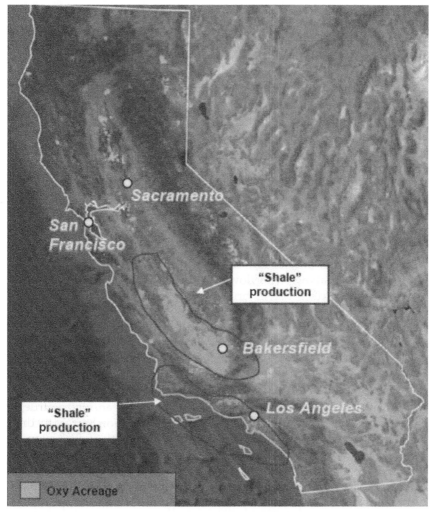

FIGURE 5.16   Monterey Shale Oil Production Area in California. *Source: Occidental Petroleum via EIA.*

and multizone fracture technology, there is increased expectation that much more of the oil can be recovered economically. The recent EIA study has said that there are 15.42 billion barrels of technically recoverable oil in the Monterey shale. Occidental Petroleum (Oxy) has reported that the cost for vertical wells ranges from $2 to $2.5 million and horizontal wells range from $5 to $7 million. They also report that the finding and development costs are between $8 and $18 per barrel depending upon the area of the basin and the quality of the rocks.

The biggest oil deposits in North America are the oil sand deposits in the Canadian province of Alberta, with some of the deposits extending over into Saskatchewan. Figure 5.17 shows the location of these deposits, which are in three areas: Athabasca, Cold Lake, and Peace River. These deposits consist of very heavy viscous oil that is essentially solid tar. The oil does not flow very well, but the shallowest deposits can be mined by digging up the ground containing the oil and processing it in retorts. They remove the sand grains and add hydrogen to the oil to make it better able to flow through pipes and be refined in refineries. The deeper deposits are contacted through wells where steam is pumped down the well to contact the oil and heat it so that it is less viscous and more able to flow into the well to be produced. This process is known as In-Situ Steam Assisted Gravity Drainage, or SAGD. This technique was developed in the 1980s by the Alberta Oil Sands Technology and Research Authority, but its use depended on the needed improvements in horizontal drilling technology. In SAGD, two horizontal wells are drilled in the oil sands, one at the bottom of the formation and another about 15 feet above it. These wells are typically drilled in groups from a central pad. In each well pair, steam is injected into the upper well and the heat melts the heavy oil, which allows it to flow into the lower well from which it is pumped to the surface. SAGD brings high oil production rates and relatively high recovery factors, sometimes more than 50% of the oil in place.

In a 2011 report entitled *Alberta's Energy Reserves 2010 & Supply/Demand Outlook 2011-20*, Alberta's Energy Resources Conservation Board (ERCB) noted that the remaining proved oil sands reserves in the Alberta province are 169 billion barrels out of an initial 177 billion barrels of proved reserves. Cumulative oil produced from the oil sands through 2010 has been 7.5 billion barrels since commercial production began. Of the remaining established reserves, mineable areas contain 34 billion barrels and the in situ SAGD areas contain 135 billion bbl. Alberta's production from mining the oil sands in 2010 was 857,000 b/d, which was higher than the 756,000 b/d produced from the in situ areas, but ERCB expects the in situ oil sands production to surpass mining production by 2015. In 2010, all the bitumen produced from mining and 11% of that produced in situ was upgraded in Alberta to yield 795,000 b/d of Synthetic Crude Oil (SCO). ERCB forecast that SCO production in Alberta will increase

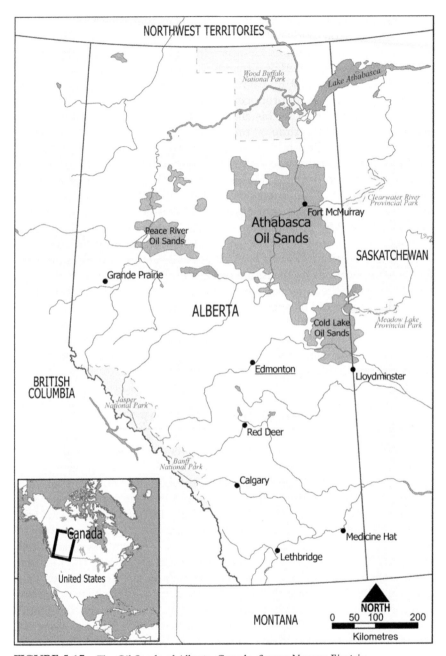

FIGURE 5.17   The Oil Sands of Alberta, Canada. *Source: Norman Einstein.*

to 1.4 million b/d by 2020. ERCB also estimates that Alberta holds 1.804 trillion barrels of oil sands initially in place and that the ultimate recovery potential is 315 billion barrels. Compare that to Saudi Arabia's oil reserves of 263 billion barrels. Alberta's oil sands production is expected to increase to 3.5 million b/d by 2020, up from 1.6 million b/d in 2010. This will make Canada one of the top oil producers in the world.

The entire discussion in this chapter has focused on both the United States and the world's crude oil production. Currently, the United States uses about 18.5 million b/d of petroleum liquids. This is made up of 6 million b/d of local crude oil production, another 3.5 million b/d of condensate and natural gas liquids (giving 9.5 million b/d of liquids), and 9 million b/d of imported crude oil and petroleum products. Obviously, the condensate and natural gas liquids are contributing significant quantities of liquids to the national supply, which is displacing significant quantities of imported crude oil. This will be important for the future.

The unconventional oil and NGLs that have been reviewed in this section have implications for the future world supply of crude. The technology that is increasing the liquids production in the United States has important implications for world oil supply. The United States is not the only country that has large unconventional resources of oil and gas. As this technology is perfected and the world needs more oil supplies, the methods of production that are being used in the United States will be exported to the rest of the world, which will add to the world supply picture.

THE WORLD PICTURE

The top 20 countries in the world in terms of proved oil reserves are shown in Table 5.1.

The number two and number three countries on this list, Venezuela and Canada, owe the majority of their reserves to their heavy oil sands. The total reported proved reserves of crude oil for the entire world is 1.3 trillion barrels, although certain peak oil theorists do not believe that these reserves are correct. They believe that some OPEC members inflate their reserves so that they can get a higher production allocation. The fact that OPEC is still restricting their members' production is evidence that there is spare capacity in the world. If the world's productive capacity was at 100% or declining, there would no longer be a need for a cartel like OPEC because cartels only exist in order to ration supply in order to keep the price at a certain level. If the members of OPEC were all producing at maximum capacity, they would have no reason to inflate their reserves so that they could get a higher allowable production. On that same note, in May 2011 an influential Saudi Prince—Al-Waleed bin Talal—said that he wants oil prices to drop so that the United States and Europe don't accelerate

TABLE 5.1 Crude Oil Proved Reserves (Billion Barrels)

1	Saudi Arabia	262.6
2	Venezuela	211.2
3	Canada	175.2
4	Iran	137.0
5	Iraq	115.0
6	Kuwait	104.0
7	United Arab Emirates	97.8
8	Russia	60.0
9	Libya	46.4
10	Nigeria	37.2
11	Kazakhstan	30.0
12	Qatar	25.4
13	United States	20.7
14	China	20.4
15	Brazil	12.9
16	Algeria	12.2
17	Mexico	10.4
18	Angola	9.5
19	Azerbaijan	7.0
20	Ecuador	6.5

Data Source: EIA.

efforts to wean themselves off his country's supply. Prince Al-Waleed is the grandson of the founding king of modern Saudi Arabia (King Saud) and is listed by Forbes as the 26th richest man in the world. He said that the oil price should be somewhere between $70 and $80 a barrel, rather than over $100 a barrel. "We don't want the West to go and find alternatives, because, clearly, the higher the price of oil goes, the more they have incentives to go and find alternatives," says Prince Al-Waleed. If Saudi Arabia were facing their "twilight in the desert," a decrease in their productive capacity as claimed by Matthew Simmons, then there would not be a need to take such an attitude. They would be pushing prices as high as possible so that they could maximize their revenue from a declining resource.

The world's top 30 producing countries are listed in Table 5.2. Production rates should have an approximate correlation to reserves; the higher

TABLE 5.2 Crude Oil Production Rates in 2010
(Thousands of Barrels/day)

1	Russia	9674
2	Saudi Arabia	8900
3	United States	5474
4	Iran	4080
5	China	4076
6	Canada	2734
7	Mexico	2621
8	Nigeria	2455
9	United Arab Emirates	2415
10	Iraq	2399
11	Kuwait	2300
12	Venezuela	2146
13	Brazil	2055
14	Angola	1939
15	Norway	1869
16	Algeria	1729
17	Libya	1650
18	Kazakhstan	1525
19	United Kingdom	1233
20	Qatar	1127
21	Azerbaijan	1035
22	Indonesia	953
23	Oman	865
24	Colombia	786
25	India	751
26	Argentina	642
27	Malaysia	554
28	Egypt	523
29	Sudan	511
30	Ecuador	486

Data Source: EIA.

the reserves, the higher the production rate. In this respect, the United States does very well. It is producing 9.65 million b/d from only 20 billion barrels of reserves. Countries such as Canada and Venezuela clearly have the capacity to increase their production, although most of their reserves are tied up in very heavy oil. This limits their production rates because the viscosity of the oil is so low. If Saudi Arabia really has 263 billion barrels of reserves, they too could increase their rates substantially and they have done this at times. Saudi Arabia, however, has elected to operate as the world's swing producer, increasing their production when the world demand increases and decreasing production when demand decreases. The Saudis would be happy to increase production if the world demanded it, but they are aware that when the price gets too high, demand decreases and the world starts looking for alternatives. This is why Prince Al-Waleed made the statements that he did. Saudi Arabia would love to sell more oil to the United States and Europe and they have the capacity to do it. Matthew Simmons' claim in *Twilight in the Desert* that the oil production of Saudi Arabia has peaked and will soon begin to decline rapidly can only make sense if Saudi Arabia is vastly overstating its reserves. Simmons does try to argue that the Saudis have less oil than they claim but their reserves would have to be about one-tenth of what they claim before they would have trouble keeping up. The United States is producing about the same amount of oil as Saudi Arabia with less than one-tenth of their reserves.

There are other countries that also have the capacity to increase their production, and one way to gauge this is to calculate their reserve life, the number of years it would take to produce their reserves if they keep producing at the current rates. To obtain this number, you divide their reserves (in barrels) by their producing rate (in barrels/year) to get their reserve life in years. Table 5.3 shows the reserve life of the top 30 countries in this category. Any country that has over 30 years of reserve life probably has the potential to increase their production. Number one on the list is Iran. Their production has been somewhat restricted by international sanctions and the lack of access to crucial technology. The United States does not appear in the top 30 on this list because the United States reserve life is only 5.7 years. That is why the catch phrase of "Drill, baby drill" is so apt for the United States. They cannot keep producing oil at its current high rates relative to its reserves unless they are continually drilling new wells and finding more oil. As soon as the United States stops drilling, its production rate plummets and its balance of payment problems gets much worse.

Even though the United States is the third largest oil producer in the world, it is also the largest oil importer in the world and has been so for a long time. Figure 5.18 shows the monthly imports of crude oil into the United States since 1973. There had been a steady increase in imports from 1983, when 4 million b/d were being imported, to August 2006 when 13.4 million b/d were imported. Since then the trend has been back down

TABLE 5.3 Reserve Life (Years) of Oil Producing Countries

1	Venezuela	269.62
2	Canada	175.61
3	Iraq	131.32
4	Kuwait	123.86
5	United Arab Emirates	110.97
6	Iran	91.99
7	Saudi Arabia	80.84
8	Libya	77.08
9	Qatar	61.71
10	Kazakhstan	53.89
11	Netherlands	42.47
12	Nigeria	41.51
13	Ecuador	36.69
14	Chad	32.57
15	Yemen	31.93
16	Bolivia	29.69
17	Sudan	26.79
18	Gabon	24.01
19	Egypt	23.05
20	Brunei	22.13
21	Australia	20.88
22	India	20.72
23	Poland	20.45
24	Trinidad and Tobago	20.31
25	Peru	20.08
26	Malaysia	19.78
27	Algeria	19.34
28	Syria	18.66
29	Romania	18.65
30	Azerbaijan	18.54

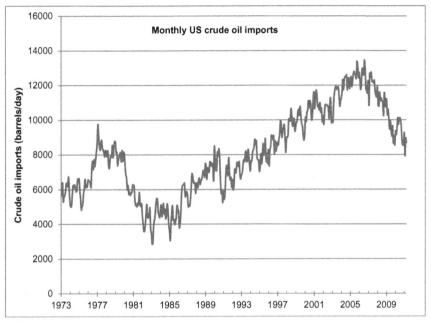

FIGURE 5.18   U.S. Monthly Crude Oil Imports. *Data Source: EIA.*

due to increased production and reduced demand in the United States. In April 2011, only 8.7 million b/d were imported. The United States does have the capacity to reduce this import bill to zero by using its potential for increased production of oil, gas, coal and electricity. This is discussed in more detail in Chapter 16. Even if that goal proves too ambitious, another option would be to reduce its crude imports to 4 million b/d and import all of this from its NAFTA partners, Canada and Mexico. This would make sense if the NAFTA partnership is to be strengthened.

THE FUTURE OF OIL

The world still has plenty of oil both in production capacity and in future reserves. Any sustained increase in the price of oil will lead to increased reserves and production. The world's oil production will not peak in the next 10 years. When it does eventually peak, it will not decline rapidly in the manner that the Hubbert Peak Oil Model predicts. The Hubbert bell-shaped production curve is only valid when there is a constant oil price and a constant technology environment; however, price and technology increases also increase the ultimate recoverable reserves, push out the day of reckoning for the peak in oil production and extend the tail of the back side of the curve.

6

Natural Gas

Natural gas is another form of hydrocarbon that is strongly related to oil. Normally, natural gas is produced from reservoirs that are separate and distinct from oil reservoirs, although there is always some gas produced form oil reservoirs as well. In the United States, as well as the rest of the world, natural gas is widely used for: space heating; to generate electricity; as an industrial fuel source; as a feedstock for petrochemicals; and, as a feedstock to make gasoline and other transportation fuels. In Chapter 5, gas production in the United States was briefly discussed as an analogy for world oil production. Even though gas production had already peaked in the United States in early 1973, it has reversed this decline and has now reached a new production peak and is still increasing. Natural gas is the third most widely used energy source in the world, just below coal. It has a very bright future with increasing importance, both in the world and in the United States.

Natural gas consists primarily of methane but can have other components. Some of these components are beneficial to its heat content such as ethane, propane, butane, and other longer chain hydrocarbons. Other components are considered impurities because they are undesirable components such as nitrogen, carbon dioxide, and hydrogen sulfide. They must be removed before the gas can be sold. Due to the fact that all oil and gas reservoirs contain residual water saturation, gas must also be dried before it can be put in a pipeline to be transported to market. Depending on the reservoir and production conditions, the produced gas can contain up to 150 pounds of water per million cubic feet of gas. The usual pipeline specifications are that gas must contain no more than 7 pounds of water per million cubic feet before it can be transported by pipeline. If this specification is not met, it leads to corrosion in the pipeline as well as water pooling in low spots in the line. Impurities such as carbon dioxide and hydrogen sulfide are acidic gasses and, if they are not removed, they also lead to corroded lines as well as other dangerous conditions.

Natural gas also tends to contain small amounts of ethane, propane, butane, and other longer chain hydrocarbons. These molecules have higher heat content than methane, so there is some benefit in letting these components remain in the gas; however, this can lead to some disadvantages which necessitate their removal. First, under certain conditions, the heavier hydrocarbons can also drop out of the gas during transportation in the pipeline and then they pool in low spots impeding the flow of the gas. They can also cause the gas to burn hotter than methane which, for certain applications, is not desired because it can lead to degradation of the equipment. Consequently, the longer chain hydrocarbons are usually reduced to meet pipeline specifications for gas.

THE WORLD PICTURE

Worldwide natural gas production has increased from 97 billion cubic feet (BCF) per day in 1970 to 309 BCF per day in 2010, as shown in Figure 6.1. During that time, the annual growth rate has been a steady 2.75%. The only decrease in supply and demand occurred due to the world financial crisis in 2008. Natural gas meets 21.4% of the world energy

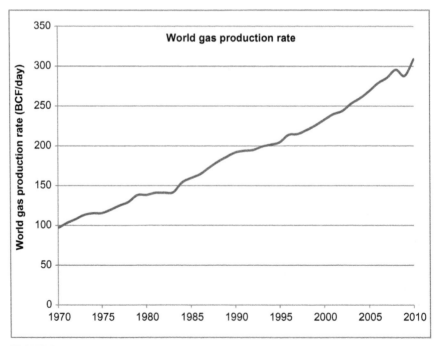

FIGURE 6.1   World gas production rate.

supply, third only to oil and coal. In the United States, it is responsible for 24.3% of the energy supply, second only to oil. The largest gas-producing countries in the world, with their daily production rate, are shown in Table 6.1. Their reserves are shown in Table 6.2 and their reserve life is shown in Table 6.3. Remember, reserve life is calculated by dividing the reserves by the daily production rate and converting the answer (given in days) to years.

From these figures, it can be seen that the United States is the largest gas producer in the world and the sixth largest in reserves. It does not appear on the list of top countries in terms of reserve life because its reserve life is less than 13 years, which puts it thirty-ninth on the world list. This list, however, is based on proved reserves and later you will see that the United States has a lot more gas that remains unproven because many more wells have to be drilled to prove these additional reserves.

As with oil reserves, all the countries on the reserve life list that have more than 25 years supply, have the capacity to increase their production if and when the world needs it. There are many more countries that have large spare gas production capacity than have spare oil production capacity. Natural gas remains an underutilized energy source. As in the United States, there is a lot of gas around the world that is still waiting to be discovered and produced.

The trade in natural gas is not as active as oil. This is because the most cost effective way to transport gas is by pipeline. This limits the gas import/export trade. There is a modest but growing liquefied natural gas (LNG) trade, but most of the international trade is still done by pipeline where possible. There are large pipelines supplying natural gas to Europe from Russia and smaller lines running from Algeria to Europe and from Canada to the United States and also between various South American countries. Most gas is produced and used locally.

LIQUEFIED NATURAL GAS

The import/export trade in natural gas is not nearly as large as the oil trade because it is more difficult to export gas. If the importing country lies close to the exporting country and they are connected by land, the gas can be transported by pipeline. Large bore pipelines exist between Russia and Europe, Algeria and Europe, Canada and the United States and between some South American countries, particularly Bolivia, Brazil, Chile, and Argentina. While oil can be readily shipped over the ocean by tanker, natural gas must first be liquefied by refrigeration to be shipped the same way. This is expensive and adds to the cost of the gas. Nevertheless, there is a growing trade in LNG.

TABLE 6.1 Largest Gas Producers in the World (BCF per Day).

1	USA	59.12
2	Russia	56.98
3	Canada	15.46
4	Iran	13.40
5	Qatar	11.29
6	Norway	10.29
7	China	9.36
8	Saudi Arabia	8.12
9	Indonesia	7.93
10	Algeria	7.78
11	Netherlands	6.82
12	Malaysia	6.43
13	Egypt	5.93
14	Uzbekistan	5.72
15	United Kingdom	5.53
16	Mexico	5.35
17	United Arab Emirates	4.94
18	India	4.92
19	Australia	4.87
20	Trinidad & Tobago	4.10
21	Turkmenistan	4.10
22	Argentina	3.88
23	Pakistan	3.82
24	Thailand	3.51
25	Nigeria	3.25
26	Kazakhstan	3.25
27	Venezuela	2.76
28	Oman	2.62
29	Bangladesh	1.94
30	Ukraine	1.79

TABLE 6.2 Largest Gas Reserves in the World (TCF).

1	**Russia**	**1580.77**
2	Iran	1045.67
3	Qatar	894.22
4	Turkmenistan	283.58
5	Saudi Arabia	283.06
6	USA	272.51
7	United Arab Emirates	212.97
8	Venezuela	192.70
9	Nigeria	186.89
10	Algeria	159.06
11	Iraq	111.87
12	Indonesia	108.40
13	Australia	103.12
14	China	99.15
15	Malaysia	84.65
16	Egypt	78.05
17	Norway	72.11
18	Kazakhstan	65.18
19	Kuwait	63.00
20	Canada	61.00
21	Uzbekistan	55.08
22	Libya	54.70
23	India	51.22
24	Azerbaijan	44.86
25	Netherlands	41.45
26	Ukraine	33.02
27	Pakistan	29.09
28	Oman	24.37
29	Vietnam	21.79
30	Romania	21.02

TABLE 6.3   Gas Reserve Life (Years).

1	**Iraq**	**2534**
2	Qatar	217
3	Iran	214
4	Venezuela	191
5	Turkmenistan	190
6	Nigeria	157
7	Kuwait	154
8	United Arab Emirates	118
9	Libya	98
10	Saudi Arabia	95
11	Azerbaijan	84
12	Yemen	78
13	Russia	76
14	Vietnam	66
15	Australia	58
16	Algeria	56
17	Kazakhstan	55
18	Romania	54
19	Ukraine	50
20	Peru	49
21	Indonesia	37
22	Malaysia	36
23	Egypt	36
24	Syria	33
25	Poland	29
26	China	29
27	Brazil	29
28	India	29
29	Myanmar	28
30	Uzbekistan	26

In order to be liquefied, natural gas must be cooled to about −259 °F at atmospheric pressure. This results in the condensation of the gas into liquid form, which takes up about one six hundredth (0.167%) of the volume of natural gas vapor. Liquefaction also has the advantage of removing the heavier hydrocarbon molecules (particularly ethane, propane, and butane) and impurities such as nitrogen, carbon dioxide, sulfur, and water. This results in LNG that is almost pure methane.

LNG is typically transported by a specialized tanker with insulated walls. It is maintained in liquid form by auto-refrigeration, a process in which the LNG is kept at its boiling point so that any heat that is conducted through the tanker walls is countered by the latent heat of vaporization, energy that is lost from the LNG as it vaporizes. The vapor is vented out of storage and then burned to provide motive power for the tanker. Once it is delivered to a port, the LNG must be vaporized again before it can be transported and burned.

There is some misplaced public concern about the safety of LNG facilities, misplaced because they actually have a remarkably good safety record over a long period of usage. In its liquid state, LNG is not explosive and cannot burn. For LNG to burn, it must first vaporize, then mix with air in the right proportions—(5-15%) and then be ignited. In the case of a leak, LNG vaporizes rapidly, turning back into a gas (methane) and mixing with air. If this mixture is within the flammable range, there is risk of ignition which, if confined, could create an explosion. If the leaking gas is vented to an unconfined space, it can burn but not explode.

There have never been any explosions or significant accidents aboard LNG tankers which have sailed more than 100 million miles without incident. In the past 65 years, there has only been one significant accident at an LNG re-vaporization facility. On October 6, 1979, at the Cove Point LNG facility in Lusby, Maryland, a pump seal failed. This allowed gas vapors to enter an electrical conduit. The explosion that followed would have been prevented by the automatic activation of a circuit breaker, but a worker had switched off the circuit breaker and the gas vapors ignited. One worker was killed, another was severely injured, and there was major damage to the building. This accident led to tougher safety codes for the operation of LNG facilities in the United States.

A more serious accident occurred on January 19, 2004 at one of the natural gas liquefaction units in Skikda, Algeria. A leak occurred in the hydrocarbon refrigerant system, forming a flammable vapor that was sucked into the inlet of the combustion fan of a steam boiler. The hydrocarbon refrigerant acted as increased fuel to the boiler, causing increased pressure within the steam-generating equipment. The rapidly rising pressure quickly exceeded the capacity of the boiler's safety valve and the steam drum ruptured, tearing apart the boiler. The flames from the ruptured boiler ignited the leaking refrigerant gas, which was confined by the equipment and structures in the area, producing an explosion and a fire. The explosion, along with the shrapnel from the ruptured steam drum, caused further damage to the piping and pressure vessels in the immediate area, leading to additional flammable fluid release. The fire took 8 h to extinguish. The explosions and fire destroyed a portion of the petrochemical complex and caused 27 deaths and injuries to 72 people. No one outside the plant was injured nor were the LNG storage tanks or the marine facilities damaged by the explosions. Even though the accident occurred in Algeria, the U.S. Federal Energy Regulatory Commission (FERC) and the U.S. Department of Energy (DOE) conducted their own investigation of the accident and a joint report was issued in April 2004. The findings of the report were that:

- There were ignition sources in the process area
- There was a lack of "typical" automatic equipment shutdown devices required by U.S. LNG design codes, and
- There was a lack of hazard detection devices.

There are currently 10 LNG unloading and re-vaporization facilities in operation in the United States (see Figure 6.2). Sixteen more have been approved for construction and, of these, two are currently under construction. Many of these were approved prior to new supplies of natural gas becoming available in the United States, and it is unlikely that all of them

FIGURE 6.2   LNG import facilities in the United States. *Source: EIA.*

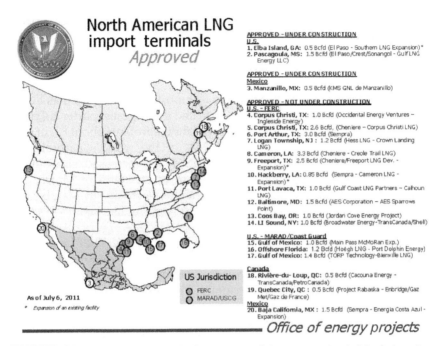

FIGURE 6.3 New LNG import facilities approved for construction in North America. *Source: U.S. Office of Energy Projects.*

will actually be built. Some of them might be converted into LNG export facilities. Figure 6.3 shows the location of these 16 facilities and 4 others located in Mexico and Canada. Due to the oversupply of the natural gas market in the United States, some LNG that is currently being delivered to the U.S. LNG terminals is now being re-exported. These re-exports occur when foreign LNG shipments are offloaded into above-ground U.S. storage tanks located on-site at the terminals. They are then subsequently reloaded onto tankers for delivery to other countries.

Due to the reduction of LNG imports into the United States, some of the import facilities are being used for temporary storage of LNG. Re-exports of foreign sourced LNG from U.S. LNG terminals exceeded 12 BCF in January 2011, equivalent to about 30% of U.S. LNG import volumes during that month, the highest volume since the start of re-export services in December 2009. There are currently three U.S. LNG terminals that have been granted federal approval to re-export LNG: Freeport in Texas, and Sabine Pass and Cameron in Louisiana.

U.S. LNG imports and deliveries from terminals to the domestic market are decreasing due to increased domestic natural gas production and low average U.S. natural gas prices that are well below the prices in other major natural gas markets with the capability to import LNG. Typically, low utilization at these terminals has created available LNG storage capacity in the terminals' storage tanks. Re-exportation of LNG lets marketers and suppliers store gas while waiting for price signals and before delivering their LNG to the higher-paying markets in Asia, Europe, and South America.

Currently, about 25–30 BCF per day is imported and exported as LNG in the world. The largest importer is Japan and the largest exporter is Qatar (discussed later in this section). 25 BCF per day represents about 8% of the world's natural gas production and 26% of the world's natural gas imports/exports. Japan's demand for LNG has increased since 2011 due to the loss of nuclear power plants in the tsunami of March 11, 2011. Japan consumed more LNG to generate electricity in March-July 2011 than over the same period in prior years. On average, Japan's electric power companies used more than 6 BCF per day (see Figure 6.4) to produce electricity during the first 5 months of 2011. Only 19 of Japan's 54 commercial nuclear facilities (or fewer than 18 GW out of a total commercial nuclear capacity of 49 GW) were in operation in the period after the tsunami. As a result, LNG consumption at power companies had increased 30% in May 2011 compared to May 2010.

Overall, imports of LNG in Japan have been trending up since 2003. As of April 2011, Japan's total LNG imports accounted for more than 10 BCF per day (see gray area in Figure 6.5) of natural gas supply, most of which are used for power generation. The average price paid for LNG in Japan in 2011 exceeds $13 per million British thermal units (BTU), or about three times the price of natural gas at the Henry Hub in the United States.

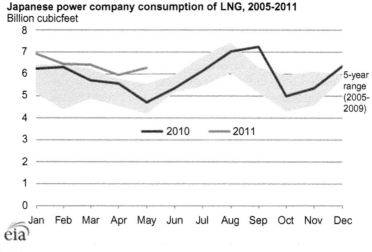

FIGURE 6.4 Japan power company consumption of LNG. *Source: U.S. Energy Information Administration based on the Japan Federation of Electric Power Companies data.*

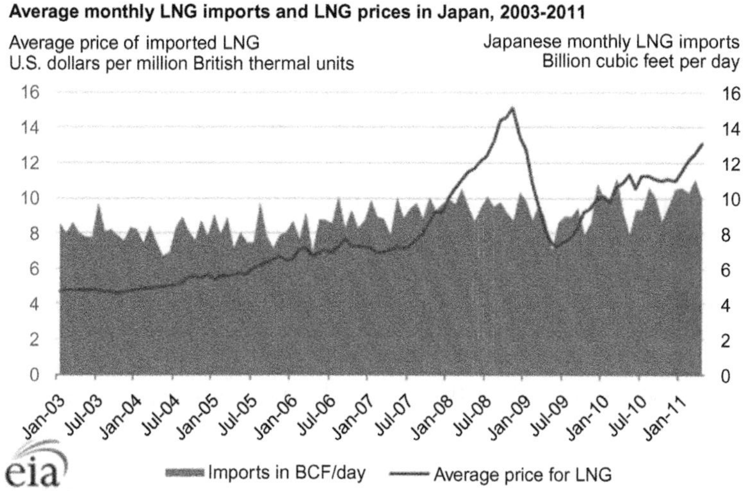

Average monthly LNG imports and LNG prices in Japan, 2003-2011

Average price of imported LNG
U.S. dollars per million British thermal units

Japanese monthly LNG imports
Billion cubic feet per day

eia

▬▬ Imports in BCF/day   ▬▬ Average price for LNG

FIGURE 6.5   LNG imports and prices into Japan. *Source: U.S. Energy Information Administration and Bloomberg.*

The price for LNG (light gray line on Figure 6.5) had peaked at $15 per million BTU in 2008, and then fallen to $7 per million BTU in 2009. Since then, the price has been trending back up and now stands at the current $13 per million BTU due to the increased demand in Japan.

Qatar, the largest LNG producer, exported 1800 BCF of LNG in 2009 (5 BCF per day), about 20% of the total global trade based on an analysis in CIA's recently released Qatar Country Analysis Brief. Its annual LNG exports are equivalent to 8% of U.S. annual marketed natural gas production. Qatar has 14% (894 trillion cubic feet (TCF)) of the world's estimated proved natural gas reserves. It began exporting LNG in 1997 and increased exports to current levels in less than 15 years. Japan, South Korea, and India accounted for 57% of Qatari LNG exports in 2009. European markets including Belgium, the United Kingdom, and Spain imported an additional 33% of Qatari LNG in 2009. North American LNG imports have been relatively low.

In 2010, Qatar exported an estimated 46 BCF of LNG directly to the United States (11% of total U.S. LNG imports) and also exported an estimated 74 BCF to the Canaport LNG terminal in Nova Scotia, Canada, most of which served as supply for New England. Very large investments in infrastructure underpin Qatar's LNG export growth. It has 13 operating LNG trains (liquefaction and purification facilities) with a total LNG capacity of 3400 BCF per year. Five of these trains were added in 2009 and 2010. The most recent addition started commercial service in November 2010 and has an annual production capacity of 380 BCF per year, currently the largest-capacity production train in existence.

NATURAL GAS PRODUCTION
IN THE UNITED STATES

The U.S. natural gas production history was discussed briefly in Chapter 5 and illustrated with production plots in Figures 5.13 and 5.14. U.S. gas production peaked in 1973 at 60 BCF per day before declining to 44 BCF per day in 1986. Since that time, the production has been increasing again and in 2010 the United States had exceeded the 1973 peak and now stands at 63 BCF per day (this only counts the dry gas produced and marketed). Moreover, on the back side of the production curve, the production rate has not followed the bell-shaped curve forecast by Hubbert's peak oil theory. Gas prices (discussed later) also increased sharply after the production peak of 1973. After this time, the shortfall in demand was partly met by imports from Canada and some limited LNG imports from Trinidad and Qatar. This is shown in Figure 6.6.

The big increase in production since 1986 has been due to coal-bed methane gas, tight gas, and shale gas. This type of gas is now known collectively as unconventional gas and it accounts for the majority of the current U.S. gas supply. Note that in Figure 6.6 it also shows that natural gas imports also started expanding in 1986 when unconventional gas began its expansion to produce a rapid increase in the natural gas supply in the

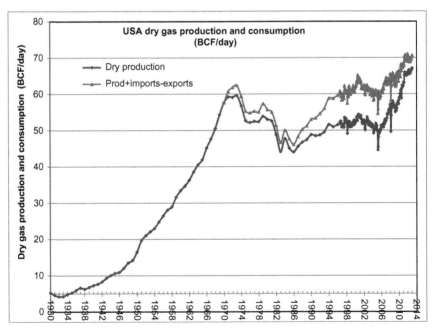

FIGURE 6.6   U.S. dry gas production and consumption (BCF per day). *Data Source: EIA.*

United States. The steepest increase in the production of gas began in 2006 when the Barnett shale in Texas became economic due to horizontal drilling and multizone fracture stimulation techniques. This also allowed many other shale gas areas to be developed in a similar manner.

The location of conventional gas fields in the lower 48 states is shown in Figure 6.7. Conventional fields have relatively high porosity and permeability, and the fields are usually developed using vertical wells. Up until 1986, all U.S. gas production came from these types of fields and they still produce a considerable amount of gas. This includes gas produced from offshore fields as shown in Figure 6.8. Most of the offshore production comes from the Gulf of Mexico.

Beginning in 1986, gas began to be produced in significant quantities from various coal-bed methane fields shown in Figure 6.9. Many types of coal have significant quantities of methane adsorbed onto the surface of the coal, which is the bane of coal miners. The methane is maintained on the surface of the coal by pressure. When the pressure is released, the gas desorbs from the coal and can flow through the coal mine or into a well. This can cause underground explosions and asphyxiation of underground miners. It was natural for gas producers to drill wells into coal seams and attempt to produce the gas through vertical wells. It took a lot of experimentation before they could get the gas to flow at commercial rates. Fracture stimulation of the coal bed is usually necessary before the well can become commercial. This technique is discussed in more detail in the next section because it is a vital technique that is common to all unconventional gas.

Coal-bed methane wells were at first quite unproductive. The wells flowed at very low rates. Even with fracture stimulation and under-reaming of the well-bore, the rates were relatively low—less than 1 million cubic feet of gas per day. Conventional gas wells tend to be much higher than that. A second problem was that many coal-bed wells produce a lot of water before any gas starts desorbing from the coal. Producers might typically spend 1 year dewatering the wells before they start producing much gas. This affects the economics significantly. Fortunately, most of the coal seams can be accessed with shallow wells which are cheap to drill and bring into production.

The largest coal-bed methane area has been the San Juan Basin field in northwest New Mexico and southern Colorado. This field, however, has become somewhat depleted and its production is now declining. At its height, the San Juan Basin field was producing about 4 BCF per day, but it is now making about 2 BCF per day, split between New Mexico and Colorado. The largest current coal-bed methane producing state is Wyoming anchored by the Powder River Basin Field, which is currently producing about 1.5 BCD per day; however, it too has begun its decline. Overall, while coal-bed methane is still contributing a robust 5 BCF per

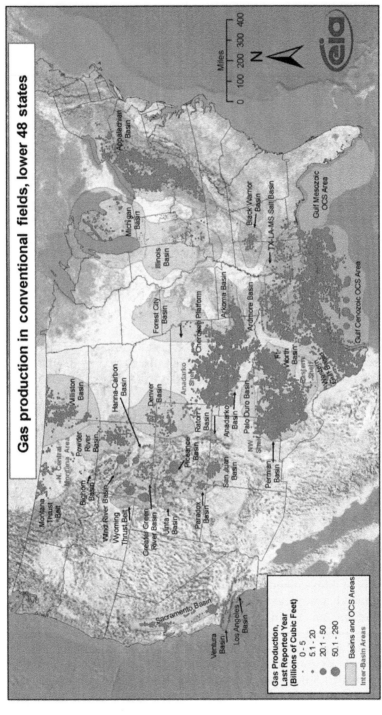

Source: Energy Information Administration based on data from HPDI, IN Geological Survey, USGS
Updated: April 8, 2009

FIGURE 6.7   Gas production in conventional gas fields (lower 48 states). *Source: EIA.*

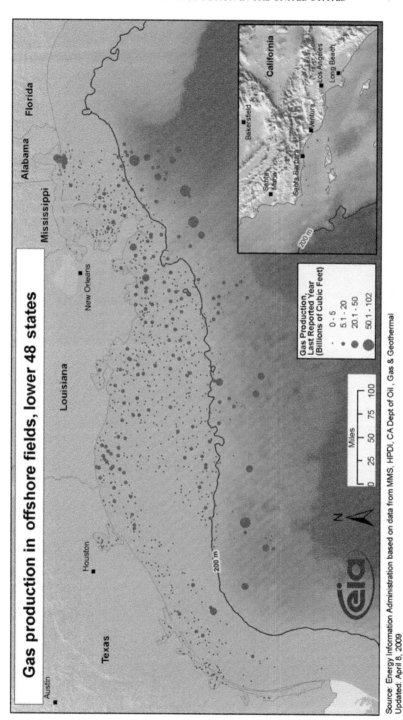

Source: Energy Information Administration based on data from MMS, HPDI, CA Dept of Oil, Gas & Geothermal
Updated: April 8, 2009

FIGURE 6.8 Gas production in the Gulf of Mexico and the West Coast of California. *Source: EIA.*

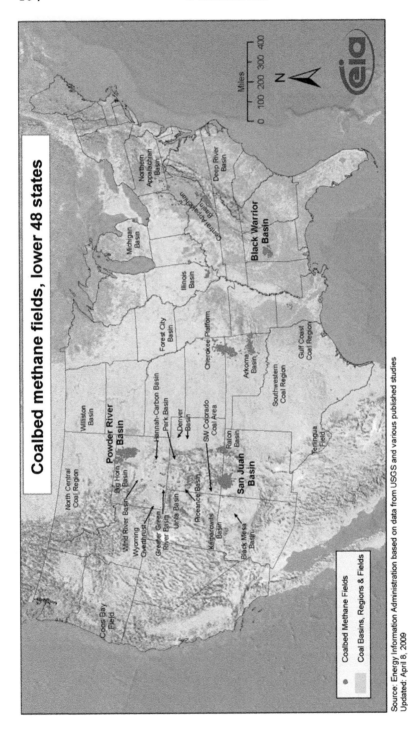

Source: Energy Information Administration based on data from USGS and various published studies
Updated: April 8, 2009

FIGURE 6.9 Gas production in coal-bed methane fields (lower 48 states). *Source: EIA.*

day to the U.S. natural gas supply, it is likely to be a decreasing supply source in the future. However, new sources of coal-bed methane are being developed in other countries, such as Australia, China and Russia. In Australia it is called coal seam gas and large supplies of this gas are being developed in the northeast state of Queensland. Most of this gas is being converted to LNG for the export market.

Another major source of unconventional gas in the United States is what has come to be called 'tight gas' Figure 6.10 shows the location of the tight gas plays in the United States. There are many of these plays involving a lot of gas. They received a large boost from the success of the Lance and Mesaverde formations in the Jonah and Pinedale fields in Wyoming, starting about 1990. Gas was initially discovered in the Pinedale Anticline in 1955, but was not successfully produced until modern drilling and hydraulic fracturing techniques were developed. The Pinedale Anticline is located within Sublette County, Wyoming, on a narrow, diagonal strip of land that stretches from just outside the Pinedale town limits running along U.S. Hwy 191 for about 30 miles southeast towards Rock Springs. While it is a large narrow anticline, the gas bearing formations are very tight. This means that the permeability of the reservoir rock containing the gas is very low (about 1-10 micro-darcies), and the gas cannot flow out of the rock at fast enough rates to be commercial. When you spend $5 million to drill and complete the well, the gas has to flow at rates higher than 1 million cubic feet of gas per day (depending on the gas price) before you start to get a return on your investment. The early Pinedale wells were flowing at less than one tenth of that rate.

In 1979, the Davis Oil Company drilled a couple of wells on a graben next to the Pinedale Anticline and produced gas from the Lance formation at low to moderate rates; however, they did not consider the rates encouraging enough to continue the exploration. The field was given the name Jonah field but it was not considered commercial. In 1990, Davis sold their wells to a small local company called McMurry Oil Company, led by an entrepreneur called Mick McMurry. He had little experience in the oil and gas business but was eager to learn. McMurry drilled a couple of additional wells and found that by fracture stimulation they could get encouraging rates from the wells. They laid some small bore gas lines down to the Opal gas hub and brought their wells into production. They drilled more wells and continued to improve their fracture stimulation techniques and gradually acquired adjoining leases around the Jonah field and the Pinedale Anticline. Eventually, they were able to prove up about 8 TCF of gas at Jonah and 40–50 TCF at Pinedale. These numbers make both these fields giant gas fields and world class by any standards. The tight reservoirs, however, had proven to be technically difficult to drill and stimulate. The reservoir has over 5000 feet of vertical pay interval, but the productive sands are discontinuous and lenticular. There were considerable technical

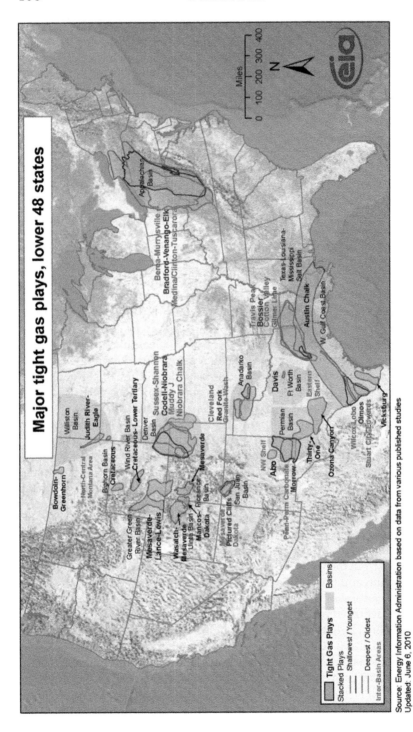

FIGURE 6.10 Gas production in tight gas plays (lower 48 states). *Source: EIA.*

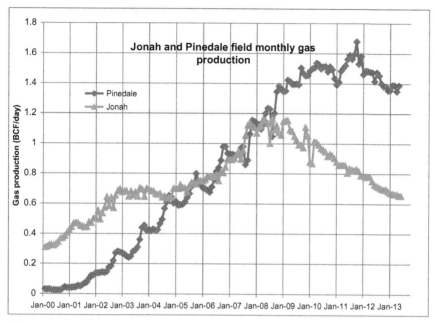

FIGURE 6.11 Pinedale and Jonah gas fields monthly gas production. *Data Source: Wyoming Oil and Gas Conservation Commission.*

difficulties to overcome, but the result was a major new source of gas for the United States.

The production data of the two fields is shown in Figure 6.11. In 2007, the combined production surpassed 2 BCF per day and they continue to be produced at these rates. Jonah's production has been declining since 2008, but this is due to a lack of drilling in the field because of low gas prices. McMurry eventually sold its operating interests in the fields, and it is now produced by a group of companies led by EnCana, Shell, Questar, and Ultra. McMurry's success opened up many tight gas bearing formations around the country, as shown in Figure 6.10. There is an estimated 500 TCF that will be eventually produced from these "tight" formations.

Meanwhile, another tight gas play was independently developing in Texas in the Barnett shale around Fort Worth and areas to the west. The pioneer here was George Mitchell and his company, Mitchell Energy Company. Mitchell has had gas-producing leases in the Fort Worth Basin since the 1950s, but he began to test wells in the Barnett shale around 1980. He was unable to make any commercial wells in the formation for the first 20 years of work. The Barnett shale was even tighter than the Lance formation that McMurry was drilling in Wyoming. The permeability of the Barnett is measured in nano-darcies and nothing like that had ever been successfully produced. Mitchell tried many different methods of stimulating the Barnett shale over a long period of time to try and coax commercial quantities of gas from the formation. Eventually, he was able to make it economic by drilling horizontal wells through the formation followed by many large fracture stimulation jobs using water-based fluids.

The average Barnett well now makes more than 2 million cubic feet of gas per day and can recover more than 5 BCF of gas per well. Mitchell eventually sold his company for about $3.5 billion. George Mitchell initiated a revolution in natural gas supply in the United States because a huge amount of gas is trapped in similar formations around the country. Eventually this technology will be exported to the rest of the world and huge new gas reserves will be produced in many other countries. It seems like the United States always leads the way on these sorts of technical innovations. This is motivated by their continuing need for energy resources to fuel their economy and maintain their culture.

Figure 6.12 shows the location of the current shale plays in the lower 48 states of the United States. Besides the Barnett, the other hot shale gas plays have been: the Haynesville in northern Louisiana; the Fayetteville in Arkansas; the Woodford in Oklahoma; the Eagle Ford in Texas; and, the Marcellus in West Virginia, western Pennsylvania, southern New York, and eastern Ohio. The biggest of these plays is the Marcellus shale, and it has barely begun to be developed. Figure 6.13 shows its location. Just like McMurry, Mitchell's success in the Barnett shale has opened up production in many other formations around the United States and Canada.

Figure 6.14 shows a table from a recent (2011) report on shale gas reserves prepared by INTEK for the U.S. Energy Information Administration (EIA). It has identified 750 TCF of unproven but technically recoverable reserves in these shale plays. The reserves are unproven because wells have not yet been drilled to prove the gas is actually there under each acre of earth where the formation is known to exist. At the same time, the numbers in Figure 6.14 may be conservative because it only shows 410 TCF of gas for the Marcellus and 43 BCF for the Barnett. In other reports, estimates have been as high as 1500 TCF for the Marcellus and 150 BCF for the Barnett. These reports may not be as reliable as the INTEK-EIA report but they are certainly more optimistic. Compare that to the listed proved reserves of gas for the entire United States of 162 TCF in 1993 and the current (2010) EIA listed proved reserves of gas of 272 TCF. The largest proved gas reserves in U.S. history were 293 TCF in 1967. These new reserves of gas are massive and are causing a paradigm shift in the supply of natural gas in the United States, which has driven the price back down to much cheaper levels.

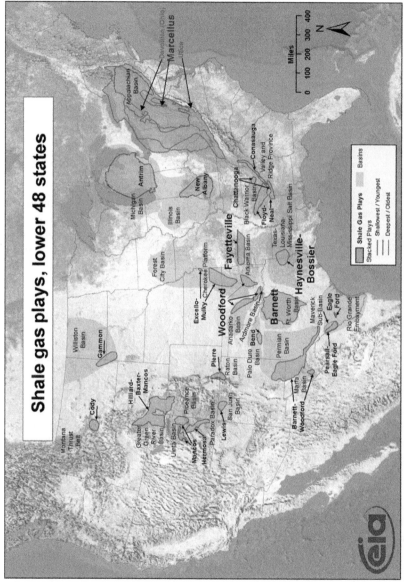

Source: Energy Information Administration based on data from various published studies.
Updated: March 10, 2010

FIGURE 6.12 American shale gas plays. *Source: EIA.*

Marcellus shale

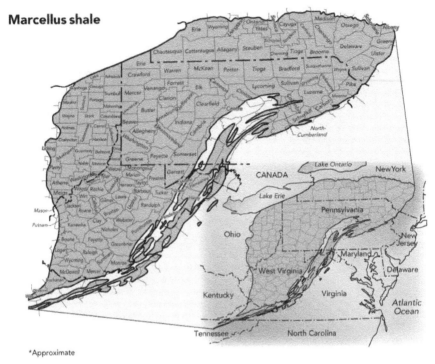

*Approximate

FIGURE 6.13   Location of the Marcellus Shale gas play. *Source: EIA.*

NATURAL GAS STORAGE

Figure 6.6 gives a picture of the annual and monthly supply of natural gas in the United States, but it does not tell the full story. The demand for natural gas is seasonal, having high demand for space heating in the winter and lower demand in the spring, fall and summer. Summer demand has been increasing because more gas is now used to generate electricity for air conditioning in the summer. Nevertheless, the highest demand for natural gas occurs in the wintertime.

Figure 6.15 shows the daily supply and demand for natural gas from January 2005 to August 2011. The light gray line is the demand curve and the gray line is the supply curve. There are clearly large seasonal variations in demand. In order to reduce the size of pipelines and also to provide a security of supply, it is more efficient to produce the gas at the same rate all year, store it close to the market (the end user) during the spring, summer and fall, and produce it from both the storage facility and the gas fields for use in the winter. For this purpose, gas is stored in underground

U.S. shale gas unproved discovered technically recoverable resources summary				
Play	Technically Recoverable Resource Gas (TCF)	Area (sq.miles)		Average EUR Gas (BCF/well)
		Leased	Unleased	
Marcellus	410.34	10,622	84,271	1.18
Big Sandy	7.40	8,675	1,994	0.33
Low Thermal Maturity	13.53	45,844		0.30
Greater Siltstone	8.46	22,914		0.19
New Albany	10.95	1,600	41,900	1.10
Antrim	19.93	12,000		0.28
Cincinnati Arch*	1.44	NA		0.12
Total Northeast	**472.05**	**101,655**	**128,272**	**0.74**
Haynesville	74.71	3,574	5,426	3.57
Eagle Ford	20.81	1,090		5.00
Floyd-Neal & Conasauga	4.37	2,429		0.90
Total Gulf Coast	**99.99**	**7,093**	**5,426**	**2.99**
Fayetteville	31.96	9,000		2.07
Woodford	22.21	4,700		2.98
Cana Woodford	5.72	688		5.20
Total Mid-Continent	**59.88**	**14,388**		**2.45**
Barnett	43.38	4,075	2,383	1.42
Barnett Woodford	32.15	2,691		3.07
Total Southwest	**75.52**	**6,766**	**2,383**	**1.85**
Hilliard-Baxter-Mancos	3.77	16,416		0.18
Lewis	11.63	7,506		1.30
Williston-Shallow Niobraran*	6.61	NA		0.45
Mancos	21.02	6,589		1.00
Total Rocky Mountain	**43.03**	**30,511**		**0.69**
Total Lower 48 U.S.	**750.38**	**160,413**	**136,081**	**1.02**

*Cincinnati Arch and Williston-Shallow Niobraran were not assessed in this report.

From the report: Review of Emerging Resources: U.S. Shale Gas and Shale Oil Plays, which was prepared by INTEK, Inc. for the U.S. Energy Information Administration (EIA)

FIGURE 6.14   U.S. Shale unproved, discovered, technically recoverable resources summary (lower 48 states). *Source: INTEK and EIA.*

reservoirs for when the need arises. There are several important reasons for storing gas in these underground facilities:

- The size of the pipelines transporting the gas are reduced
- It allows the gas fields to be produced at a constant rate
- It provides a security of supply when interruptions occur, and
- It allows for trading and balancing of supply from different sources.

The fourth item bears some explanation. In the past, U.S. gas contracts were often written with "take or pay" provisions. This meant that, if the utilities buying the gas did not need all the gas in the contract, it had to produce and store the gas in the storage reservoirs or pay penalties. Take or pay contracts are less common in today's market, but with the deregulation of the U.S. gas industry that resulted from the FERC Order

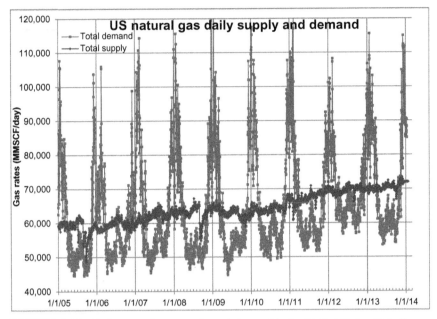

FIGURE 6.15   Daily natural gas supply and demand in the United States. *Data Source: Bentek.*

No. 636 (issued on April 8, 1992), new reasons for storing gas became evident. As the price became more volatile, gas buyers tried to buy the gas at low prices and store it for later use or resale at higher prices. In this environment, the demand for gas storage facilities increased. Under current circumstances, the gas does not need to be stored close to the end-user but can be stored anywhere along the pipeline, particularly near pipeline interchanges.

The operators of underground storage facilities are usually pipeline companies and local distribution companies, but there are some independent storage service providers. There are about 80 companies that currently operate the nearly 400 underground storage facilities in the lower 48 states. The location of these storage facilities are shown in Figure 6.16. There it can be seen that there are three types of reservoirs used to store gas: depleted oil and gas fields; salt caverns; and, salt-water aquifers. If a storage facility serves interstate commerce, it is subject to the jurisdiction of the FERC; otherwise, it is regulated by the state authority.

Operators of storage facilities are not usually the owners of the gas held in storage. Most working gas held in storage facilities is held under lease with traders, gas distribution companies, and even end users who buy the gas before they need to use it. The United States has the most efficient gas storage system in the world. Figure 6.17 shows the total amount of gas

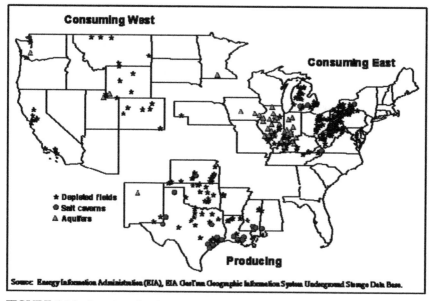

FIGURE 6.16   Location of underground storage facilities in the U.S. lower 48. *Source: EIA.*

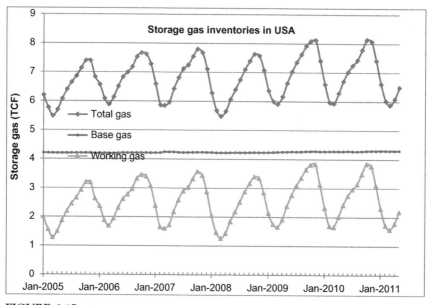

FIGURE 6.17   Storage gas inventories in the United States. *Data Source: EIA.*

held in storage over the past 6.5 years. As can be seen, the total gas in storage at any one time is divided up into the base gas and the working gas. The base gas (also called the cushion gas) is the volume of gas intended as permanent inventory in a storage reservoir needed to maintain adequate pressure and flow rates throughout the winter withdrawal season. Working gas is the volume of gas in the reservoir above the level of base gas that is available to be withdrawn in the winter. Working gas capacity is equal to the total gas storage capacity minus the base gas. The specification of what is base gas and what is working gas is somewhat arbitrary in that during a crisis of supply some of the base gas could be withdrawn to satisfy an emergency. Usually, the base gas is left in the storage reservoir permanently and is never withdrawn. The working gas is injected into the storage reservoirs in the spring, summer and fall and produced again in the wintertime. This provides a tremendous back-up and security of supply to the U.S. gas consumer and keeps the whole supply chain moving along in a most reliable manner.

In Figure 6.17, it can be seen that the base gas remains relatively constant at about 4.3 TCF. The working gas varies from about 1.25 TCF up to 3.8 TCF. The total gas in storage varies from 5.55 TCF to 8.1 TCF. There is always more than 5.5 TCF gas in storage, even at the end of the winter season when the storage gas is relatively depleted.

To illustrate the importance of gas storage, one only has to look at the effects a single accident had in a natural gas facility in the Australian state of Victoria in September 1998. The largest city in the state of Victoria is Melbourne, a bustling city of 4 million people—the second largest in Australia. The city of Melbourne and the entire state of Victoria is supplied with natural gas from the offshore oil and gas fields in the Bass Strait, south of the state, and operated by a subsidiary of Exxon. The gas from the fields is processed at three gas plants at Longford (near the city of Sale) and then is piped to Melbourne and to other cities further afield. Natural gas is the primary heating source for 80% of Victorian households, 50% of the commercial enterprises and 25% of industry.

On Friday, September 25′ 1998, a vessel ruptured at one of the gas plants leading to several explosions and fires. These fires and leaks continued at the plant until the last fires were extinguished 2 days later on Sunday, September 27. The winter months in Australia are June, July, and August. Even though the winters in Melbourne are not as fierce as in North America, the weather was still somewhat cool and the loss of the gas supply left a lot of people without heat.

As a result of the fires and explosions, all three gas plants were shut down. As Victoria is not connected to any gas storage facilities, supplies of gas to all domestic, commercial and industrial consumers in metropolitan Melbourne and in several areas beyond were shut down. Most Victorian gas consumers were left without gas for 19 days. The accident and subsequent loss of supply is considered to have been one of Victoria's worst economic disasters. It is estimated that 1.5 million households and 90,000 businesses were affected. In addition to directly affecting the daily lives of 4.5 million people for almost 3 weeks, the estimated cost of the accident to the Victorian economy was put at $1.5 billion. Industrial sectors particularly affected included the car industry, plastics production, food and drink manufacturers and the hospitality sector. Tens of thousands of workers were temporarily unemployed and several manufacturing industries were forced to shut down for a month. The disaster spawned the largest class action suit in Australian legal history, with 10,000 litigants signed on. Other impacts included temporarily curtailed production of some basic consumables including bread, milk, and other dairy products. Supply lines of basic consumables had to be brought in from out-of-state sources to overcome local shortfalls. This disruption would not have occurred if there was a large gas storage facility located near Melbourne. Gas could be withdrawn from the storage facility until the regular supply resumed.

THE PRICE OF GAS

In Figure 5.14 of Chapter 5, it was shown that the average annual price of natural gas in USA had been driven up by the reduced supply of natural gas after its production peaked in 1973. In recent years, the price of natural gas has been quite volatile. As with any commodity, the price of gas is governed by supply and demand. When demand outstrips supply, the price rises; when the supply outstrips demand, the price falls. Due to the seasonal variations in demand, it is not a simple task to measure when supply and demand are out of balance. There are also the follow-up questions as to the factors that affect supply and demand. Figure 6.15 showed that demand in particular is always rising and falling relative to supply and in the last section it was shown that the industry has learned to balance these variations using natural gas storage. These factors can mask a supply/demand imbalance. The price of gas in the United States is usually quoted in dollars per million BTU, abbreviated $/MMBTU. The primary source of the price data is the New York Mercantile Exchange, abbreviated NYMEX.

Gas was previously sold under long-term fixed price contracts. This was beneficial to both buyer and seller because both knew exactly what price was being paid for the gas over the entire life of the contract. These contracts were also necessary to finance field development. Bankers would not lend money on the development of a gas field unless the long-term contract was in place, which assured the banker that money would be there to repay the loan. These contracts had take or pay provisions whereby the buyer (a large well financed utility or a large industrial user) would be required to pay for the gas whether they took delivery of it or not. While much of the risk lay with the buyers, they had a fixed price that removed the risk of cost increases. Over time these types of contracts fell out of favor because, after spot prices began to rise in the late 1970s, gas producers found that being locked into long-term low price contracts removed any upside potential for their product. Moreover, when the industry was deregulated and made more open by FERC Order 636 in 1992, it created the conditions that allowed for contracts and sales at the market price for gas.

Figure 6.18 shows the daily spot price for natural gas since January 1, 1997. The spot price is the price paid for daily deliveries of gas at the prevailing price. The data shows that the spot price can spike rapidly when shortages develop. The majority of gas is not actually sold at the spot price or in long-term fixed price contracts. The majority of gas is now sold under futures contracts. A futures contract is a promise to deliver gas during a certain month at a particular price. The futures price for four forward months is shown in Figure 6.19. These show the same volatilities and price spikes as the spot price trends; however, the spikes are not quite as severe. On February 25, 2003, for example, the spot price peaked at $18.48 per

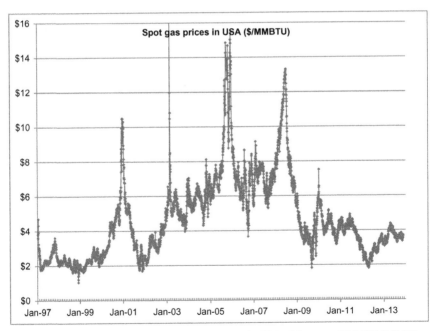

FIGURE 6.18   Daily spot price of natural gas in the United States on NYMEX ($/Million BTU). *Data Source: EIA.*

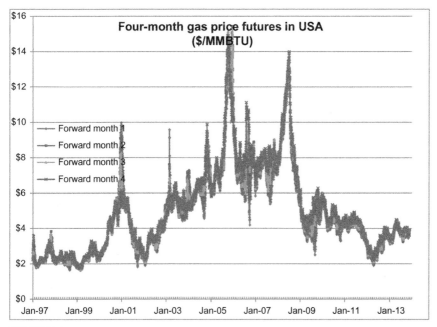

FIGURE 6.19 Four forward months of gas futures prices in the United States on NYMEX. *Data Source: EIA.*

MMBTU from a price of $5.88 per MMBTU on February 14. By March 5, 2003 the price was back down to $7.80 per MMBTU, and it continued declining from there. This brief spike was caused by a reduction in supply and an increase in demand during a winter blast of cold weather. The forward months futures prices also spiked, but at $9.58 for the one-month futures contract and $6.58 for the two-month futures contract. The three-month and four-month contract did not react at all because this was a short-term temporary situation.

Examination of Figure 6.19 indicates that the four forward months futures prices usually follow each other across long-term trends. When the one-month price increases, the two-, three-, and four-month future prices also increase, and vice versa. While this is often the case, it is not always true, as will be seen later.

The question remains as to what exactly causes these spikes. Why does the price run up and what causes it to return to prespike prices? Figure 6.20 shows the spike that occurred in the winter of December 2000 to January 2001. A severely cold winter caused a larger than usual demand for gas for heating, and the price shot up. There was no real shortage of gas because the extra demand was filled from storage gas. Traders, however, recognized that when a larger than usual supply comes from storage gas, this gas must eventually be replaced and more gas must

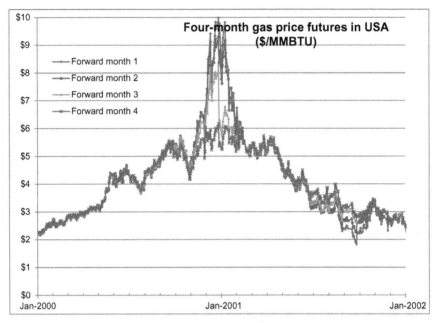

FIGURE 6.20   Four forward months of gas futures prices in the United States on NYMEX. *Data Source: EIA.*

eventually be produced to replace the gas being used by the cold winter. This pushes the price up.

Throughout the year 2000, the price of gas had been rising gradually and the four future month's prices tracked each other very closely. The market was saying that if the price is rising now, these conditions will still exist in four months and the price for four months in the future will also rise by the same amount. When the price spike came in December 2000, the four-month price separated from the one-month price because the market perceived that the conditions pushing the price up in December may still exist in January but probably would be over by April when the cold winter had receded. Indeed by April 2001, the price of gas was declining and the four future month's prices declined in lockstep with one another. By September 2001, the future months were separating again. While low demand in September may justify the price continuing to fall, the four-month future price was for December 2001 when another cold winter might be looming. There were no price spikes in the winters of January 2000 or January 2002 because these winter were normal and did not result in any demand for gas above and beyond the normal high winter demand.

In September 2005, another huge price spike occurred, but this one was not caused by unusually cold winter weather. To illustrate what happened, Figure 6.21 shows the one-month forward price along with the

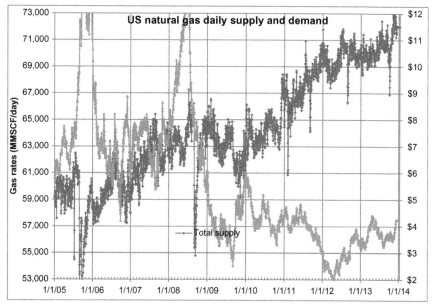

FIGURE 6.21 U.S. natural gas futures price versus daily supply. *Data Source: Bentek and EIA.*

daily supply curve from Bentek's daily production data. In September 2006 two huge hurricanes tore through the Gulf of Mexico: Hurricane Katrina and Hurricane Rita. In addition to destroying New Orleans, they also shut down most of the oil and gas production in the Gulf of Mexico for several months. The total U.S. gas supply dropped from about 60 BCF per day to about 52 BCF per day. This created a temporary shortage of gas for about 2 months. When the offshore platforms returned to production, the price returned to normal levels. This was a price spike caused by a temporary supply interruption.

A similar supply interruption was caused by two more Gulf of Mexico hurricanes in September 2008 when Hurricanes Gustav and Ike shut in another 10 BCF per day. This time the supply interruption did not cause a price spike and the reason is rather interesting. In the summer of 2008, a gas price spike developed due to rising demand. The robust economy of 2007 and early 2008 produced rising gas demand, which caused the price of gas to run up to $14 per MMBTU. These high prices then caused the demand to soften and the price began to decrease again. By September 2008, the world financial crisis—precipitated by bad subprime loan defaults—suddenly hit the market and the price of gas continued to decrease rapidly. When the hurricanes hit at the same time as the financial crisis, the market ignored the supply interruption because there was a strong perception that demand was going to be greatly decreased by a weakening world economy. The supply disruption had no effect on the gas price.

To keep track of the balance between supply and demand, traders primarily use the weekly gas storage numbers released by the EIA of the U.S. government. These can be just as difficult to interpret because they also change with the season. A more effective method is to plot the moving average of the supply and demand. If 52-week or 1-year moving averages of supply and demand are compared, you can see over the past year if the supply or the demand has been higher and whether the price should be increased or decreased. The peaks and valleys of the working gas in storage are useful information, but these only occur twice per year. Each week (on Thursdays at 10:30 am EST) when the storage data is released, traders compare the storage levels to the same week of the previous 5 years and try to adjust the price according to their interpretation of the data.

Figure 6.22 shows the 52-week moving average of supply and demand along with the weekly working gas in storage values. In 2007 and early 2008, the average demand was rising faster than the average supply. This resulted in the storage gas in early 2008 being reduced to less than 1.5 TCF, its lowest level in many years. This was a signal for the price to be pushed up in the spring and summer of 2008. When the world financial crisis hit in late 2008, demand was indeed scaled back. The storage low of early 2009 was above 1.6 TCF, and the storage high of November 2009 was above 3.8

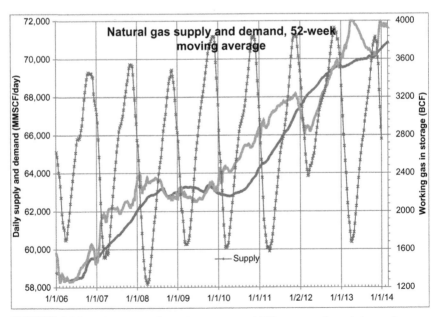

FIGURE 6.22 Fifty-two-week moving average of U.S. gas supply and demand versus working gas in storage. *Data Source: Bentek and EIA.*

TCF, its highest value ever. These were signals for the gas price to be pushed down to $2.51 per MMBTU on September 9, 2009, its lowest price since March 2002.

When the Gulf of Mexico production was restored after the damage and devastation caused by Hurricanes Katrina and Rita in September 2005, the supply of gas in the United States received a major boost with the success of the shale gas revolution. The large increases in supply from January 2006 to the present time, seen in Figure 6.21, are due to the Barnett shale production, followed by the Fayetteville in Arkansas, the Woodford in Oklahoma, the Haynesville in northern Louisiana, and more recently by the Eagle Ford in Texas and the Marcellus in Pennsylvania.

The other factor that affects the natural gas supply is the number of gas wells that are being drilled. Most gas wells produce at their highest rate at the beginning of their productive lives. Very soon after a gas well goes on production, its production rate begins to decrease as the pressure in the reservoir decreases. If the industry is not continuously drilling new wells, the gas supply will naturally decrease. Traders watch the drilling rig report to see how many wells are being drilled and put into production. This is an indication of how much the gas production rate will be increasing or decreasing in the future. Most wells that are drilled these days are going to be successful producers so the rig report is a strong indicator of future supply.

STATE PRODUCTION AND TRANSPORTATION OF NATURAL GAS IN USA

The United States has the most efficient and widespread natural gas pipeline network in the world. As shown in Figure 6.23, the gray lines represent interstate pipelines and the light gray represent intrastate pipelines. This network is continually being revised as production in some locations declines and new fields are brought into production. With the paradigm shift to unconventional gas in the past 5 years, a large reorientation of the pipeline network has occurred. Nevertheless, with such a dense and ubiquitous network in place, it is relatively easy and rapid to respond to new production in most places. Figure 6.23 does not give an accurate picture of the amount of gas moving in the pipelines because the diameter of the pipelines is not shown. The larger the diameter, the greater the amount of gas that can flow through the pipeline for a given pressure gradient in the pipeline. Figure 6.24 shows the corridors where the gas is currently flowing. The width of the corridors indicates the volume of gas flowing through the pipeline network. Further details about the directions and volumes of flows can be found on the EIA website.

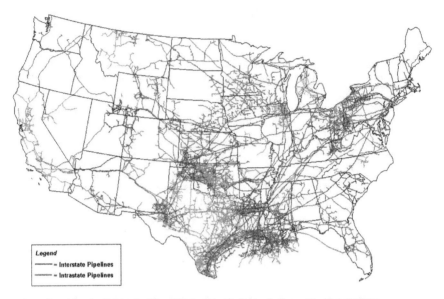

Source: Energy Information Administration, Office of Oil & Gas, Natural Gas Division, Gas Transportation Information System

FIGURE 6.23   U.S. natural gas pipeline network (2009). *Source: EIA.*

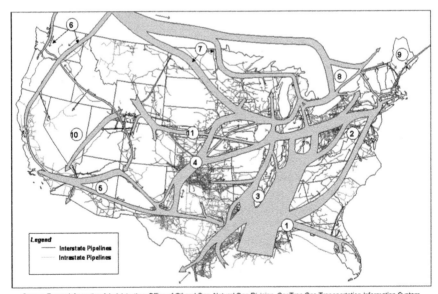

Source: Energy Information Administration, Office of Oil and Gas, Natural Gas Division, GasTran Gas Transportation Information System.

The EIA has determined that the informational map displays here do not raise security concerns, based on the application of the Federal Geographic Data Committee's *Guidelines for Providing Appropriate Access to Geospatial Data in Response to Security Concerns.*

FIGURE 6.24   U.S. natural gas pipeline network showing transportation corridors. *Source: EIA.*

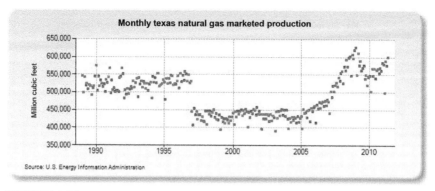

FIGURE 6.25 Monthly Texas natural gas marketed production. *Source: EIA.*

The largest gas-producing state is Texas (see production in Figure 6.25), and this is reflected in the network of pipelines that crisscross the state including the offshore Gulf of Mexico area, where a great deal of gas is produced. The second-largest gas-producing state is Louisiana, which overtook Wyoming for second place in 2010 due to large increases in its shale gas production from the Haynesville formation. Its production history is shown in Figure 6.26. The Federal Offshore Gulf of Mexico natural gas production, which has been steadily declining in recent years, is shown in Figure 6.27. The Federal Offshore Gulf of Mexico production has only been taken into account since 1997. Prior to that, the offshore Gulf of Mexico production was counted in state figures. That is why Texas and Louisiana show a decrease in production in 1997. The third-, fourth-, and fifth-largest gas-producing states are Wyoming, Oklahoma, and New Mexico, whose production volumes are shown in Figures 6.28–6.30, respectively.

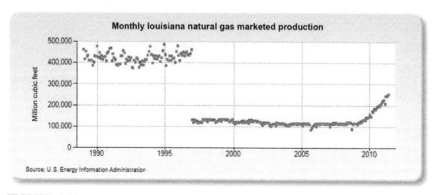

FIGURE 6.26 Monthly Louisiana natural gas marketed production. *Source: EIA.*

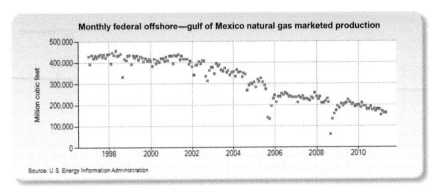

FIGURE 6.27 Monthly federal offshore Gulf of Mexico natural gas marketed production. *Source: EIA.*

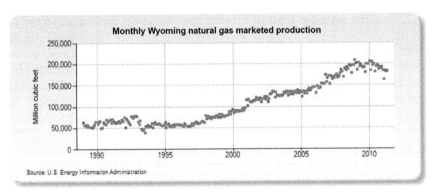

FIGURE 6.28 Monthly Wyoming natural gas marketed production. *Source: EIA.*

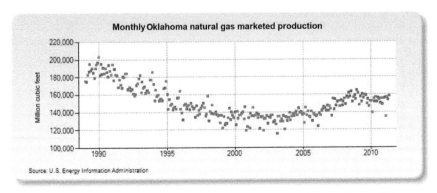

FIGURE 6.29 Monthly Oklahoma natural gas marketed production. *Source: EIA.*

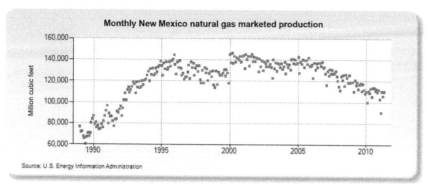

FIGURE 6.30 Monthly New Mexico natural gas marketed production. *Source: EIA.*

FRACTURE STIMULATION TECHNIQUES

All of the unconventional gas now being produced in the United States has been made economic by the hydraulic fracturing process (commonly known as "fracking"), in which water, chemicals and proppant materials are pumped into the well to release the gas trapped in low permeability formations by creating fractures in the rock and allowing natural gas to flow from the rock into the well. This process is illustrated in Figure 6.31.

After the water-based fluid creates the fracture, the proppant is pumped down the well transported by a carrier fluid. The purpose of the proppant is to prop open the fracture so that it does not close up again, which allows the gas to flow out of the formation through the fracture. Most commonly the proppant is sand, which is sometimes coated with resin. The proppant can also be aluminum bauxite or special ceramics, which are mainly used when the weight of the overburden rock is very high and creates high stresses. When the overburden stress is too high, the regular sand particles get crushed in the fracture and the fracture closes up. Ceramic proppants are more expensive than sand, so they are only used when sand will not do the job. The carrier fluid used to transport the proppant down into the fracture is often borate-based cross-linked gel polymers. These polymers are used because they can carry a lot of proppant out into the formation. The more proppant placed in the fracture, the better it works. After a certain period of exposure to the higher temperatures in the formation, the chemical bonds in the cross-linked gel polymer break down and the polymer becomes more fluid, which allows it to flow back out of the formation, leaving the proppant behind. Cross-linked gel polymers are widely used as carrier fluids in hydro-fracturing operations.

Most of the shale gas-fracturing operations use a so-called slick-water fracture fluid. It is called slick-water because gel polymers are not used,

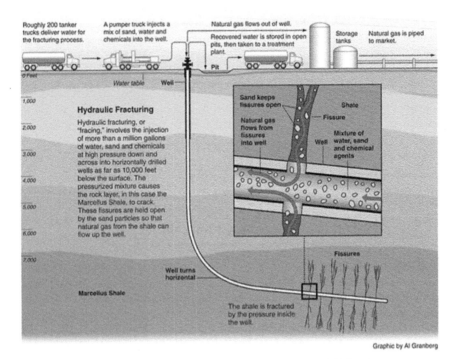

FIGURE 6.31   The hydraulic fracturing process. *Source: EIA and Al Granberg.*

but chemicals are added to the water-based fracturing fluids to reduce the water's viscosity to make it flow more readily. Slick water cannot carry proppant as effectively as the gels, but it seems to be able to create a broader network of induced fractures in the shale, which makes the shale more productive. Slick water fracks are also cheaper as well as being more effective in some shale formations.

One of the environmental concerns about fracturing operations is that some of the chemicals used can contaminate freshwater formations. This should not happen because the shale formations are usually widely separated from the freshwater aquifers by thousands of feet of rock. Accidents can occur, however, and fracturing fluid containing chemicals could be spilled on the ground and then washed into aquifer recharge areas. As of the current moment, there have not been any such incidents and there have not been any water supplies contaminated by fracturing fluids. Despite this, regulatory authorities are now requiring fracturing service companies to identify the chemicals being used in their operations so that if water wells do end up contaminated, the contamination can be traced.

LIQUID TRANSPORTATION FUELS FROM NATURAL GAS

Natural gas can be used to manufacture petrochemicals such as methanol, ammonia, fertilizer, and polymers; however, these are relatively small users of the gas reserves with limited markets. Liquid fuels represent a much larger market. Moreover, they can be moved in existing pipelines or product tankers and even blended with existing crude oil or product streams. No special contractual arrangements are required for their sale in most markets.

The technology to convert gas to liquids (GTL) is well known, and new technology is being developed and applied on a regular basis. The technology is scalable, allowing design optimization and application to smaller gas deposits. The key influences on their competitiveness are the cost of capital, operating costs of the plant, feedstock gas costs and the ability to achieve high utilization rates in production. As a generalization, however, GTL has not been competitive against conventional oil production unless the gas has a low opportunity value and is not readily transported. When oil is selling for close to $100 per barrel, the economics become much more attractive, as will be seen later.

GTL not only adds value, but is capable of producing products that could be sold or blended into refinery stock as superior products with fewer pollutants for which there is growing demand. Reflecting GTL's origins as a gas, GTL processes produce diesel fuel with an energy density comparable to conventional diesel, but with a higher cetane number permitting a superior performance engine design. Another "problem" emission associated with diesel fuel is particulate matter, which is composed of unburnt carbon and aromatics and compounds of sulfur. Fine particulates are associated with respiratory problems, while certain complex aromatics have been found to be carcinogenic. Low sulfur content leads to significant reductions in particulate matter that is generated during combustion, and the low aromatic content reduces the toxicity of the particulate matter. This has resulted in a worldwide trend towards the reduction of sulfur and aromatics in fuel.

It is technically feasible to synthesize almost any hydrocarbon from any other. In the past five decades, several processes have been developed to synthesize liquid hydrocarbons from natural gas. Indirect conversion to liquids can be carried out via Fischer-Tropsch (F-T) synthesis or via methanol. The discovery of F-T chemistry in Germany dates back to the 1920s, and its development has been for strategic rather than economic reasons, as in Germany during World War II and in South Africa during the apartheid era. Mobil developed the "M-gasoline" process to make gasoline from methanol, and this was implemented in 1985 in a large integrated methanol-to-gasoline (MTG) plant in New Zealand. The New Zealand plant was a technical success, but produced gasoline at costs above $30 per barrel. This was higher than the cost of refining gasoline from crude oil in the 1980's and 1990's. The plant was eventually converted to methanol-only production.

The first step in the GTL process is to convert the natural gas to syngas (which consists of hydrogen and carbon monoxide) by partial oxidation, steam reforming or a combination of the two processes. The key variable is the hydrogen to carbon monoxide ratio with a 2:1 ratio recommended for F-T synthesis. Steam reforming is carried out in a fired heater with catalyst-filled tubes that produce a syngas with at least 5:1 hydrogen to carbon monoxide ratio. To adjust the ratio, hydrogen can be removed by a membrane or pressure swing adsorption system. This process is widely used if the surplus hydrogen is used in a petroleum refinery or for the manufacture of ammonia in an adjoining plant. The partial oxidation route provides the desired 2:1 H:C ratio and is the preferred route in isolation of other needs.

Conversion of the syngas to liquid hydrocarbon is a chain growth reaction of carbon monoxide and hydrogen on the surface of a heterogeneous catalyst. The catalyst is either iron- or cobalt-based, and the reaction is highly exothermic. The temperature, pressure and catalyst determine whether light or heavy products are produced. For example, at 330 °C mostly gasoline and olefins are produced whereas at 180-250 °C mostly diesel and waxes are produced. Similar technologies are used to make liquid transportation fuels from coal, and more details on these technologies are given in Chapter 13. It is slightly more difficult to make syngas from coal, but once created, the syngas is converted into gasoline and diesel fuels using the F-T process in exactly the same manner.

The South African company, Sasol, is a technology supplier that was originally formed to provide synthetic petroleum products to South Africa during the apartheid era. The firm has built a series of F-T coal-to-oil plants and is one of the world's most experienced synthetic fuel organizations. It is now marketing a natural GTL technology. It has developed the world's largest synthetic fuel project, the Mossgas complex at Mossel Bay in South Africa. This project was commissioned in 1993 and produces 25,000 barrels per day. To increase the proportion of higher-molecular-weight hydrocarbons, Sasol has modified its Arge reactor to operate at higher pressures. It has commercialized four reactor types, with the slurry phase distillate process being the most recent. Its products are more olefinic than those from the fixed bed reactors and are hydrogenated to straight chain paraffins. Its slurry phase distillate converts natural gas into liquid fuels, most notably superior-quality diesel, using technology developed from the conventional Arge tubular fixed-bed reactor technology. The resultant diesel is suitable as a premium blending component for standard diesel grades from conventional crude oil refineries.

Blended with lower grade diesels, it makes it easier to comply with the stringent specifications being set for transport fuels in North America and Europe.

Its other technology uses the Sasol Advanced Synthol (SAS) reactor to produce mainly light olefins and gasoline fractions. Sasol has developed high performance cobalt-based and iron-based catalysts for these processes. The company claims a single module of the Sasol slurry phase distillate plant converts 100 million cubic feet per day (110 TJ per day) of natural gas into 10,000 barrels per day of liquid transport fuels and can be built at a capital cost of about US$250 million. This cost equates to a cost per daily barrel of capacity of about US$25,000, including utilities, off-site facilities, and infrastructure units. If priced at $5 per MMBTU, the gas amounts to a feedstock cost of US$50 per barrel of product. The fixed and variable operating costs (including labor, maintenance, and catalyst) are estimated at a further US$5 per barrel of product, thereby resulting in a direct cash cost of production of about US$55 per barrel (excluding depreciation). Most of this cost is for the gas feedstock.

In June 1999, Chevron and Sasol agreed to an alliance to create ventures using Sasol's F-T technology. The two companies have conducted a feasibility study to build a GTL plant in Nigeria that was to begin operating in 2013. This joint venture (Chevron and Sasol) has also partnered with the Nigerian National Petroleum Corporation to build the GTL plant, which is expected to convert 325 million cubic feet of natural gas per day into 33,000 barrels of liquids, principally synthetic diesel. When completed, the plant is expected to export the diesel fuel product to Europe.

Shell has carried out research and development (R&D) since the late 1940s on the conversion of natural gas, leading to the development of the Shell middle distillate synthesis (SMDS) route—a modified F-T process. SMDS focuses on maximizing yields of middle distillates, notably kerosene and diesel and bunker fuel. Shell built a 12,000 barrel per day GTL plant in 1993 in Bintulu, Malaysia. The process consists of three steps: the production of syngas with a $H_2:CO$ ratio of 2:1; syngas conversion to high-molecular-weight hydrocarbons via F-T using a high performance catalyst; and hydro-cracking and hydro-isomerization to maximize the middle distillate yield. Shell is investing US$6 billion in GTL over 10 years with four plants. It announced in October 2000 that there was an agreement with the Egyptian government to build a 75,000 barrel per day facility in Egypt and a similar plant to be built in Trinidad & Tobago. In April 2001, it announced interest for plants in Australia, Argentina and Malaysia at 75,000 barrels per day costing US$1.6 billion.

Exxon has developed a commercial F-T system from natural gas feedstock. Exxon claims its slurry design reactor and proprietary catalyst systems result in high productivity and selectivity along with significant economy of scale benefits. Exxon employs a three-step process: fluid bed synthesis gas generation by catalytic partial oxidation; slurry phase F-T synthesis; and fixed bed product upgrade by hydro-isomerization. The process can be adjusted to produce a range of products. More recently, Exxon has developed a new chemical method based on the F-T process to synthesize diesel fuel from natural gas. Exxon claims that better catalysts and improved oxygen-extraction technologies have reduced the capital cost of the process and is actively marketing the process internationally. Made from gas, the high-molecular-weight liquid GTL products can be hydro-cracked in a simple low-pressure process to produce naphtha, kerosene, and diesel that is virtually free of sulfur and aromatics. These derivative fuels are therefore potentially more valuable, notably in the United States, Europe, and Japan because of their high environmental standards.

The Syntroleum Corporation in the United States is marketing an alternative natural-gas-to-diesel technology based on the F-T process. It claims to be competitive with a lower capital cost due to the redesign of the reactor. Instead of oxygen, an air-based auto-thermal reforming process is used for synthesis gas preparation to eliminate the capital expense of an air separation plant. This type of catalyst results in high yield. It claims to be able to produce synthetic crude at around $50 per barrel. The syncrude can be further subjected to hydro-cracking and fractionation to produce a diesel/naphtha/kerosene range at the user's discretion. The company indicates its process has a capital cost of around $13,000 per daily barrel of diesel for a 25,000 barrel per day facility and an operating cost between $3.50 and $5.70 per barrel.

The thermal efficiency of the Syntroleum process is reported to be about 60%, implying a requirement of about 90 million cubic feet (85 TJ) per day of dry gas for a 25,000 barrel per day facility costing $350 million. These figures indicate a unit cost of less than $50 per barrel ($8 per GJ) of diesel fuel. The Syntroleum Corporation now also licenses its proprietary process for converting natural gas into other synthetic crude oils and transportation fuels. In February 2000, Syntroleum Corporation revealed its plans to construct a 10,000 barrel per day (requiring 130 TJ per day or 800,000 tons per year of gas) GTL plant for the state of Western Australia. This project did not get built. The project planned to produce synthetic specialty lubricating oils, naphtha, normal paraffin, and drilling fluids and was estimated to cost US$500 million, generating sales of around US$200 million per year.

On August 1, 2011, China Petroleum & Chemical Corporation (Sinopec) and Syntroleum Corporation announced the opening of the Sinopec/Syntroleum Demonstration Facility (SDF) located in Zhenhai, China. It is an 80 barrel per day facility utilizing the Syntroleum-Sinopec F-T technology for the conversion of coal, asphalt and petroleum coke into high-value synthetic petrochemical feedstocks.

Rentech Inc. in Denver, Colorado, has been developing an F-T process using a molten wax slurry reactor and precipitated iron catalyst to convert gasses and solid carbon-bearing material into straight chain hydrocarbon liquids. In their process, long straight chain hydrocarbons are drawn off as liquid heavy wax while the shorter chain hydrocarbons are withdrawn as overhead vapors and condensed to soft wax, diesel fuel, and naphtha. It is promoted as suitable for remote and associated gas fields as well as sub-pipeline quality gas. It currently has a 10 barrel per day demonstration plant operating in Commerce City, Colorado.

There are several methanol-based routes to manufacture gasoline from syngas. The principal one is Mobil's MTG process based on their ZSM-5 zeolite catalyst, which was commercialized in 1985 in the GTL plant now owned by Methanex in New Zealand. Commercial applications of the MTG process are now anticipating using a fluid bed reactor because of its higher efficiency and lower capital cost.

The use of GTL for chemical and energy production is forecast to advance rapidly with increasing pressure on the energy industry from governments, environmental organizations, and the public to reduce pollution, including the gaseous and particulate emissions traditionally associated with conventional petroleum-fueled and diesel-fueled vehicles. The commercial success of GTL technology has not yet been fully established and returns from GTL projects will depend on projections of market prices for petroleum products and presumed price premiums for the environmental advantages of GTL-produced fuels. Unit production costs will reflect: the cost of the feedstock gas; the capital cost of the plants; marketability of by-products such as heat, water, and other chemicals (e.g., excess hydrogen, nitrogen, or carbon dioxide); the availability of infrastructure; and the quality of the local workforce. Clearly, the feedstock gas cost will have an influence as it may vary widely depending on alternative applications. As one indication, based on current efficiencies, a change in the cost of gas feedstock of $0.50 per million BTU (or per one gigajoule GJ) would shift the synthetic crude oil price by around $5 per barrel. This is predicated on the relationship that the processes require about 10 MMBTU (10.5 GJ) of gas to produce 1 bbl of fuel with variations depending on scale, quality of output and variable production costs traded off against capital costs.

Shell estimates that a GTL plant processing 600 million standard cubic feet (0.7 TJ) of gas per day would cost 60% more than an LNG plant, but the more readily used end-products make GTL cheaper than LNG. According to Shell, a 75,000 barrel per day plant would cost around US $1.6 billion.

Capital costs for GTL projects currently tend to be in a range of double that of refineries, between $20,000 and $30,000 per daily barrel of capacity (compared with refinery costs of $12,000-$14,000 per daily barrel). The cost of GTL-produced fuel could also vary by approximately $1.50 per barrel with a shift of $5000 in capital cost. Estimates of the crude oil prices necessary to allow positive economic returns from a GTL project vary widely, with optimistic estimates ranging as low as $14-$16 per barrel plus fuel costs.

Presently there are only four GTL facilities producing synthetic petroleum liquids at more than a demonstration level: the Mossgas Plant (South Africa), with an output capacity of 23,000 barrels per day; the Shell Bintulu plant in Malaysia at 20,000 barrels per day; and the MTG project in New Zealand, which is now making only methanol. The fourth and largest GTL project is the recently completed Shell GTL plant in Qatar. It uses 1.6 BCF per day of gas to produce about 140,000 barrels per day of liquid fuels. The joint project in Nigeria of Chevron, Sasol and Nigerian National Petroleum Corporation to build a 30,000 barrel per day plant costing $1 billion using the Sasol Slurry Phase Distillate process was expected to begin operations in 2013. The Nigeria project would benefit from the infrastructure already in place for nearby oil and gas production and export facilities.

THE FUTURE OF NATURAL GAS

The future for natural gas is very bright, particularly in the United States, which is already the world's largest producer and consumer of natural gas. With the large deposits of shale gas that have recently been made economic, there are abundant new supplies to be produced. Currently, the United States is producing about 68 BCF per day of total gas, netting 63 BCF per day of marketed dry gas as well as a large amount of condensate and natural gas liquids. This has allowed the United States to reduce its volume of imported oil from a high of 13.4 million barrel per day in August 2006 to the current (in 2014) 5 million barrel per day. The United States has the capacity to double its current gas production to as much as 136 BCF per day and convert the additional gas to electricity or to liquid transportation fuels, displacing much more imported oil in the process. If all of the additional 68 BCF per day of gas was converted into liquid transportation fuels, it would amount to 7 million barrel per day less imported oil. If the gas was used to generate electricity instead and the electricity was used to power electric vehicles, a similar volume of crude oil could be displaced. If the 136 BCF per day were maintained for the next 30 years, it would require reserves of 1500 TCF of gas. There is every expectation that the oil and gas industry can prove up that volume of gas.

In order to increase the gas production rate by 68 BCF per day, about 34,000 new wells would have to be drilled with the assumption that each well would have an average initial rate of 2 million cubic feet per day. After that, an additional 20,000 wells per year would have to be drilled to keep the gas production rate at a constant 136 BCF per day. It would be a significant logistical task to drill that many wells each year, but the oil and gas industry is capable of doing it. Shale gas and tight gas wells are not cheap to drill and the industry would not drill the wells if the gas markets were not developed and a minimum gas price was not certain. An estimate is that a gas price of $5 per MMBTU would be required and at this price the industry could produce transportation fuels for about $70 per barrel. The current (2014) wholesale price for gasoline is $2.85 per gal, which is equivalent to $120 per barrel.

There are many uses for natural gas, which are summarized in Figure 6.32. It is most widely used for space heating, but Figure 6.32 shows that it is also a very versatile energy source that can be adapted to a wide variety of uses.

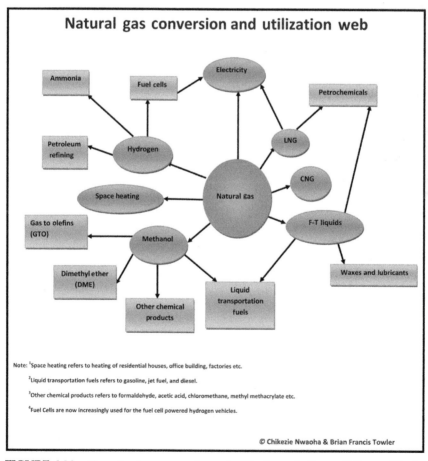

Natural gas conversion and utilization web

Note: [1]Space heating refers to heating of residential houses, office building, factories etc.

[2]Liquid transportation fuels refers to gasoline, jet fuel, and diesel.

[3]Other chemical products refers to formaldehyde, acetic acid, chloromethane, methyl methacrylate etc.

[4]Fuel Cells are now increasingly used for the fuel cell powered hydrogen vehicles.

© Chikezie Nwaoha & Brian Francis Towler

FIGURE 6.32   The natural gas conversion and utilization web. *Source: Chikeze and Towler 2013.*

Nuclear Energy

Nuclear power had its genesis in Einstein's theory of relativity which was first published as the special theory of relativity in 1905, although that is not actually what led scientists to generate nuclear power by splitting the atom. When heavy elements such as uranium (the heaviest naturally occurring element) are split into other elements using nuclear fission, the resulting elements are collectively slightly lighter than the original uranium atom. According to Einstein's equation, the extra mass is converted into energy. Conversely, when light elements such as hydrogen (the lightest element) are fused into heavier elements (two hydrogen atoms fuse into helium), using nuclear fusion, they end up slightly lighter than the original hydrogen atoms used to create them. Again, according to Einstein's equation, the extra mass is converted into energy.

The energy released in each process (fission and fusion) is stored as the binding energy of the protons and neutrons in the nucleus of the atom. It may seem paradoxical that the fission of heavy elements and the fusion of lighter elements both release energy. This occurs because the energy required to bind neutrons and protons together to form atoms goes through a maximum near the middle elements of Iron and Nickel, as shown in the attached plot. Elements heavier than iron release energy when they are split by nuclear fission; elements lighter than iron release energy when they are fused together by nuclear fusion.

When scientists were probing the structure of the atom in the first half of the twentieth century they began using neutrons to bombard the nucleus. These experiments inevitably led to the splitting of uranium atoms into two roughly equal size pieces. The potential for splitting uranium atoms had been implied by work done with radium by Frederic and Irene Joliot-Curie (the daughter and son-in-law of Pierre and Marie Curie) in 1935. Enrico Fermi's group in Italy was probably the first to split uranium atoms in 1937, but they struggled to understand the products of their uranium experiments. Fermi proved to himself and the world that the products were not any element in the range from element number 82 (lead) to element number 91 (protactinium). So he surmised that his experiments were creating trans-uranium elements, with an atomic number higher than 92. It was not until Otto Hahn and Fritz Strassman repeated the experiments in Germany in 1938 that they were able to prove that the uranium atoms were being split approximately in two and one of the products was barium. The Germans were so surprised and unsure of their results that they prevailed upon a former colleague, Lise Meitner, to examine their results and try to come up with an alternative explanation. Meitner had recently fled Nazi persecution in Germany. Instead of proposing an alternative explanation, she and her nephew (Otto Frisch) developed the theory for splitting the uranium atom into two using Einstein's theory and equation and also predicted the energy that would be generated. Hungarian Leo Szilard, a former protégé and scientific collaborator with Albert Einstein, had previously patented the nuclear chain reaction process, surmising that any nuclear process that could generate neutrons could create a chain reaction that might be used to create a huge release of energy. He called the process "an atomic bomb." When Szilard heard about the Hahn-Strassman experiment and saw it duplicated he realized that the splitting of Uranium-235 (U-235) with neutrons was exactly the process that could do this. Hahn called the process "nuclear fission."

The first public demonstrations of the fission process were made in 1945 when a uranium fission bomb was dropped on Hiroshima in Japan and a plutonium fission bomb was dropped on Nagasaki, Japan. The results were horrifying and catastrophic, but this resulted in the surrender of Japan and the end of World War II. After World War II, the even more explosive hydrogen fusion bombs were developed by both the United States and Russia. These hydrogen fusion weapons were never deployed or used in any war; however, their explosive power and rapid release of energy have been demonstrated in test firings both above ground and underground. It can be seen from Figure 7.1 that a fusion process releases more energy than a fission process.

After World War II and the development of the hydrogen bomb, scientists turned their attention to harnessing this energy for the generation of electricity. The first plants that were developed utilized the easier to initiate and control uranium fission process. A typical nuclear fission power plant design is shown in Figure 7.2.

How does a nuclear power plant work? It is a relatively simple process. In the core of the reactor in the containment building, U-235 is bombarded with neutrons that split the uranium atom into two smaller atoms and also releases extra neutrons. These extra neutrons bombard other uranium atoms, splitting them and creating a chain reaction that sustains the reaction as long as the uranium fuel is present. The chain reaction can be

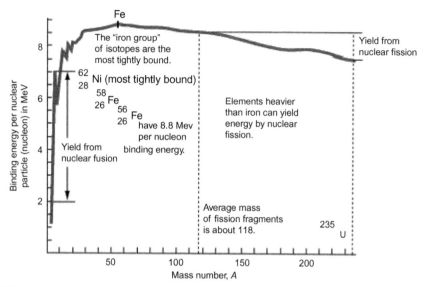

FIGURE 7.1   Binding energy of the elements as a function of mass number.

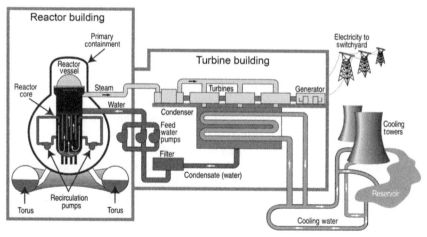

FIGURE 7.2   Typical nuclear power plant design.

controlled or stopped by the insertion of rods into the reaction chamber that absorbs some or all of the extra neutrons. These rods are usually made of boron or graphite. When the U-235 atoms are split, the extra binding energy is released and is transferred as heat to a water jacket surrounding the reaction core. The water absorbs the heat, raising its temperature, and vaporizes it into steam at high pressure. The steam is then fed to a steam turbine which generates the electricity.

When nuclear power plants were being designed in the 1950s, some nuclear engineers predicted that, with the deployment of nuclear power, electricity would become so cheap it would be too cheap to meter. The reason they came to this conclusion was that the amount of energy that could be extracted by splitting uranium atoms was enormous and the cost of the fuel relative to the energy generated was very small. Electricity, however, did not become too cheap to meter because the capital and operating costs of the nuclear power plants was so large that the electricity still had to be metered to pay back those large costs. The same arguments can be made about wind power and solar power. The sun and wind are free and cannot be metered, but in Chapter 9 you will see that wind power is still more expensive than fossil fuel power stations. The harnessing of solar energy discussed in Chapter 8 is so expensive due to high capital costs that it may never be more than a niche player.

Nuclear energy grew rapidly in the 1960-1975 period in countries such as the United States, France, Japan and Germany. It ran into problems in the 1970s due to the public concern raised by the radioactive waste it generated. These concerns were fueled by serious accidents at the Three Mile Island plant in Pennsylvania in 1979 and the Chernobyl disaster in the Ukraine in 1986. This suppressed its further expansion. In recent years, public perception had begun to change as concerns about atmospheric carbon dioxide levels lead to a renewed interest in nuclear power. The Fukushima nuclear power plant accident due to the earthquake and tsunami in northern Japan on March 11, 2011 has once again, however, focused attention on nuclear plant safety.

Most nuclear power is used to generate electricity. The United States and Russia do have some ships and submarines that are nuclear powered, but this represents a tiny fraction of the energy generated from nuclear reactions. The 29 countries with the largest nuclear power generation are shown in Table 7.1. The United States is far and away the - largest user of nuclear energy with 841 TWh. France is the second largest with 411 TWh, and after that Japan, Russia, South Korea, and Germany round out the top six (only these six countries generate more than 130 TWh/year). According to the percentage of electricity generated by nuclear power shown in Table 7.2, France is number 1 with 76%, while the United States—which generates a healthy 20% of its electricity from nuclear power—is number 16 in that respect. For the entire world, the percentage of electricity generated by nuclear power is 13.42%.

The United States currently has 104 nuclear reactors. An aerial view of two of these reactors at the Diablo Nuclear Power Plant in Diablo Canyon, California, is shown in Figure 7.3.

TABLE 7.1 2009 Nuclear Energy Consumption (TWh)

1	USA	840.78
2	France	410.53
3	Japan	274.64
4	Russian Federation	163.58
5	South Korea	147.77
6	Germany	134.90
7	Canada	89.81
8	Ukraine	82.16
9	China	70.13
10	United Kingdom	69.19
11	Spain	52.89
12	Sweden	52.63
13	Belgium & Luxembourg	47.22
14	Taiwan	41.57
15	Switzerland	27.52
16	Czech Republic	27.11
17	Finland	23.65
18	India	16.82
19	Bulgaria	15.57
20	Hungary	15.42
21	Slovakia	14.08
22	Brazil	13.00
23	South Africa	12.12
24	Romania	11.76
25	Lithuania	10.85
26	Mexico	9.59
27	Argentina	7.99
28	Netherlands	4.23
29	Pakistan	2.85

TABLE 7.2 Percentage of Electricity Generated by Nuclear Power

1	France	75.69%
2	Lithuania	70.68%
3	Belgium & Luxembourg	55.88%
4	Slovakia	53.72%
5	Ukraine	47.52%
6	Hungary	42.75%
7	Switzerland	40.11%
8	Sweden	38.95%
9	Bulgaria	36.14%
10	Czech Republic	33.59%
11	Finland	33.04%
12	South Korea	32.42%
13	Japan	24.63%
14	Germany	22.60%
15	Romania	20.46%
16	USA	20.26%
17	United Kingdom	18.61%
18	Spain	18.33%
19	Taiwan	18.10%
20	Russian Federation	16.47%
21	Canada	14.16%
22	Argentina	6.45%
23	South Africa	4.67%
24	Netherlands	3.77%
25	Mexico	3.72%
26	Pakistan	3.07%
27	Brazil	2.78%
28	India	1.93%
29	China	1.88%

FIGURE 7.3 Diablo Canyon Nuclear Power Plant, California. *Source: Emdot, 2007.*

THE URANIUM FUEL CYCLE

All current nuclear power plants use U-235 as its fuel. There are a variety of uranium-based fuels that could be used, but U-235 is the simplest. The 235 refers to the atomic weight of the uranium atom that is used. Every uranium atom has 92 protons in the atomic nucleus and a varying number of neutrons. The atomic weight is collectively the total number of protons and neutrons in the atomic nucleus. There are three isotopes of uranium that exist in nature: U-234, U-235, and U-238. U-235 is easier to split than the other two isotopes because of the odd number of neutrons in the atom. When a neutron is fired at the atom with enough speed (kinetic energy), the U-235 atom absorbs the neutron and momentarily becomes U-236. The kinetic energy of the absorbed neutron causes the U-236 nucleus to begin vibrating and then it splits into two atoms of roughly equal size. Depending on the conditions, the products of the fission process are varied and range from Krypton, Strontium, Zirconium, Technetium to Cesium and Barium. When uranium is mined, it is recovered as uranium oxide (U_3O_8). Only about 0.7% of the uranium atoms in the U_3O_8 are U-235 atoms. Most of the other 99.3% of the uranium atoms are U-238. There are also a few U-234 atoms. The purified uranium oxide (U_3O_8) that is recovered from the mining operation is called yellowcake because of its color. It should be noted that some so-called yellowcake is actually brownish in color.

The next step in the cycle involves the conversion of the yellowcake into uranium hexafluoride (UF_6) gas. This step is required to separate the uranium atoms into the three isotopes: U-234, U-235, and U-238. The U_3O_8 molecules are first chemically reacted into uranium hexafluoride (UF_6) from which the individual U-235 and U-238 isotopes can be separated. This gas is sent to an enrichment plant where the isotope separation takes place. The United States currently has one operating enrichment plant, which uses a diffusion process to separate the isotopes. The slightly smaller and lighter U-235 atoms travel slightly faster than the U-238 atoms and are forced to diffuse through a porous membrane where they are collected and concentrated. The final product is still mostly U-238 but the concentration of U-235 has been raised from 0.7% to about 5%, and this mixture is now called enriched uranium fuel. Note that it is still molecular UF_6.

There are other processes that are used for enrichment, and newer ones are being proposed and developed. Another enrichment technique that is used elsewhere is the gas centrifuge process, where the UF_6 gas is spun at high speed in a series of cylinders which separates the lighter 235-UF_6 and heavier 238-UF_6 atoms based on their different atomic masses. The lighter U-235 travels faster than the U-238, causing it to separate. Other enrichment technologies that are currently being developed use laser technology to do the separation and concentration. One is called atomic vapor laser isotope separation and another is molecular laser isotope separation. These laser-based enrichment processes can achieve higher initial U-235 enrichment (isotope concentration) factors than the membrane or centrifuge processes and are capable of operating at higher throughput rates.

The next step in the production of nuclear fuel takes place at one of the five U.S. fuel fabrication facilities. Here, the enriched UF_6 gas is reacted to form a black uranium dioxide powder (UO_2). The powder is compressed into small pellets and sealed into metal fuel rods, which are long metal tubes about 1 cm in diameter. The fuel rods are then bundled together to make up a fuel assembly (see Figure 7.4). Depending on the reactor type, there are about 179-264 fuel rods in each fuel assembly and a typical reactor core holds 121-193 fuel assemblies.

These fuel rods can be handled safely without precautions because the uranium is only slightly radioactive and essentially all radiation is contained within the metal tubes. These fuel assemblies last up to 6 years and about one-third of the assemblies are changed out every 2 years.

The reactor core itself is a cylindrical arrangement of the fuel bundles, about 12 ft in diameter and 14 ft high. It is encased in a steel pressure vessel. The core has essentially no moving parts except for a small number of

FIGURE 7.4 A nuclear fuel rod assembly. *Source: Commissariat à l'Énergie Atomique.*

control rods that can be inserted to regulate the reaction. These control rods, typically made of boron or graphite, are designed to absorb some of the neutrons that are released by fission so that the reaction can be slowed or stopped as necessary.

Following their use in the reactor, the fuel assembly becomes highly radioactive and must be removed and stored under water for several years. Even though the fission reaction has stopped, the spent fuel continues to give off heat from the decay of radioactive elements that were created when the uranium atoms were split apart. The water cools the fuel and at the same time absorbs the radiation. As of 2002, there were over 165,000 spent fuel assemblies stored in about 70 interim storage pools throughout the United States. After cooling for several years, the spent fuel is moved to a concrete container for on-site storage.

About 80% of U-235 is consumed before the fuel rods are considered spent. This means that less than 4% of the uranium loaded into the reactor is consumed in the nuclear reactions. The rest of the uranium, which is mostly U-238, remains unchanged. Chemical processing of the spent fuel material to recover the remaining portion of fissionable products for use in fresh fuel assemblies is technically feasible. Some countries, such as France, reprocess spent nuclear fuel, but it is not permitted in the United States.

Eventually, it is intended that the spent nuclear fuel will be stored in a permanent underground repository. The United States designed and built the Yucca Mountain repository in Nevada for the purpose of storing this spent nuclear fuel but so far it has not been permitted for this purpose. The spent fuel continues to be stored on site at the nuclear reactors. This spent fuel, containing mostly U-238, could be reprocessed to enrich the U-235 or could be used as fuel in fast breeder reactors, which are discussed in the next section.

FAST BREEDER REACTORS

One of the problems with current nuclear reactor technology is that only some of the U-235 is consumed in the reactor and the U-235 represents less than 1% of the uranium atoms. The supply of uranium is limited, and nuclear power cannot be expanded much unless this limited uranium supply is increased. There are many other nuclear fission processes that rectify this situation. The fast breeder reactor is one such design that falls into the general class of breeder reactors. These reactors actually generate more fissile material than they use and can use a broader array of nuclear fuels and so they do not run into the supply limitation. Moreover, fast breeder reactors actually utilize the much more abundant U-238 atoms in its fuel process. Other breeder reactors can use thorium, which is three times more abundant than uranium.

Several fast breeders have been built and tested but, because their fission process is more complicated and their capital costs higher, they have not been as economically successful as standard U-235 nuclear reactors. If nuclear power is to be expanded, however, this technology has to be further developed in some form. In addition to the fast breeder reactor, other types of breeders are currently being developed. India, for example, is currently developing a thermal breeder reactor for commercial use. The Indian reactor uses thorium as its fuel. India has about one-third of the world's reserves of thorium, so it makes economic sense for India to develop a breeder reactor that utilizes thorium.

Fast breeder reactors convert U-238 to Plutonium-239 (PU-239) under the action of fast moving neutrons. The PU-239 is fissile, whereas the U-238 is not. Due to the complexities in the handling of the fast moving neutrons, the standard nuclear reactors cannot be used.

One variation on the fast breeder reactor is the liquid metal fast breeder reactor shown in Figure 7.5. In this type of reactor, a liquid metal is used as the primary coolant. This liquid metal captures heat directly from the nuclear reactor core, then exchanges heat with water in the steam cycle. The steam is used to generate electricity in a steam turbine. The liquid metal is used as the primary coolant because water tends to slow down the fast neutrons needed for the nuclear reaction. Sodium, mercury, and lead have been used as the circulating liquid metal. In this reactor, even though the process "breeds" fissile material, the U-238 is actually consumed

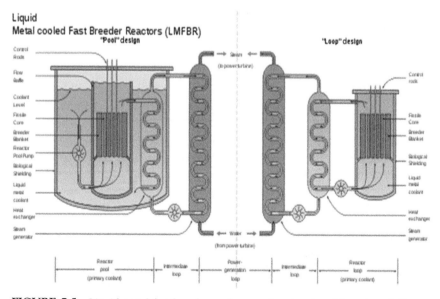

FIGURE 7.5 Liquid metal fast breeder nuclear reactors (pool and loop types). *Source: Graeve Moore.*

as fuel. Due to the fact that U-238 is much more abundant than U-235, the limited uranium supply is not a problem for nuclear expansion.

France built two fast breeders called the phenix and the superphenix and tested them extensively before shutting them down in 1996. The cleanup cost for the superphenix cost $1 billion. Germany spent €3.6 billion and 19 years to build a fast breeder reactor called the SNR-300, completing it in 1985. It sat idle for about 7 years before it was scrapped for political reasons without ever having run. The important point to note is that breeder technology is necessary if nuclear power is to be expanded.

Thorium-232 (Th-232) can be used in both fast breeder reactors and in thermal breeder reactors. It is first converted to U-233 according to the following breeder reaction:

$$^1n + {}^{232}\text{Th} \rightarrow {}^{233}\text{Th} \rightarrow {}^{233}\text{Pa} \rightarrow {}^{233}\text{U}$$

In this reaction, the neutron bombards the nucleus of the Th-232, which then becomes Th-233. This then transmutes into 233-Protactinium (Pa-233) via beta decay. The Pa-233 undergoes a second beta decay to transmute into U-233, which is fissile in a similar manner to U-235. There are several advantages to using Th-232 as the fuel, particularly in thermal breeder reactors:

- It is more abundant than uranium
- Mined thorium consists of a single isotope (Th-232)
- It has a superior neutron usage
- It has favorable physical and chemical properties (higher melting point, higher thermal conductivity, lower coefficient of thermal expansion, and more stable oxides) which improve reactor performance, and
- It produces less toxic and shorter lived trans-uranium waste products that improve waste repository performance.

URANIUM RESERVES AND SOURCES

Most of the uranium used in U.S. nuclear power reactors comes from foreign suppliers. Only about 14% comes from U.S. mined uranium. The operators of U.S. nuclear power reactors purchased and used a total of 50 million pounds of U_3O_8 during 2009, at a weighted-average price of $45.86 per pound. The 2009 total was a decrease of 7% compared with the 2008 total of 53 million pounds. 14% of the U_3O_8 delivered in 2009 was of U.S. origin at a weighted-average price of $48.92 per pound. Foreign-origin uranium accounted for the remaining 86% of deliveries at an average price of $45.35 per pound. Australian-origin and Canadian-origin uranium together accounted for 40% of the 50 million pounds. Uranium originating in the former Soviet Union (Kazakhstan, Russia, and

Uzbekistan) accounted for 29% and the remaining 17% originated in Brazil, Czech Republic, Namibia, Niger, and South Africa.

The nuclear power plants purchased uranium of several different types: Uranium oxide (U_3O_8) concentrate was 69% of the deliveries in 2009, and natural UF_6 and enriched uranium were 31%. During 2009, 17% of the uranium was purchased under spot contracts at a weighted-average price of $46.45 per pound. The remaining 83% was purchased under long-term contracts at an average price of $45.74 per pound.

The U.S. Energy Information Administration last updated its estimates of U.S. uranium reserves for year-end 2008. This represented the first revision of their estimates since 2004. The update is based mainly on analysis of company annual reports rather than geological analysis. At the end of 2008, their estimate of U.S. uranium reserves totaled 1227 million pounds of U_3O_8 at a maximum forward cost of up to $100 per pound of U_3O_8. At up to $50 per pound, U_3O_8, estimated reserves were 539 million pounds. Based on average 1999-2008 consumption levels, uranium reserves available at up to $100 per pound of U_3O_8 represented approximately 23 years' worth of demand, while uranium reserves at up to $50 per pound of U_3O_8 represented about 10 years' worth of demand; however, because domestic U.S. uranium production supplies only 14% on average of U.S. requirements for nuclear fuel, the effective years' supply of domestic uranium reserves is actually seven times higher than these numbers.

In 2008, Wyoming led the nation in total uranium reserves in both the $50 and $100 per pound U_3O_8 categories, with New Mexico second. Taken together, these two states constituted about two-thirds of the estimated reserves in the country available at up to $100 per pound U_3O_8 and three-quarters of the reserves available at less than $50 per pound U_3O_8. By the mining method, uranium reserves in underground mines constituted 39% of the available product at up to $100 per pound of U_3O_8. This is followed by open-pit mining at 31% and *in situ* leaching (ISL) at 29%. At $50 per pound U_3O_8, however, uranium available through ISL was about 40% of the total reserves, somewhat higher than uranium in underground and open-pit mines in that cost category. ISL is the dominant mining method for U.S. production today.

Up until 2004, the average price being paid for uranium supplies was about $10 per pound. After that, the price increased steadily to $45 per pound due to increased demand and diminishing supplies. It briefly spiked at over $100 per pound in late 2007 and early 2008, but it then settled out at $45 per pound. As the price goes up, the supply increases because more uranium deposits become economic to mine. The low price discouraged the development of mines in the United States, but as the price has now increased, new U.S. supplies are being developed.

U.S. mines produced 4.1 million pounds of U_3O_8 in 2009, 7% more than in 2008. Fourteen underground mines produced ore-containing uranium during 2009, four more than 2008. Four ISL mining operations produced solutions containing uranium in 2009, two less than during 2008. Overall, during part or all of 2009, there were 18 U.S. mines that produced uranium to be processed into uranium concentrate. The total production of U.S. uranium concentrate (yellowcake) in 2009 was 3.7 million pounds, 5% below the 2008 level, from one U.S. mill (White Mesa Mill) and four ISL plants (Alta Mesa Project, Crow Butte Operation, Kingsville Dome, and Smith Ranch-Highland Operation). Shipments of uranium concentrate from these facilities were 3.6 million pounds in 2009, 12% below the 2008 level. The annual production of uranium from U.S. mines for the last 17 years is shown in Figure 7.6. Figure 7.7 shows the comparison between the annual production and annual shipments of uranium yellowcake from mines in the United States.

The United States has the capacity and reserves to increase its indigenous supply of uranium from U.S. sources but, at the moment, the nuclear power industry is highly dependent on foreign suppliers to fuel its nuclear power plants. This situation does not represent as large a balance of payments problem as does the large oil import bill because the 45 to 50 million pounds of yellowcake that it is currently importing only costs about $2.5 billion per year. It does, however, represent a significant security of supply issue, particularly if the United States wants to expand its nuclear power production. It is unlikely that the United States can

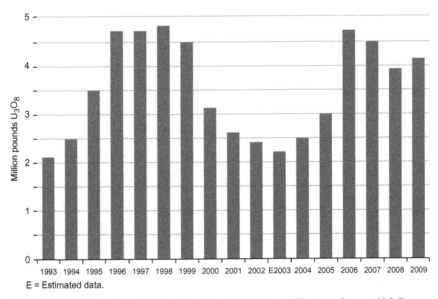

FIGURE 7.6 Annual production of uranium (U_3O_8) from U.S. mines. *Sources: U.S. Energy Information Administration: 1993-2002-Uranium Industry Annual 2002 (May 2003), Table H1 and Table 2. 2003-2009-Form EIA-821A, "Domestic Uranium Production Report" (2003-2009).*

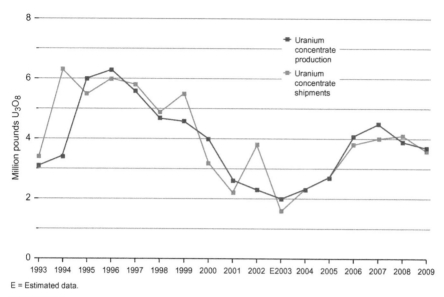

E = Estimated data.

FIGURE 7.7 Comparison of annual production and shipments of uranium (U_3O_8) from U.S. mines. *Sources: U.S. Energy Information Administration: 1993-2002-Uranium Industry Annual 2002 (May 2003), Table H1 and Table 2. 2003-2009-Form EIA-821A, "Domestic Uranium Production Report" (2003-2009).*

increase its production of uranium enough to become self-sufficient in uranium. There is also some uncertainty about how long the foreign supplies will continue, particularly if nuclear power production was increased. This situation would be rectified if fast breeder reactors were commercialized.

The world supply of uranium is dominated by Kazakhstan, Australia, and Canada, with significant production coming from Russia, Uzbekistan, Namibia, and Niger. The world resources are sufficient for the current usage. If the price increases due to shortages, more dilute uranium deposits could be produced. The world price of uranium did spike in 2007-2008 due to strong demand, and it is likely to do so again if nuclear power demand picks up. The world reserves of uranium are shown in Table 7.3. These reserves are a function of the current world price, which is $45 per pound. At this price, Australia dominates in world reserves holding 1.673 million tons, which is 31% of the world's known economic reserves of 5.404 million tons. The current usage rate of uranium is 53,663 tons/year, so the 5.404 million tons of reserves represent 100.7 years of reserve life at today's usage rates. Based on the U.S. reserve estimates above, it could be projected that if the price of uranium doubled (which would not substantially affect the economics of today's reactors), the amount of uranium reserves would also double. This would represent

TABLE 7.3   World Uranium Reserves 2009

	Tons U	% of world	EJ
Australia	1,673,000	31%	292.9
Kazakhstan	651,000	12%	114.0
Canada	485,000	9%	84.9
Russia	480,000	9%	84.0
South Africa	295,000	5%	51.6
Namibia	284,000	5%	49.7
Brazil	279,000	5%	48.8
Niger	272,000	5%	47.6
USA	207,000	4%	36.2
China	171,000	3%	29.9
Jordan	112,000	2%	19.6
Uzbekistan	111,000	2%	19.4
Ukraine	105,000	2%	18.4
India	80,000	1.50%	14.0
Mongolia	49,000	1%	8.6
Other	150,000	3%	26.3
World total	5,404,000		946.0

200 years of supply at today's usage rate. If fast breeder reactors were commercialized, the fuel supply from uranium would be increased by 150-fold. Clearly, the capacity to increase the supply of nuclear power exists, even with the current technology; however, substantial increases would require the commercialization of fast breeder technology.

Table 7.3 conversions require some explanation. A 1-GW nuclear power plant requires approximately 200 short tons of uranium per year. In 1 year, the 1-GW plant can generate 8760 GWh of electricity. This equates to 43.8 GWh/ton, 48.63 GWh/ton, or 175.06 TJ/ton. This conversion factor has been used to convert the reserves of uranium into exajoules in Table 7.3. The current total world reserve of uranium is 946 EJ, of which 293 EJ are located in Australia, even though Australia produces substantially less than Kazakhstan and Canada. World uranium production for the past 8 years is shown in Table 7.4. In 2010, 53,663 tons of uranium was produced, representing 63,285 tons of U_3O_8 (yellowcake). In the past 4 years, production increases of 9% per year have been seen.

TABLE 7.4 World Uranium Production (Tons)

Country	2003	2004	2005	2006	2007	2008	2009	2010
Kazakhstan	3300	3719	4357	5279	6637	8521	14,020	17,803
Canada	10,457	11,597	11,628	9862	9476	9000	10,173	9783
Australia	7572	8982	9516	7593	8611	8430	7982	5900
Namibia	2036	3038	3147	3067	2879	4366	4626	4496
Niger	3143	3282	3093	3434	3153	3032	3243	4198
Russia	3150	3200	3431	3262	3413	3521	3564	3562
Uzbekistan	1598	2016	2300	2260	2320	2338	2429	2400
USA	779	878	1039	1672	1654	1430	1453	1660
Ukraine (est)	800	800	800	800	846	800	840	850
China (est)	750	750	750	750	712	769	750	827
Malawi							104	670
South Africa	758	755	674	534	539	655	563	583
India (est)	230	230	230	177	270	271	290	400
Czech Repub.	452	412	408	359	306	263	258	254
Brazil	310	300	110	190	299	330	345	148
Romania (est)	90	90	90	90	77	77	75	77
Pakistan (est)	45	45	45	45	45	45	50	45
France	0	7	7	5	4	5	8	7
Germany	104	77	94	65	41	0	0	0
Total world	35,574	40,178	41,719	39,444	41,282	43,853	50,772	53,663
Tons U_3O_8	41,944	47,382	49,199	46,516	48,683	51,716	59,875	63,285

NUCLEAR POWER PLANT SAFETY

One of the battles that nuclear power has had to fight is the public perception that it is too risky to be used. Nuclear power generates radioactive waste. If this waste escapes into the environment, the consequences can be severe and include such things as radiation sickness, cancer, and death. The nuclear industry has a remarkably good safety record, but the consequences of a nuclear accident can be so catastrophic and the effects so long lasting that the standards of safety must be extremely high. Society cannot tolerate a mishap that would not only kill many people but would also render large areas of land uninhabitable. There have been a few incidents where public safety was put at risk but three accidents in particular stand out as fueling the fears of the public about expanded use of nuclear power.

THE THREE MILE ISLAND INCIDENT

The accident at the Three Mile Island Unit 2 (TMI-2) nuclear power plant near Middletown, Pennsylvania on March 28, 1979, caused no deaths or injuries to plant workers or members of the public. It was, however, the first serious nuclear accident and is still considered the most serious in U.S. nuclear power plant operating history. It also ushered in changes in nuclear reactor operations and accident response. Because the nuclear industry learned lessons from the event, the industry has been made much safer.

The accident began about 4:00 a.m. on March 28, 1979, when the main feed water pumps broke down and shut down the steam circulation to the turbines. The steam turbine and the main reactor shut down automatically, but reactor fuel was still generating heat and so the pressure in the primary cooling system began to increase. In order to prevent that pressure from becoming excessive, a pressure relief valve opened automatically. The valve should have closed again when the pressure decreased by a certain amount, but it did not. As a result, cooling water leaked out of the open valve and caused the core of the reactor to overheat. The plant did not have instruments that showed that the valve remained open or that showed the level of coolant in the core. Because adequate cooling was not available, the nuclear fuel overheated and the metal tubes which hold the nuclear fuel pellets ruptured and the fuel pellets began to melt. About one-half of the core melted during the early stages of the accident. Although the TMI-2 plant suffered a severe core meltdown, it did not lead to a breach of the walls of the containment building and it did not release large quantities of radiation to the environment.

Detailed studies of the radiological consequences of the accident have been conducted by many state and federal agencies. Estimates are that the average dose to about 2 million people in the area was only about 1 mrem. This is a very low radiation exposure, less than a single chest X-ray. There were no deaths or injuries and no evidence of cancer increases in the area.

THE CHERNOBYL DISASTER

On April 26, 1986, an accident occurred at Unit 4 of the nuclear power station at Chernobyl in the Ukraine in the former USSR. An aerial view of the plant taken in 1997, 11 years after the accident, is shown in Figure 7.8. The accident was apparently caused by a sudden power surge to the reactor controls. As a consequence, the reactor caught fire and exploded and released massive amounts of radioactive material into the environment.

FIGURE 7.8   Aerial view of the Chernobyl Power Plant. *Source: NASA/Jerry M. Linenger,*
1997.

This radioactive material drifted downwind and was detected in Sweden
within hours. Boron and sand were poured on the reactor from the air and
then the damaged unit was encased in concrete. The three other units
of the Chernobyl nuclear power station were subsequently restarted.
The Soviet nuclear power authorities presented an initial report on the
accident at an International Atomic Energy Agency meeting in Vienna,
Austria, in August 1986.

After the accident, access to the area in a 30-km (18-mile) radius around
the plant was closed, except for persons requiring official access to the
plant and to the immediate area for evaluating and dealing with the consequences of the accident and operation of the undamaged units. 116,000
people were evacuated from the most heavily contaminated areas in 1986
and another 230,000 people in subsequent years. Pripyat, the town near
Chernobyl where most of the workers at the plant lived before the

accident, was evacuated several days after the accident because of radiological contamination. It was included in the 30-km exclusion zone around the plant and is still closed to all but those with authorized access.

The Chernobyl accident caused many severe radiation effects almost immediately. Among the 600 workers present on the site at the time of the accident, two died within hours of the reactor explosion and 134 received high radiation doses and suffered from acute radiation sickness. Of these, 28 workers died in the first 4 months after the accident. Another 200,000 recovery workers involved in the initial cleanup work of 1986-1987 were exposed to high doses of radiation, between 0.01 and 0.50 grays. The gray is the SI unit for absorbed radiation, which is defined as one joule of deposited energy per kilogram of tissue. To assess the risk of radiation, the absorbed dose is multiplied by the relative biological effectiveness of the radiation to get the biological dose equivalent in rems or sieverts (1 Sv is equal to 100 rem or 100,000 mrem). As time went on, the number of workers involved in cleanup activities at Chernobyl rose to 600,000, although only a small fraction of these workers was exposed to dangerous levels of radiation. Both groups of cleanup and recovery workers may still become ill because of their radiation exposure so their health is still being monitored.

The Chernobyl accident also resulted in widespread contamination in the areas of Belarus, the Russian Federation, and the Ukraine—areas inhabited by millions of residents. Radiation exposure to residents evacuated from areas heavily contaminated by radioactive material from the Chernobyl accident also has been a concern. Average doses to Ukrainian and Belarusian evacuees were 17 and 31 mSv, respectively. Individual exposures ranged from a low of 0.1 to 380 mSv. The majority of the 5 million residents living in contaminated areas, however, received very small radiation doses, which are comparable to natural background levels, which are 1 mSv (or 100 mrem) per year.

The health of these residents has also been monitored since 1986 and to date there is no strong evidence for radiation-induced increases of leukemia or other cancers (other than thyroid cancer). An exception is a large number of children who in 1986 received substantial radiation doses to their thyroid after drinking milk contaminated with radioactive iodine. To date, about 4000 thyroid cancer cases have been detected among these children. Although 99% of these children were successfully treated, nine children in the three countries died from thyroid cancer. No other evidence of any effect on the number of adverse pregnancy outcomes, delivery complications, stillbirths or overall health of children has been observed among the families living in the most contaminated areas.

Apart from the increase in thyroid cancer after childhood exposure, no increase in overall cancer or noncancer diseases have been observed that can be attributed to the Chernobyl accident and exposure to radiation.

Nevertheless, it is estimated that approximately 4000 radiation-related cancer deaths may eventually be attributed to the Chernobyl accident over the lifetime of the 200,000 emergency workers, 116,000 evacuees and 270,000 residents living in the most contaminated areas. This estimate is somewhat lower than initial speculations that radiation exposure would claim up to 100,000 lives.

THE FUKUSHIMA-DAIICHI INCIDENT

The Fukushima-Daiichi nuclear incident was caused by a huge earthquake (9.0 on the Richter scale) off the northeast coast of the main Japanese Island of Honshu on March 11, 2011. The earthquake spawned a giant tsunami that created waves up to 128 ft tall in some areas adjacent to the earthquake. The waves were estimated at 45 ft tall when it swamped the Fukushima-Daiichi nuclear power plant, 120 miles away, about 1 h after the earthquake. The power plant is on the northeast coast of Honshu island about 120 miles southwest of the epicenter of the earthquake, which was located about 80 miles offshore from the city of Sendai.

There are 6 reactors at the Fukushima-Daiichi nuclear power plant, but at the time of the quake, reactors 4, 5 and 6 had been shut down for maintenance. Reactors 1, 2 and 3 were operating, but they shut down automatically after the earthquake, with emergency electrical generators starting up to run the water pumps needed to cool the reactor cores. The plant was protected by a seawall designed to withstand a 19-ft tsunami, but when the 45-ft wave arrived the entire plant was flooded, knocking out the generators and the cooling water pumps. This caused a partial core meltdown in the operating reactors 1, 2 and 3. Explosions and severe damage occurred, and radiation leaked out into the surrounding countryside. The entire area, which had been badly damaged by the earthquake and tsunami, had to be evacuated to within 30 km of the plant. Figure 7.9 shows the cumulative radiation contamination as of April 29, 2011. No deaths or severe illnesses can be attributed to the accident, but the incident is ongoing at the time of publication.

NUCLEAR FUSION REACTORS

It is clear from the discussion at the beginning of the chapter that nuclear fusion processes release much more energy than the fission processes; for example, the hydrogen bomb is far more potent than the uranium or plutonium bombs that were dropped on Hiroshima and Nagasaki in 1945. It would be expected, therefore, that a nuclear fusion reactor would be much more efficient and generate much more energy than a

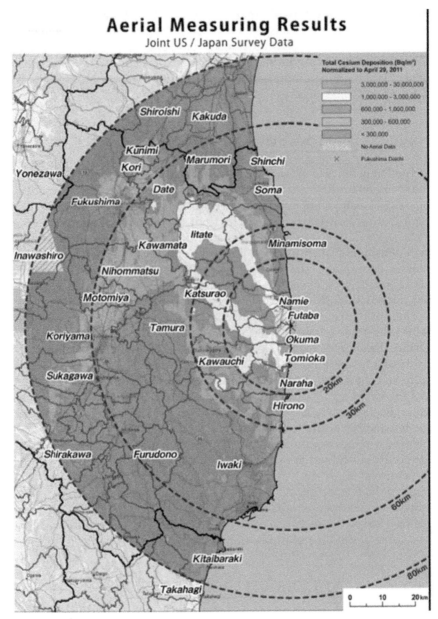

FIGURE 7.9 Map of deposition of Radio-Cesium (sum of Cs-134 and Cs-137) for the land area Within 80 km of the Fukushima-Daiichi Plant, as reported by the Japanese Authorities.

nuclear fission reactor. This was the expectation from nuclear scientists. Consequently, a lot of time and effort has been devoted to the development of a fusion reactor. Surprisingly, this effort has failed. In fact, so far no one has been able to build a fusion reactor that even produces more energy than it consumes. The hydrogen bomb clearly puts out more energy than it consumes, but for a nuclear power reactor, the reaction must be controlled so that the energy can be harnessed in an orderly fashion. To initiate the fusion reaction in a controlled environment, very high pressures and temperatures must be generated. This takes more energy than can be recovered from the resulting fusion process. In a hydrogen bomb, the nuclear fusion reaction is initiated by detonating a fission bomb. This technique cannot be used in a nuclear fusion reactor.

In the fusion process, two hydrogen atoms are fused into a helium atom; however, standard hydrogen (which is more than 99.8% of the hydrogen that exists on Earth) has just one proton and no neutrons in its nucleus, whereas a standard helium atom has two protons and two neutrons in its nucleus. The simplest way to create helium using nuclear fusion would suggest using an isotope of hydrogen called deuterium. Deuterium is a hydrogen atom that has one proton and one neutron in its nucleus. 0.0156% of the water molecules in seawater contain deuterium. Due to the large volume of seawater on Earth, there is also a large amount of deuterium that can be made available to fuel fusion reactors on a commercial scale.

This type of water containing deuterium is called heavy water, and it can be separated from seawater using the standard diffusion or centrifuge processes described earlier. The deuterium is liberated from the water using electrolysis to produce pure deuterium. When two deuterium atoms are fused with each other, however, one of the neutrons escapes and light helium (He-3) is created. A second possibility is one of the two deuterium atoms sheds a neutron that is captured by the other to create tritium (which is hydrogen with two neutrons and one proton in its nucleus). This is shown in the following two reactions:

$$^2D + {}^2D \rightarrow {}^3T + {}^1H$$

$$^2D + {}^2D \rightarrow {}^3He + {}^1n$$

It turns out that there is an easier fusion process to initiate and that is to fuse the two main isotopes of hydrogen, deuterium, and tritium, which produces standard helium plus one neutron.

$$^2D + {}^3T \rightarrow {}^4He + n$$

Deuterium is twice as heavy as standard hydrogen (^1H) which makes the separation from seawater rather easy compared to the uranium enrichment process. Tritium (^3H or ^3T) is also an isotope of hydrogen, but it does not occur naturally, so it is usually manufactured from lithium using the following nuclear reactions:

$$^1n + {}^6\text{Li} \rightarrow {}^3\text{T} + {}^4\text{He}$$

$$^1n + {}^7\text{Li} \rightarrow {}^3\text{T} + {}^4\text{He} + {}^1n$$

The reactant neutron can be supplied by the deuterium-tritium (D-T) fusion reaction shown above. The reaction with ^6Li is exothermic, providing a small energy gain for the reactor. The reaction with ^7Li is endothermic, but does not consume the neutron. At least some ^7Li reactions are necessary because some neutrons are lost by reactions with other elements. Most reactor designs use the naturally occurring mix of lithium isotopes (^7Li and ^6Li).

The D-T nuclear reaction is the easiest fusion reaction to initiate, but it has some disadvantages.

- It produces neutrons that result in increased radioactivity within the reactor structure
- Only about 20% of the fusion energy yield appears in the form of charged particles, which limits the extent to which direct energy conversion techniques might be applied
- The tritium supply currently depends on lithium resources, which are less abundant than deuterium resources, and
- Tritium is radioactive and hard to handle.

The neutron flux in a commercial D-T fusion reactor has been estimated to be more than 100 times the neutron flux of current fission power reactors. In a commercial reactor, the neutrons produced could be used to react with lithium in order to create the tritium. This also channels the energy of the neutrons into the lithium. This reaction protects the reactor envelope from the neutron flux.

Though more difficult to initiate than the D-T reaction, the Deuterium-Deuterium (D-D) nuclear fusion reaction has certain advantages. The D-D reaction has two branches as shown above. The optimum energy for the initiation of the D-D reaction is 15 keV, which is slightly higher than the optimum for the D-T reaction. The first reaction does not produce neutrons, but it does produce tritium. Most of the tritium produced will be consumed in the D-T reaction before leaving the reactor, which reduces the tritium handling required, but also means that more neutrons are produced and that some of these are very energetic. The neutron produced from the second branch of the D-D reaction has an energy of only 2.45 MeV (0.393 pJ), whereas the neutron from the D-T reaction has an energy of 14.1 MeV (2.26 pJ), resulting in more isotope products and more material damage than occurs in the D-D reaction. Assuming complete tritium burn-up, the reduction in the fraction of fusion energy carried by neutrons is only about 18%, so the primary advantage of the D-D fuel cycle is that tritium production is not required. Other advantages are independence from limitations of lithium resources and less energetic neutron production.

Current research technology uses a magnetic torus to confine the deuterium fuel as plasma in a small enough space to create the reaction. Lasers have also been proposed to confine the plasma fuel.

The largest fusion reactor experiment so far has been the Joint European Torus (JET) in Culham, England. In 1997, JET produced a momentary peak of 16 MW of fusion power which was only 65% of the input power, with the output fusion power of 10 MW being sustained for just half a second. The next generation fusion reactor, ITER, is currently under construction in Cadarache, France. It is designed to produce 10 times more fusion power than the power put into the plasma, but it will only operate for a few minutes at a time. After 60 years of unsuccessful attempts to develop a commercial fusion reactor, research continues but there is considerable skepticism about its commercial viability.

There have been claims over the years of simpler cheaper methods of creating fusion reactions, but no commercial process has been demonstrated. The most famous of these experiments was the cold fusion claims of Martin Fleischman and Stanley Pons at the University of Utah in 1989. Pons and Fleishman announced results of an experiment where they electrolyzed heavy water using a palladium electrode. They measured more heat coming from the electrolysis chamber than was being input by the electrolysis electrodes and surmised that they had created cold fusion in the palladium electrode. Subsequent scrutiny discounted this as being caused by a fusion reaction. Nevertheless, some researchers still believe that cheap cold fusion processes are a possibility.

THE FUTURE OF NUCLEAR POWER

Because of the limited supply of uranium, it might seem that with current commercial technology, nuclear power might not be able to play an expanded role in the future energy resources of the world or of the United States. A doubling in the uranium price, however, is likely to lead to a 200-year supply, which would allow expansion of nuclear power to triple what it is now. Moreover, if public concerns about safety can be addressed, it is also feasible that breeder reactor technology does show promise for even more greatly increased nuclear power resources to be used. Breeder reactors can be based on both uranium (U-238) and thorium (Th-232) fuels, and there are sufficient quantities of these fuels to supply a lot of breeder reactor power for a very long time. Until the technology is further developed and commercial breeder reactors are built, however, nuclear power may remain at today's levels.

8

Solar Power

There is immense public interest in developing solar power for widespread use, but there are several major problems with this development. First, the sun does not shine all the time and so it cannot be a reliable source without reliable energy storage technology. Second, and more importantly, the cost is still 10 times the cost of other energy sources and so it is not cost competitive. The public likes solar energy because they perceive it to be cheap and clean with no environmental impacts. Surprisingly, this is not true; it does have environmental impacts that are not fully appreciated. For instance, would people be happy if we had to cover the Mojave Desert with solar collectors or if we cut down 50 million trees so they would not shade the solar collectors on our roofs? Just as hydroelectric power changes the ecology of the rivers that provide the power, solar collectors change the ecology of the areas where they are deployed.

In some respects, most of the energy used on earth is solar energy. The only exceptions are nuclear power and geothermal energy. Nuclear power originated in the creation of uranium in a supernova hundreds of millions of years ago, while geothermal energy comes from radioactive processes deep in the earth's core. All fossil energy originally came from solar energy generated millions of years ago. Wind energy and hydroelectric energy also depend directly on solar activity. The sun shining on the earth creates the hot spots that cause the wind to blow. Solar activity also evaporates the water from the oceans to cause the rain to fall at higher elevations, where it is captured as potential energy and converted into electrical energy.

Every day the sun beams an average of 60,000 EJ of energy directly to the earth. If it did not receive this energy, the earth would die. Figure 8.1 illustrates what happens to this energy: 6% is reflected by the atmosphere; 20% is reflected by the clouds; 4% is reflected by the earth's surface; 16% is absorbed by the atmosphere; 3% is absorbed by the clouds; and 51% is absorbed by the land and the oceans. Very little of the energy is actually captured. Most of what is absorbed is radiated back into space. Nevertheless, solar energy plays a crucial role in controlling the temperature of the

The Future of Energy
http://dx.doi.org/10.1016/B978-0-12-801027-3.00008-7

161

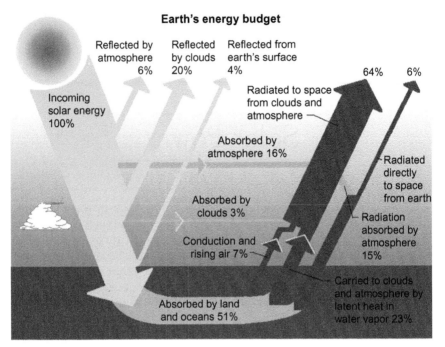

FIGURE 8.1 The capture and disbursement of solar energy by the earth. *Source: NASA.*

earth. When solar insolation increases, the earth warms up and when it decreases the earth cools down. Solar insolation has the most significant impact on the earth's temperature, much larger than all the greenhouse gases combined. A very small part of the sun's energy is captured by plants and animals on the earth and used to support the growth of living organisms on the planet.

The total solar energy that reaches the upper atmosphere is 1367 W/m². Of this, about 697 W/m² reaches the earth's surface and is absorbed by the land and the oceans. The other 670 W/m² is intercepted by the atmosphere and reflected or absorbed and reradiated back into space. Not all of this energy is in a form that can be captured by photovoltaic (PV) technology. By the time it reaches the earth's surface, the usable solar energy received averages about 300 W/m². The diameter of the earth is 12,750 km (7920 miles). The surface area of the earth is 510 Tm², of which 149 Tm² is land and the other 361 Tm² is ocean. The usable solar energy hitting the land area is 300×149 TW $= 45$ PW. The world uses about 500 EJ/year, which is $500/(365 \times 24 \times 3600)$ EW $= 0.016$ PW. A tiny fraction of the usable solar energy that is hitting the land mass of the earth could power the entire world. Put another way, if the 45 PW hitting the land was captured with 10% efficiency, it would take 0.35% of the land area to be used to capture enough solar energy to power the entire world. Of course, it is not as simple as that. Solar energy is very expensive. It is intermittent, unreliable and requires exotic materials. There are simply not enough of these materials to do the job.

Solar power technology attempts to capture some solar energy and convert it to electricity to power a portion of our lifestyle. Some solar energy can also be captured as heat energy and used to heat our houses, buildings and hot water. Capturing the vast energy of the sun has been the dream of scientists for millennia. Unfortunately, it is not an easy thing to do economically. Part of the reason for this is that solar energy is very diffuse and it costs a lot to concentrate it into usable forms.

There are two main methods of directly capturing sunlight to be used as an energy source for electricity generation. The first is through the use of semiconductors to convert sunlight directly into electricity. This method is known as photovoltaic (PV) solar energy. The second method involves using mirrors to concentrate sunlight onto pipes or towers containing liquid material that is used to heat steam to drive a turbine to make electricity. This method goes by the name of concentrated solar power (CSP). Figure 8.2 shows one such facility while Figure 8.3 shows the schematic of a different design.

The basic PV installation, converting sunlight into direct current (DC) electricity, outputs DC electricity, which is used to charge a battery, which is then used to power DC loads such as lights or heating elements. The

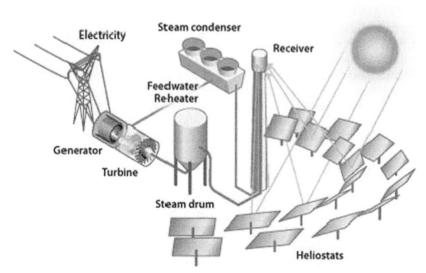

FIGURE 8.2 Diagram of a Solar Power Tower thermal power plant. *Source: US DOE.*

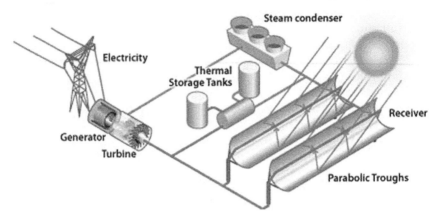

FIGURE 8.3 Schematic of the parabolic trough solar power at the SEGS Plants in California. *Source: US-DOE.*

electrical grid, however, is alternating current (AC), which can be used to power lights and heating elements but also to run electric motors that drive fans and compressors. DC motors are more difficult and expensive to run as electrical motors, so it is easier to convert DC into AC using an inverter.

Figure 8.4 shows the growth in solar energy in the United States over the past 15 years. The production was very flat up until 2008, when a sharp rise occurred. Nevertheless, the solar power production is still very small. Figure 8.5 shows the monthly production over the 2 years from January 2009 to March 2011. This indicates that solar energy production decreases dramatically in the wintertime. The short winter days and the lower angle of the winter sun have a large effect on this, but later analyses suggest that the decrease is much larger than expected. In places such as Tucson, Arizona, and the Mojave Desert in California where the Solar Energy Generating Systems (SEGS) CSP plants are located, the decrease in winter solar radiation is not nearly as much as Figure 8.4 would suggest. The explanation for this is that some of the larger solar energy plants are deliberately shut down through the winter because they have recently been operated as daily summer peaking units.

PHOTOVOLTAICS

Solar PV cells work on the principle of the photoelectric effect. When a photon of light strikes a solar cell, its energy is transferred to an electron in the semiconductor material in the top layer. If enough energy is absorbed

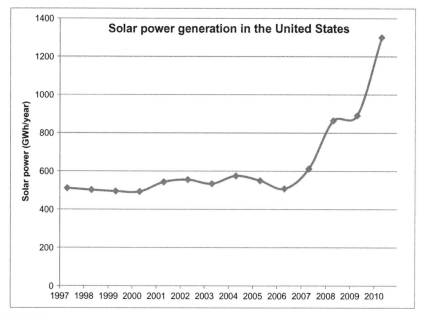

FIGURE 8.4   Annual solar power generated in the United States. *Data Source: EIA.*

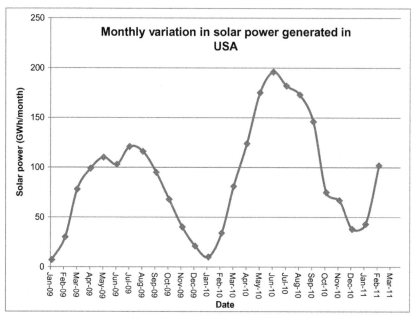

FIGURE 8.5   Monthly variation in solar power generated in the United States. *Data Source: EIA.*

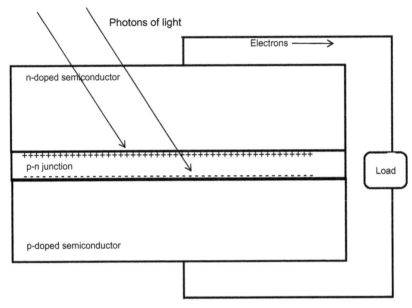

FIGURE 8.6 Operation of a photovoltaic cell.

by the electron, it can escape from its normal position in the atom. In this process, the electron creates an empty spot where the electron used to be, which is termed the electron "hole." Each photon of light with the right level of energy will free one electron and create one hole. The semiconductors and the junction between them create an electrical field that is strong enough to push the freed electron through the material and around an electric circuit to create an electrical current. This is illustrated in Figure 8.6.

Father and son scientists Antoine César Becquerel and Alexandre Edmond Becquerel discovered this PV effect in 1839, when Alexandre was only 19 (Antoine was 51). Alexandre, under the supervision of his father, was experimenting with a beaker containing an electrolyte and two metal electrodes and noticed that certain metals and solutions could produce small electric currents when exposed to light. They reported this discovery to the scientific community, but could not find a practical use for it. The Becquerels were part of a very famous family of French scientists, and Alexandre's son, Henri, became even more famous than his father and grandfather when he codiscovered radioactivity along with Pierre and Marie Curie. Henri Becquerel was awarded the Nobel Prize in physics in 1903. The SI unit of radioactivity is named after him.

In 1877, American Charles Fritts improved on the Becquerels' discovery using selenium as the semiconductor, separated by a very thin layer of gold. Fritts was able to generate electricity with an efficiency of about 1%. Albert Einstein's landmark paper on the photoelectric effect in 1905

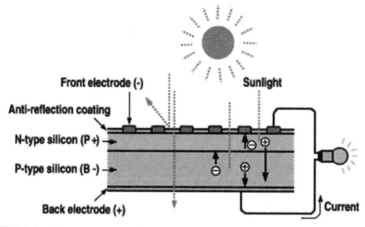

FIGURE 8.7 The photovoltaic effect

provided the theoretical framework that began to explain the PV phenomenon, which led to further improvements. By 1935, both selenium and copper oxide semiconductors were being used as light sensors in PV cells for use in photographic photometers. The first solar cell based on a semiconductor made from silicon was developed by Bell Laboratories' Russell Ohl in 1941, and this led to substantially increased conversion efficiency. It was further improved and commercialized by Bell Laboratories in 1954. The Bell cells have since been widely used in satellite technology and space exploration.

The standard solar cell consists of three primary layers (see Figures 8.6 and 8.7). The top layer is made of a positively charged, electron-attracting, n-type semiconductor, while the bottom layer is made of a negatively charged, hole-attracting, p-type semiconductor. In between these layers is a doped semiconductive material called the p-n junction, which is a very thin single crystal layer that diffuses from p-type to n-type. An electric force field is created by the depletion of the electrons and holes at the p-n junction. This electric field provides the voltage needed to force the electrons to jump across the p-n junction and move through the semiconductors and around the loaded electric circuit (see Figure 8.6). The electron flow represents the electrical current driven by the voltage of the cell's electric field. The product of the generated current (in amperes) and the generated voltage (measured in volts) constitutes the electrical power in watts. The power in watts multiplied by time in seconds gives the energy generated in joules.

The solar cell also has a conducting layer on the top and bottom of the cell to collect the electrical current flowing into and out of the cell. The contact layer on the top face of the cell, where light enters, is designed to let light in; at the same time, it also conducts the electrons out. It is composed of a metal wire conductor arranged in a grid pattern, with enough wire to conduct the electrons away, but also with enough space between the wires to allow the light sensitive semiconductor to have access to the light source. The electrical contact layer on the bottom of the cell covers the entire bottom surface of the cell because it does not have to be transparent to the light. Covering the surface wire grid and the n-type semiconductor is an antireflective coating to reduce reflection losses and hold the wire grid in place.

The most common material used in solar cells has been single crystal silicon. Commercial solar cells made from single crystal silicon are currently limited to about 20% efficiency because they are most sensitive to infrared light. Solar radiation in this region of the electromagnetic spectrum is lower in energy than violet and ultraviolet light. The best research lab efficiency has been at the University of New South Wales (Australia) where their single crystal silicon solar cells have achieved efficiencies up to 25% (see Figure 8.8).

Polycrystalline solar cells are made by a casting process in which hot liquid silicon is poured into a mold and cooled, then sliced into very thin wafers. This process makes the solar cells significantly cheaper to produce than single crystal cells, but their commercial efficiency has been less than 15% due to internal resistance at the boundaries of the silicon crystals. The best research lab efficiency has been 20.4% (see Figure 8.8).

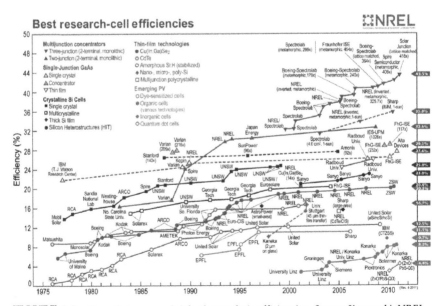

FIGURE 8.8   Growth in research lab photovoltaic efficiencies. *Source: Kazmerski, NREL.*

Amorphous silicon solar cells are made using silane (SiH_4) gas to deposit silicon onto a transparent medium. This type of solar cell can be applied as a thin film to low-cost substrates such as glass or plastic. Thin film cells have a number of advantages: easier deposition and assembly; the ability to be deposited on inexpensive substrates; the ease of mass production; and high suitability to large applications. Given that amorphous silicon cells have no crystal structure at all, their efficiencies have been less than 10% due to internal energy losses. United Solar, however, has recently been able to achieve efficiencies of 12.5% with these types of solar cells.

A number of other materials can also be used to make solar cells and currently research programs have been investigating gallium arsenide, copper indium diselenide, copper gallium diselenide, and cadmium telluride. Newer, high-tech solar cells have also achieved improved energy conversion efficiency by incorporating two or more layers of different materials with different wavelength sensitivities. Top layers are designed to absorb higher energy photons while allowing lower energy photons through to be absorbed by the layers beneath. Double-junction cells are already commercially available and triple-junction cells have achieved efficiencies up to 43.5% in the laboratory using concentrators. In this process, large lenses are used to concentrate the light onto a small solar cell, greatly increasing their efficiency.

CONCENTRATED SOLAR POWER

CSP installations use mirrors to concentrate sunlight onto pipes containing a working fluid such as steam or Therminol. The working fluid is heated by the concentrated solar energy which drives a heat engine such as a steam turbine, which generates electricity.

The total installed capacity of CSP plants in the world in 2011 was about 1.5 GW, mostly located in the United States and Spain. While Spain has 582 MW, the United States has 507 MW of capacity. 1.5 GW is about the size of one medium size power plant burning coal or natural gas, so the installed capacity is still very small. There is also another 17.5 GW of CSP projects under development worldwide, of which 8.7 GW is in the United States, 4.5 GW in Spain and 2.5 GW in China.

There are four common CSP forms in use: parabolic troughs, dish stirlings, linear fresnel reflectors, and solar power towers. These systems usually use tracking systems to allow the concentrating mirrors to follow the sun across the sky and maintain the maximum energy. The different designs produce different operating temperatures and varying thermodynamic efficiencies due to the differences in the way that they track the sun and focus light.

The most common type of CSP is the *parabolic trough*, which consists of a linear parabolic reflector that concentrates light onto a pipe containing the working fluid positioned along the reflector's focal line. The parabolic reflector follows the sun during the daylight hours by single axis tracking. The SEGS plants in California, Nevada Solar One near Boulder City, Nevada and Plataforma Solar de Almería's SSPS-DCS plant in Spain use this technology.

Figure 8.3 illustrates the parabolic trough system used at the SEGS Plants in California. This is the largest system of solar power plants in the United States. The SEGS units use patented synthetic oil (called Therminol) as the working fluid in the pipes at the focal point of the parabolic units. There are nine SEGS units, but the ownership was split when the original owners, LUZ International, went bankrupt in 1991. Seven of the units, SEGS III-IX, are now co-owned and operated by NextEra Energy (formerly FPLEnergy). SEGS I and II were sold to Sunray Energy Incorporated, but are now owned and operated by Cogentrix Solar Services. The working fluid used in these plants, Therminol, is a synthetic oil and as such is highly flammable. This led to a disaster in February 1999, when a 900,000 gallon Therminol storage tank exploded at the SEGS II plant in Daggett, CA and caused extensive damage.

Currently, the SEGS units are operated as peaking units, which means that they do not operate all the time. On hot summer days when the generating capacity of the units is at its peak, the plants are turned on to satisfy the increased demand for electricity to run air conditioners in California. Table 8.1 details some of the history of operation from 1996-2002 when the units were operated year round.

The data shows a capacity factor of 21% on average during this time. This, however, is somewhat misleading because seven of the units (III-IX) were operated at night and on cloudy days by supplementing the steam turbines with natural gas supplied energy. The true capacity factor from solar energy is probably closer to the 13%, seen at the SEGS I and II units.

A *Dish-Stirling System* consists of a stand-alone parabolic reflector that concentrates sunlight onto a receiver positioned at the reflector's focal point (Figure 8.9). The dish reflector tracks the sun along two axes. The working fluid in the receiver is heated up to the optimal temperature and then used by the Stirling engine to generate power. Dish engine systems eliminate the need to transfer heat to a boiler by placing a Stirling engine at the focal point. The University of Nevada Las Vegas, the Australian National University in Canberra and the Sandia National Labs have pilot projects using this technology. A commercial application of this technology has been approved for construction by Stirling Energy Systems Solar Two, LLC (SES Solar Two, LLC) in Imperial County, California.

The California Energy Commission provided the following details on this project:

The proposed Imperial Valley Solar/SES Solar Two project would be a nominal 750-megawatt (MW) Stirling engine project, with construction planned to begin either late 2009 or early 2010. Although construction would take approximately 40 months to complete, renewable power would be available to the grid as each 60-unit group is completed. The primary equipment for the generating facility would include the approximately 30,000, 25-kilowatt solar dish Stirling systems (referred to as

TABLE 8.1   Operational Data for the Nine SEGS Solar Power Plants

Plant	Year Built	Location	Net Turbine Capacity (MW)	Field Area (m²)	Oil Temperature (°C)	Gross Solar Production of Electricity average 1998-2002 (MWh)	Gross Solar Production of Electricity Capacity Factor (%)
SEGS I	1984	Daggett	14	82,960	307	16,500	13.45
SEGS II	1985	Daggett	30	165,376	316	32,500	12.37
SEGS III	1986	Kramer Jct.	30	230,300	349	68,555	26.09
SEGS IV	1986	Kramer Jct.	30	230,300	349	68,278	25.98
SEGS V	1987	Kramer Jct.	30	250,500	349	72,879	27.73
SEGS VI	1988	Kramer Jct.	30	188,000	391	67,758	25.78
SEGS VII	1988	Kramer Jct.	30	194,280	391	65,048	24.75
SEGS VIII	1989	Harper Lake	80	464,340	391	137,990	19.69
SEGS IX	1990	Harper Lake	80	483,960	391	125,036	17.84
Totals			354	2,290,016		654,544	21.11

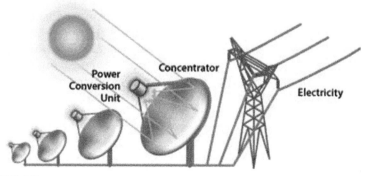

FIGURE 8.9   Solar dish power plant

SunCatchers), their associated equipment and systems, and their support infrastructure. Each SunCatcher consists of a solar receiver heat exchanger and a closed-cycle, high-efficiency Solar Stirling Engine specifically designed to convert solar power to rotary power then driving an electrical generator to produce grid-quality electricity. The 6500 acre project site is located on approximately 6,140 acres of federal land managed by the Bureau of Land Management (BLM) and approximately 360 acres of privately owned land. The site is approximately 100 miles east of San Diego, 14 miles west of El Centro, and approximately 4 miles east of Ocotillo, California.

The project will be constructed in two phases. Phase I of the project will consist of up to 12,000 SunCatchers configured in 200 1.5-MW solar groups of 60 SunCatchers per group and have a net nominal generating capacity of 300 MW. Phase II will add approximately 18,000 SunCatchers, expanding the project to a total of approximately 30,000 SunCatchers configured in 500–1.5-MW solar groups with a total net generating capacity of 750 MW.

The Applicant has applied for a ROW grant for the Project Site from the Bureau of Land Management (BLM) California Desert District. Although the Project is phased, it is being analyzed in this Application for Certification as if all phases will be operational at the same time.

The project would include the construction of a new 230-kV substation approximately in the center of the project site, and would also be connected to the SDG&E Imperial Valley Substation via an approximate 10.3-mile, double-circuit, 230-kV transmission line. Other than this interconnection transmission line, no new transmission lines or off-site substations would be required for the 300-MW Phase I construction. The full Phase II expansion of the project will require the construction of the 500-kV Sunrise Powerlink transmission line project proposed by SDG&E. Within the Project boundary, Phase I requires approximately 2,600 acres and Phase II requires approximately 3,500 acres. The total area required for both phases, including the area for the operation and administration building, the maintenance building, and the substation building, is approximately 6,500 acres. The 230-kV transmission line required for Phase I would parallel the Southwest Powerlink transmission line within the designated right-of-way (ROW). A water supply pipeline for the project would be built on the approved Union Pacific Railroad ROW. Since the proposed project does not have a steam cycle, the primary water use would be for mirror washing.

A *solar power tower* (Figure 8.2) consists of an array of dual axis tracking reflectors (heliostats) that concentrate light on a central receiver on the top of a tower which contains the working fluid. This fluid is heated to the optimal temperature and then used as a heat source for power generation in a turbine. The Solar Two in Daggett, California and the Planta Solar 10 (PS10) in Sanlucar la Mayor, Spain (see Figure 8.2) use this technology. eSolar's 5-MW Sierra SunTower located in Lancaster, California is the only CSP tower facility currently operating in North America, but Rice Solar Energy recently had a 150 MW solar tower unit approved for construction in Riverside County in California. The California Energy Commission provided the following details on this project:

Rice Solar Energy, LLC, (RSE) a wholly owned subsidiary of SolarReserve, LLC, proposes to construct, own, and operate the Rice Solar Energy Project (RSEP or project). The RSEP will be a solar generating facility located on a privately owned site in unincorporated eastern Riverside County, California. The proposed project will be capable of producing approximately 450,000 megawatt hours (MWh) of renewable energy annually, with a nominal net generating capacity of 150 megawatts (MW).

The RSEP will be located in an unincorporated area of eastern Riverside County, California. Land surrounding the project site consists mostly of undeveloped open desert that is owned by the federal government and managed by the U.S. Bureau of Land Management (BLM).

The proposed facility will use concentrating solar power (CSP) technology, with a central receiver tower and an integrated thermal storage system. The RSEP's technology generates power from sunlight by focusing energy from a field of sun-tracking mirrors called heliostats onto a central receiver. Liquid salt (The salt is a mixture of sodium nitrate, a common ingredient in fertilizer, and potassium nitrate, a fertilizer and food additive. These mineral products will be mixed onsite as received directly from mines in solid crystallized form and used without additives or further processing other than mixing and heating.), which has viscosity and appearance similar to water when melted, is circulated through tubes in the receiver, collecting the energy gathered from the sun. The heated salt is then routed to an insulated storage tank where it can be stored with minimal energy losses. When electricity is to be generated, the hot salt is routed to heat exchangers (or steam generation system). The steam is then used to generate electricity in a conventional steam turbine cycle. After exiting the steam generation system, the salt is sent to the cold salt thermal storage tank and the cycle is repeated. The salt storage technology was demonstrated successfully at the U.S. Department of Energy-sponsored 10-MW Solar Two project near Barstow, California, in the 1990s.

According to the applicant, this unique CSP technology offers several important benefits. Because liquid salt has highly efficient heat transfer and storage properties, it is used as the heat transfer medium in the cycle. Natural gas heating is therefore not required for startup or for operating stability during routine cloud cover. Second, the stored energy in the salt can be extracted upon demand and produce electricity even when there is no sunlight. Finally, the output from the RSEP will produce a stable electricity supply, compensating for potential impacts on the electricity grid from other intermittent energy sources having less predictable operating characteristics.

The solar facility will have the following key elements:

- A large circular field of mirrors (heliostats) that reflect the sun's energy onto a central receiver tower
- A conventional steam turbine generator to produce electricity
- Insulated tanks to store the hot and cold liquid salt heat transfer fluid, and
- An air-cooled condenser (ACC) to eliminate water consumption for cooling the steam turbine exhaust

There is also another 370 MW solar tower installation being planned for the Ivanpah Dry Lake, in San Bernardino County, California, on federal land managed by the Bureau of Land Management. The owners are Solar Partners LLC. The details on this project are as follows:

The proposed project includes three solar concentrating thermal power plants, based on distributed power tower and heliostat mirror technology, in which heliostat (mirror) fields focus solar energy on power tower receivers near the center of each heliostat array. Each 100-MW site would require approximately 850-acres (or 1.3 square miles) and would have three tower receivers and arrays; the 200-MW site would require approximately 1,600-acres (or 2.5 square miles) and would have 4 tower receivers and arrays. The total area required for all three phases would including the administration building/operations and maintenance building and substation and be approximately 3,400-acres (or 5.3 square miles). Given that the three plants would be developed in concert, the proposed solar plant projects would share the common facilities mentioned above to include access roads, and the reconductored transmission lines for all three phases. Construction of the entire project is anticipated to begin in the first quarter of 2009, with construction being completed in the last quarter of 2012.

In each solar plant, one Rankine-cycle reheat steam turbine receives live steam from the solar boilers and reheat steam from one solar reheater located in the power block at the top of its own tower. The reheat tower would be located adjacent to the turbine. Additional heliostats would be located outside the power block perimeter road, focusing on the reheat tower. Final design layout locations are still being developed. The solar field and power generation equipment would be started each morning after sunrise and insolation build-up, and shut down in the evening when insolation drops below the level required to keep the turbine online.

Each plant also includes a partial-load natural gas-fired steam boiler, which would be used for thermal input to the turbine during the morning start-up cycle to assist the plant in coming up to operating temperature more quickly. The boiler would also be operated during transient cloudy conditions, in order to maintain the turbine on-line and ready to resume production from solar thermal input, after the clouds pass. After the clouds pass and solar thermal input resumes, the turbine would be returned to full solar production. Each plant uses an air-cooled condenser or "dry cooling," to minimize water usage in the site's desert environment. Water consumption would therefore, be mainly to provide water for washing heliostats. Auxiliary equipment at each plant includes feed water heaters, a deaerator, an emergency diesel generator, and a diesel fire pump.

Electricity would be produced by each plant's Solar Receiver Boiler and the steam turbine generator. The heliostat mirrors would be arranged around each solar receiver boiler. Each mirror tracks the sun throughout the day and reflects the solar energy to the receiver boiler. The heliostats would be 7.2-feet high by 10.5-feet wide (2.20-meters by 3.20-meters) yielding a reflecting surface of 75.6 square feet (7.04 square meters). They would be arranged in arcs around the solar boiler towers asymmetrically.

Each solar development phase would include:

- A natural gas-fired start-up boiler to provide heat for plant start-up and during temporary cloud cover
- An air-cooled condenser or "dry cooling," to minimize water usage in the site's desert environment
- One Rankine-cycle reheat steam turbine that receives live steam from the solar boilers and reheat steam from one solar reheater located in the power block at the top of its own tower adjacent to the turbine
- A raw water tank with a 250,000 gallon capacity; 100,000 gallons to be used for the plant and the remainder to be reserved for fire water

- A small onsite wastewater plant located in the power block that treats wastewater from domestic waste streams such as showers and toilets, and
- Auxiliary equipment including feed water heaters, a deaerator, an emergency diesel generator, and a diesel fire pump.

SOLAR EFFICIENCY OF CSP INSTALLATIONS

The theoretical efficiency, η, of the conversion to electricity of the solar energy received can be determined from the theory of thermal radiation and from Carnot's principle deduced from the second law of thermodynamics (see Chapter 3). In these systems, the solar insolation is first converted into heat and collected by the receiver with efficiency, $\eta_{receiver}$. The heat is then converted into work with the thermodynamic efficiency of which the maximum possible is given by the Carnot efficiency η_{carnot}. Hence, the total efficiency, $\eta = \eta_{receiver} \times \eta_{carnot}$.

For a solar receiver providing the heat source at temperature T_H and the heat engine rejecting heat to a heat sink at temperature T_L (the atmosphere is usually the heat sink, so let $T_L = 300\,°K$), the Carnot efficiency was shown in Chapter 3 and appendix A to be:

$$\eta_{carnot} = 1 - \frac{T_L}{T_H} \tag{8.1}$$

The receiver efficiency can be calculated from the Stefan-Boltzmann law for thermal radiation, but to simplify the equations you must also assume the following: perfect optics ($\eta_{optics} = 1$); the collecting and reradiating areas are equal; and, the absorptivity and emissivity of the collector and receiver are perfect. The receiver efficiency can then be calculated from:

$$\eta_{receiver} = 1 - \frac{\sigma T_H^4}{IC} \tag{8.2}$$

Hence the total theoretical efficiency, η, is given by:

$$\eta = \left(1 - \frac{\sigma T_H^4}{IC}\right)\left(1 - \frac{T_L}{T_H}\right) \tag{8.3}$$

Where $\sigma = $ Stefan-Boltzmann constant $= 56.704\,nW/(m^2K^4)$. For $I = $ solar insolation, let us assume an optimistic $500\,W/m^2$; and $C = $ concentrator magnification. From Equation 8.3, it is obvious that the efficiency is a nonlinear function of the receiver temperature, T_H. The higher the receiver temperature, the higher the Carnot efficiency; however, the receiver efficiency will be lower. The efficiency actually becomes zero at two points: the temperature where the Carnot efficiency is zero and the temperature where the receiver efficiency is zero. The Carnot efficiency is zero when $T_L = T_H$. This occurs at night when the sun is not shining and the receiver source temperature reduces to the atmospheric sink temperature. The receiver efficiency reduces to zero when $\sigma T_H^4 / IC = 1$. This occurs when the receiver has become so hot that all the solar energy that it is receiving is being reradiated back into the atmosphere. This is the maximum temperature possible for the receiver because all of the thermal energy the receiver is getting is being reradiated and none can be transmitted to the heat engine. At this temperature, the receiver efficiency reduces to zero. This is illustrated in Figure 8.10, which shows the theoretical efficiency as a function of the receiver temperature, T_H, and the concentrator magnification, C. The higher the concentrator magnification (C), the higher the maximum efficiency that occurs at some optimum temperature, T_{opt}. If the receiver is allowed to exceed this temperature, the efficiency of energy conversion will decline again because the receiver has become too hot and is losing too much energy from reradiation.

The maximum achievable receiver temperature, when the efficiency is zero, is given by:

$$T_{max} = \left(\frac{IC}{\sigma}\right)^{0.25} \tag{8.4}$$

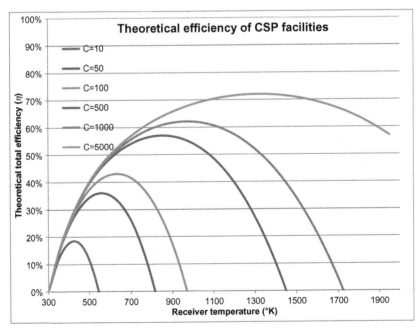

FIGURE 8.10   Theoretical efficiency of CSP facilities.

The optimal operating temperature is determined by differentiating Equation 8.3 with respect to the operating temperature, T_H, and setting the derivative to zero.

This leads to the following equation for the optimal temperature, T_{opt}:

$$T_{opt}^5 - 0.75 T_L T_{opt}^4 - \frac{T_L IC}{4\sigma} = 0 \qquad (8.5)$$

This equation (8.5) can be solved numerically to find the optimal temperature, T_{opt}. A plot of the maximum temperature (Equation 8.4) and the optimum operating temperature (Equation 8.5) as a function of the magnification parameter, C, is shown on a semilog scale in Figure 8.11. Figure 8.10 implies that large solar efficiencies are feasible using these CSP facilities. In practice, however, these theoretical efficiencies are not achievable because of other inefficiencies in the optics, the absorptivity, the emissivity and the thermal inefficiencies of the heat engine. In fact, the thermodynamic efficiency of the heat engine is usually about 40% and the receiver efficiency is less than this, giving efficiencies in the range of less than 20%. Equation 8.3 and Figure 8.10 demonstrate the concept of the optimal operating temperature for CSP plants. In practice, Figure 8.10 can be used as a guide to search for the actual optimal temperature. Note that the data on the SEGS plants in Table 8.1 shows the working fluid is Therminol and the operating temperature is 391 °C (664 °K). The

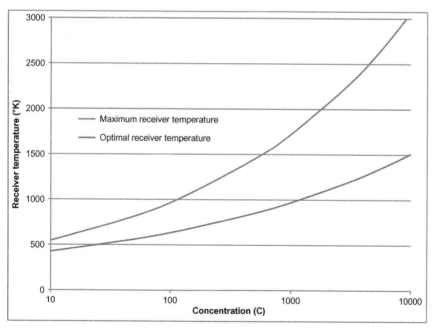

FIGURE 8.11   Maximum and optimum receiver temperature in a CSP solar plant.

magnification parameter (C) is not given for these plants, but using the analysis above, an optimum temperature of 664 °K is consistent with $C = 130$.

SELECTED LOCATION STUDIES

The U.S. NREL also records solar radiation data that allows one to determine the electricity generated by solar cells at locations around the United States and the world. Solar cells do not generate much electricity at large latitudes (northern or southern) in the wintertime because of the shortness of the days and the low angle of the sun. Much higher generation can be achieved at the same locations in the summertime. Locations at the equator can achieve relatively constant generation all year round, only being diminished on cloudy or rainy days. The larger the latitude, the larger the discrepancy will be between winter and summer generation. Of course, there is also a diurnal variation every day. Solar cells generate electricity only when the sun is shining and not at all at nighttime.

The next series of graphs shows the electricity generated from a PV system at selected locations using solar cells as a function of the month of the year. The assumed PV system is a single silicon crystal array with a collecting area of 35 m^2 (377 ft^2). This corresponds to a PV array that can theoretically generate 4 kW under optimal conditions. It also assumes that the solar radiation is converted to DC with a conversion efficiency of 11% and is then converted to AC with an efficiency of 77%. The overall efficiency of solar conversion is approximately 8%. You can also assume that the solar arrays track the sun with a dual axis system. A dual axis system is the most efficient way to maximize the solar energy that is harnessed. Fixed angle systems would harness lower energy amounts than those shown here.

Figures 8.12 through 8.17 show the results of the analysis for six cites in the western hemisphere arranged in order from Whitehorse (the capital city of the Yukon Territory in Canada) in the north, to Belem, Brazil, in the south. Whitehorse, which has latitude of 60.7°N, has very short days of low angle sun in the winter and so generates very little electricity in the winter months. In the long days of summer, it can generate more solar electricity than the tropical cities of Fort-de-France on the Caribbean island of Martinique, or Belem, which is situated at the mouth of the Amazon River in Brazil. Fort-de-France, at a latitude of 14.6°N and Belem, 1.4° south of the equator, both have relatively constant daily generating capacity all year round, but are more affected by the wet tropical climate than the length of the day or the solar radiation received. Table 8.2 summarizes the total annual electricity generated in the six

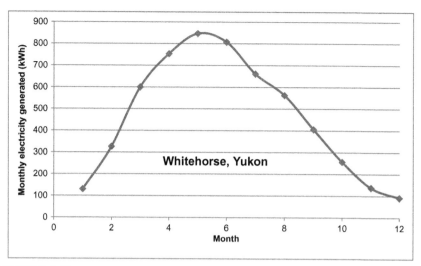

FIGURE 8.12 Annual electricity generated in Whitehorse, YK, with 4-kW sun-tracking solar panel.

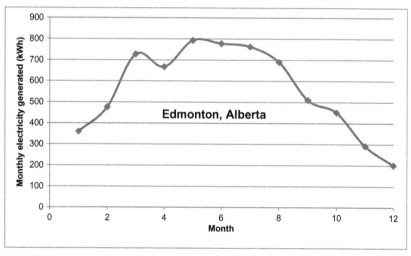

FIGURE 8.13 Annual electricity generated in Edmonton, AB, with 4-kW sun-tracking solar panel.

locations along with the other city parameters. Note that of these six cities, Tucson is the best place to generate solar-powered electricity, partly because of its more southerly latitude but also because of its dry climate.

Looking at Figures 8.15-8.17, it is not easy to explain the monthly variation seen in the U.S. solar energy production shown in Figure 8.4. Solar

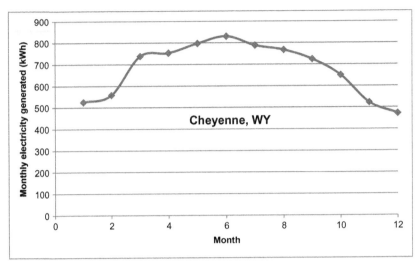

FIGURE 8.14   Annual electricity generated in Cheyenne, WY, with 4-kW sun-tracking solar panel.

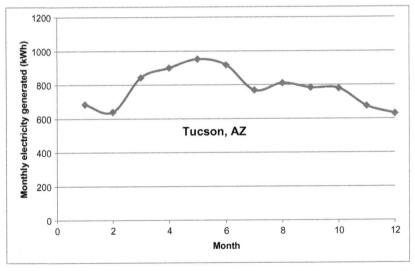

FIGURE 8.15   Annual electricity generated in Tucson, AZ, with 4-kW sun-tracking solar panel.

production in the United States in the winter declines to a small fraction of the summer production value. Given that most of this production is coming from the southwest of the United States suggests that the winter decline should not be so large. The reason is that the large SEGS units in California are now used as peaking units in the summer and are largely

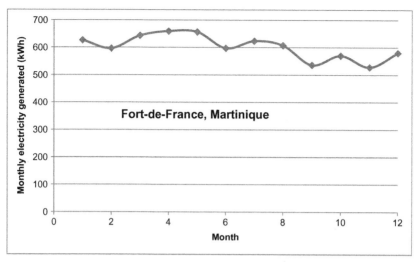

FIGURE 8.16 Annual electricity generated in Fort-de-France with 4-kW sun-tracking solar panel.

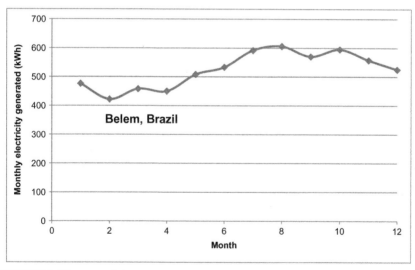

FIGURE 8.17 Annual electricity generated in Belem, Brazil, with 4-kW sun-tracking solar panel.

shut down in the winter. Table 8.2 also shows the average solar radiation received in each of the six cities by a dual axis sun tracking PV module in W/m^2. If single axis systems or fixed tilt systems are used, the solar energy captured will be less than this.

TABLE 8.2   Comparison of Six Cities for Photovoltaic Electricity Generation

City	Latitude, °N	Longitude, °W	Elevation, ft	Average Solar Radiation, W/m²	Total Annual AC Energy, kWh
Whitehorse, YK	60.72	135.07	2306	210	5578
Edmonton, AB	53.31	113.58	2346	254	6708
Cheyenne, WY	41.15	104.82	6142	307	8125
Tucson, AZ	32.12	110.93	2556	384	9383
Fort-de-France, M	14.6	61.01	13	295	7221
Belem, BR	−1.38	48.48	52	261	6292

ENVIRONMENTAL ISSUES

Solar energy is perceived as being environmentally friendly. It burns no fuel, it has minimal operating costs and there are no emissions. It is therefore perceived as having no impact on the environment. Because of the principle that I adopted in Chapter 1, this cannot be true. There are some environmental impacts in the manufacture and construction of the PV materials for PV installations. The same is true for solar thermal units. These impacts can be discounted as far less than the impacts of a coal burning plant; however, the big impact of a large solar installation (whether PV or CSP) is that it requires a lot of land. A 1-GW CSP facility requires at least 6000 acres of land while a 1-GW PV facility requires over 12,000 acres of land. Compare this to a 1-GW coal-fired or nuclear power plant, which requires about 600 acres of land. A 1-GW natural gas facility requires less than 300 acres. For solar installations, unlike wind power, land has to be dedicated to the harness of the energy. Moreover, if deployed in pristine environments, the ecology of that environment will be changed forever. Plants and animals that previously relied on the sun's warmth and energy will be covered over and die. Because of the large land area required, there is also a visual impact. No matter where they are placed, solar installations will be noticed and will alter the visual landscape, the so called view-shed.

While solar power's contribution to the energy mix remains small, these impacts will not be noticed. Solar power is likely to remain small because of the economic issues discussed in the next section. If some large breakthrough were made to make solar power more economic and the reach of solar power exploded into very large installations, the environmental impact would become more apparent. In that circumstance, just as people complain now about hydroelectric installations and are starting to complain about wind farms, so too will there be significant objections to solar power if it ever became successful. This is because you cannot extract energy from the environment without having an impact on the environment. Energy is part of the environment. These environmental impacts are covered in more detail in Chapter 4.

ECONOMICS

At its summit on April 4–7, 2011 in New York, Bloomberg New Energy Finance estimated that the capital cost of solar PV electricity generated from large, solar PV installations is $1.80 per watt and further predicted that the price would decline to $1.50 per watt by the end of 2011. From this, the installed cost can be estimated at $3-$4 per watt, or at least $3000 per kilowatt, allowing for installation costs. The capital cost of CSP plants is about the same as this. If a capital cost of $3000 for a 1-kW panel is amortized over 20 years at an interest rate of 5%, the payment would be $240 per year. In 1 year the 1-kW solar panel can generate about 876 kWh of electricity, assuming that over 1year the panel can generate electricity at about 10% of the maximum possible. This allows for the fact that the maximum possible electricity is generated only at noon in the middle of summer. On average on winter days, it only generates 50% of the electricity that it does on summer days. The cost of the electricity, assuming no fuel costs and negligible operating costs, is $240/876, or 27.4c per kWh. This is more than 10 times the cost of electricity generated from coal, nuclear, natural gas or hydroelectric and 5–8 times the cost of wind power. It also does not take into account any land costs, maintenance costs or electricity storage costs. CSP facilities produce power at about the same cost as PV systems. CSP systems were previously a little cheaper than PV systems but the costs of both have now equalized due to improvements in the manufacture of PV systems.

To avoid storage costs, most solar power generators prefer to sell any excess power they generate when the sun is at its peak back to the utility grid. The utilities do not like to buy this power because they have no control over the quantity and quality of the power they are getting. The utility has to balance the power available on the grid according to the demand. When demand is higher, they have to switch in more generating capacity; when demand reduces, they have to switch off the units. Solar (and wind) power that does not have energy storage facilities has to be added to the grid when it is generated. The utility has no control over how much they are getting or when or if it contains voltage fluctuations.

Political pressure has forced the utilities to buy back all the excess power that solar generators produce. As long as this power remains small, they can use their other generators to balance the load. The question remains: what tariff should the utility pay to the solar generator?

The utility can buy electricity from other generators for the wholesale cost (which is often less than 2c/kWh), which they then sell to the consumer for the retail cost (which is 7c-12c per kWh) depending on the location and the customer size. The average cost for residential customers is now 12c/kWh, while large industrial customers pay about 7c/kWh. This is not an unreasonable mark-up because the electric utility supplies the distribution network which has to be capitalized and maintained. They also must include a reasonable profit in their cost to the consumer. If they are forced to buy back electricity from the consumer, they would like to pay the same <3c/kWh cost that their other suppliers charge. Initially, however, they were forced to allow the solar generators to treat the electrical grid as a storage bin, putting power in when they had excess and taking it out again when they needed it. Whether or not an individual consumer is able to exactly balance their peaks and valleys, this arrangement effectively means that the solar power generators are selling the electricity to the utility for the retail cost. This represents an additional subsidy to the solar power generators. The utility has to make up this extra cost by raising its retail charges to all consumers. As long as solar energy remains small, this is not a big imposition. If for some reason the solar generation capacity was increased substantially, this would raise electricity rates for all consumers. Even this sweet deal, however, is an insufficient subsidy to make solar power cost effective because solar energy costs more than retail costs to generate. Consequently, many U.S. states and many countries have adopted special "feed-in tariffs" for solar generators. These rates vary widely from about 20 c/kWh up to 80 c/kWh. In the Australian state of Victoria, the rate paid to solar generators for small systems (up to 5 kW), the feed-in-tariff, is A60 c/kWh.

In 2011, the Australian dollar was about equal to the U.S. dollar. In the Canadian province of Ontario, the Renewable Energy Standard Offer Program, introduced in 2006 with the passage of the Green Energy Act, allowed residential homeowners in Ontario with small roof-top systems (less than 10 kW) to sell the energy they produce back to the grid at C42¢ per kWh, while buying power from the grid at C8¢ per kWh. Obviously this C42 c/kWh rate was still too low to make it worthwhile because in March 2009 the feed-in-rate for small solar panel installations was increased to C80¢/kWh. This gives one an idea of the cost of electricity from small roof-top systems—it is 30-40 times the cost of conventional power. The additional cost is subsidized by the electricity consumer with increased retail rates.

The Andosol-I Solar Energy plant, a relatively large 50-MW parabolic trough plant in Spain, receives a feed-in-tariff of €0.27/kWh (38.6 c/kWh) from the utilities in Europe. This plant cost €300 million ($425 million) to build so its capital cost can be also amortized at 5% over 20 years to give a payment of $21.25 million/year. Assuming a capacity factor of 15%, the 50-MW plant should be able to generate 65.7 GWh/year. Dividing the $21.25 million/year into the 67.5 million kWh/year gives 32.3c/kWh. This is the estimated cost of the electricity, and so they can make a profit from the feed-in-tariff of 38.6 c/kWh. This is, however, 20 times the cost of other electric sources.

THE FUTURE OF SOLAR ENERGY

Solar energy has started to grow but until the costs are substantially reduced it will remain a very small component of the world and the U. S. energy picture. Even though the cost has been substantially reduced over the past 50 years, it still remains a very expensive energy source. The PV effect has been known since 1839 and, despite extensive research efforts since then, solar power is still very expensive. It does not seem likely that further research efforts will make the breakthroughs that will lead to the commercialization of solar power in the foreseeable future. There is a lot of solar power waiting to be harnessed if such a break-through could be made; however, even if there was a way to make solar power viable, it is to be expected that this cannot be done without having a significant impact on the environment.

Wind Energy

Wind energy is relatively cost competitive with other energy sources, but it has the same reliability issues as solar energy. The wind does not blow all the time and cannot be turned up when the electricity demand increases. As its use becomes more widespread, the environmental impacts are also coming to the forefront. Wind energy creates noise and visual pollution and health effects such that people object to living next to them or placing them where they can be seen. Advocates claim that wind power has the lowest external costs when considering human health and ecological impacts, building and crop damage, global warming and loss of amenities when compared to coal, oil, gas, biomass, nuclear, hydro and solar power.

Energy derived from wind power consumes no fuel and consequently emits no air pollution. But is there any environmental impact during the manufacture of the turbines? A study by Lenzen and Munksgaard of the University of Sydney (Australia) and the Danish Institute for Local Governmental Studies reported that the energy consumed in manufacturing and transporting the materials used to build a wind power plant is paid back within a few months.

There are reports of bird and bat mortality at wind turbines, as there are around other artificial structures. Overall, the ecological impact of wind turbines on these species is not significant; however, depending on the particular site, prevention and mitigation of wildlife fatalities can affect the siting and operation of the turbines. There are also claims about the health impacts that may occur as a result of the noise pollution generated from the wind turbines. These issues are discussed in more detail later in this chapter.

WIND ENERGY STATISTICS

Figure 9.1 shows the statistics of electricity generation from wind power in the United States. In early 2011, the United States was generating about 10 TWh/month of electricity from wind out of a total electricity generation that ranges from a low of 305 TWh/month in November to a high

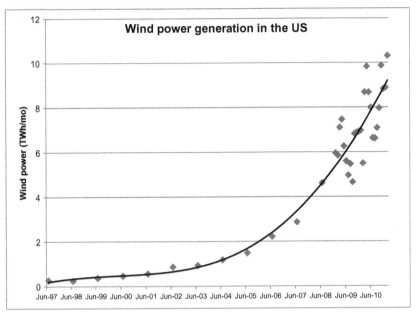

FIGURE 9.1   Electricity Generation from Wind Power in the United States.

of 410 TWh/month in July (when air conditioners are running full blast). Currently, wind power ranges from 2.5% to 3% of the total electricity demand. This has increased substantially from about 0.1% of the total electricity demand in 1997 and is still increasing as shown in the attached plot in Figure 9.1. Though wind power is rising and outperforms solar, biomass (wood), geothermal and petroleum as electricity fuel sources, it still trails coal, gas, nuclear and hydroelectric by a large margin. By contrast, coal comprises 45% of the electricity generation market, natural gas is 25%, nuclear is 20% and hydroelectric is 6%. There is no doubt, however, that wind power is gaining in importance.

The installed capacity of U.S. wind generators is currently 46 GW (2011 estimate). The total generating capacity is thus 33.3 TWh/mo. The 10 TWh/mo actually generated represents a capacity factor of 30%. This capacity factor has also been rising rapidly, up from 22.5% in 2005 and 23.5% in 2008. Current design strategies are aimed at maximizing this capacity factor by minimizing the generator size on each turbine, as explained in the next section.

Even though the installed capacity of wind power in the United States has been rising rapidly, it still trails Europe by a wide margin and was recently overtaken by China, which has been installing wind generators at a faster rate than the United States. In 2010, Europe had an installed capacity of 84 GW and was expected to reach 100-GW capacity by 2012.

In 2010, China had an installed capacity of 42 GW and was expected to reach 70-GW capacity by 2012.

PHYSICS AND ENGINEERING OF WIND POWER

A wind turbine converts the kinetic energy of the wind into electrical energy. The elements of a wind turbine are shown in Figure 9.2. Wind power has also been used to grind grain (the traditional windmill) and pump water, but the large application in today's energy environment is electricity generation. This is a relatively new application.

The equation for kinetic energy is:

$$E = \tfrac{1}{2}mv^2 \tag{9.1}$$

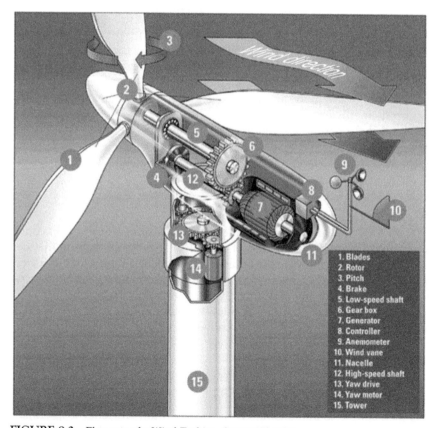

FIGURE 9.2 Elements of a Wind Turbine. *Source: US DOE.*

Where

m = mass = density × volume = ρV

v = velocity of the wind

E = energy

The volume (V) can be calculated from the area (perpendicular to the flow) by the length. The length is equal to the velocity by time. The area perpendicular to the flow is given by:

$$A = \frac{1}{4} \pi D^2$$

Where D is the diameter of the turbine blades, ρ is the density of the air, t is the time, and V is the volume of the air

Using the first law of thermodynamics, the energy extracted by the turbine is given by the kinetic energy of the air flowing into the turbine blades minus the kinetic energy of the wind flowing out of the turbine blades minus any frictional energy lost in the moving parts of the turbine and the generator:

$$E = \frac{1}{2} m v_{in}^2 - \frac{1}{2} m v_{out}^2 - \text{frictional losses} \tag{9.2}$$

Where v_{in} is the velocity of the air flowing into the turbine, v_{out} is the velocity of the air flowing out of the turbineThe velocity of the wind leaving the turbine, v_{out}, cannot be zero because this would stop the air flow and no new air could flow into the turbine. In practice, the energy lost to friction and the energy lost to the air flowing back out of the turbine are expressed as turbine inefficiency.

Equation 9.2 can be written in terms of the wind velocity, v, of the air flowing into the turbine and the energy conversion efficiency, η, as follows:

$$E = \frac{1}{2} \eta m v^2 \tag{9.3}$$

Where η is the turbine efficiency

Even if there were no frictional losses, the turbine efficiency (η) could not be more than 59.3% because of the velocity of the air leaving the turbine. This limit is known as the Betz limit, after Albert Betz who published this result in 1920. It is relatively easy to derive this limit using the principles of conservation of energy, the conservation of momentum, the conservation of mass and the well known Bernoulli's equation for fluid flow. Using these principles, Betz showed that the theoretical efficiency of a frictionless turbine is given by:

$$\eta = \frac{1}{2} \left[1 - \frac{v_{out}}{v_{in}} \right] \left[1 + \frac{v_{out}}{v_{in}} \right]^2 \tag{9.4}$$

It is then simple to demonstrate that the maximum efficiency is $\eta = 16/27$ ($\approx 59.3\%$), which occurs when $v_{out}/v_{in} = 1/3$. This means that if you could design a wind turbine such that the velocity of the wind flowing out of the turbine was one-third of the velocity of the wind flowing into the turbine, and the turbine had no frictional losses, the efficiency (η) of the turbine would be 59.3%, which is the absolute maximum that could be achieved. In practice, the value of η for large commercial turbines ranges from 40% to 45%.

Converting the mass to a density $\times$ volume (ρV) and the volume to area $\times$ velocity $\times$ time ($\frac{1}{4}\pi D^2 vt$) leads to Equation 9.5

$$E = \eta \rho \pi D^2 t v^3 / 8 \tag{9.5}$$

Power is the rate of energy conversion, calculated using energy (E) divided by time (t). The t can be eliminated from Equation 9.5 by writing the equation in terms of power (P):

$$P = \eta \rho \pi D^2 v^3 / 8 \tag{9.6}$$

The density of air varies according to the temperature, humidity, and pressure. At higher elevations, the pressure and temperature decrease, as does the density. At sea level with an atmospheric pressure of 1075 millibars (mb), a temperature of 30 °C (86 °F), and a dew point temperature of 20 °C (68 °F), the density of air is 1.225 kg/m³. At an elevation of 2062 m (6762 ft., e.g., Rawlins, WY) with an atmospheric pressure of 844 mb, a temperature of 20 °C (68 °F), and a dew point temperature of 0 °C (32 °F), the density of air is 1.000 kg/m³. With an assumed turbine efficiency of 40% at sea level, Equation 9.6 becomes (using SI units):

$$P = 0.1924 D^2 v^3 \tag{9.7}$$

With an assumed turbine efficiency of 40% at an elevation of 6762 ft., where the density of air is 1 kg/m³, the equation (in SI units) would be:

$$P = 0.157 D^2 v^3 \tag{9.8}$$

Notice that the power that can be delivered by the turbine is proportional to the cubic power of the wind speed.

As an example, the Suzlon S88 wind turbine has blades with a diameter of 88 m (288 ft.) and is rated to operate in winds up to 25 m/s (56 mph). The turbine and generator sit on a 79-m tower. If you are wondering about the geometry of an 88-m diameter turbine sitting on a 79-m tower, just remember that the blades only extend down 44 m from the top of the tower and 44 m up above the tower. Above the maximum allowable wind speed of 25 m/s, the turbine is shut down to avoid blade and tower damage. The rated wind speed is 14 m/s (31.3 mph). This is the speed where the turbine is guaranteed to produce the rated power. This rated wind speed is slightly above the actual wind speed where maximum power is achieved.

Using those numbers, including the maximum allowable wind speed of 25 m/s in Equation 9.6, the power output of the turbine can be calculated:

$$P = 0.1924\,D^2 v^3 = 0.1924{*}88^{2}{*}25^3 = 23{,}283{,}000\,\text{W} = 23.3\ \text{MW}$$

If you look up the specifications of the Suzlon S88, you will find that the generator attached to the turbine can only generate 2.1 MW. This is because the turbine is not regularly exposed to 56-mph winds. There is no use putting a very expensive 23-MW electrical generator on the turbine if the winds speeds are usually not fast enough to generate that power. To generate 2.1 MW, the wind speeds would have to be 11.2 m/s (25 mph). If the wind speed is less than 11.2 m/s, the turbine will generate less than the rated 2.1 MW. If the Suzlon S88 was fitted with a generator twice the size (4.2 MW), it would only harness maybe 10% more energy (depending on the wind profile) and its capacity factor would decrease substantially.

The power generated by the Suzlon S88, calculated from Equation 9.7, is shown in Figure 9.3. Compare this with the published power curve for the S88 shown in Figure 9.4. The power generated increases with a cubic exponent up to the wind speed of 11.2 m/s (25 mph). From 11.2 to 25 m/s, the power generated remains constant at 2.1 MW. To achieve this constant power output, the pitch of the blades is feathered to reduce the efficiency further and keep the power at 2.1 MW. At 25 m/s (56 mph), the turbine is

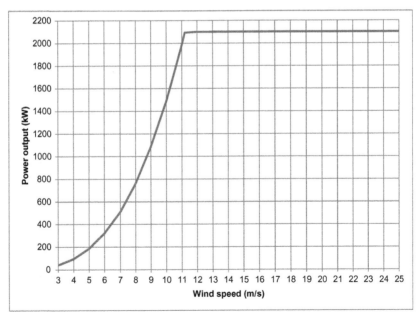

FIGURE 9.3 Power Output of the Suzlon S88 Wind Turbine as a Function of Wind Speed, Calculated from Equation 9.7.

S88-2.1 MW: Wind speed (M/S)

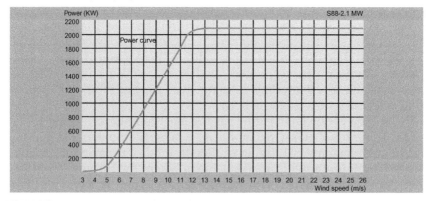

FIGURE 9.4 Power Output of the Suzlon S88 Wind Turbine as Published by Suzlon.

shut down to avoid damage that might be caused by high winds. It is allowed to restart when the wind speed has decreased significantly.

In any given area, the wind speed tends to follow a Rayleigh distribution in terms of the time spent at each wind speed. Without going into the equations for a Rayleigh distribution, this means that if you plotted the number of hours that the wind blows at a particular wind speed versus the wind speed, it would look like Figure 9.5. The plot in Figure 9.5 has

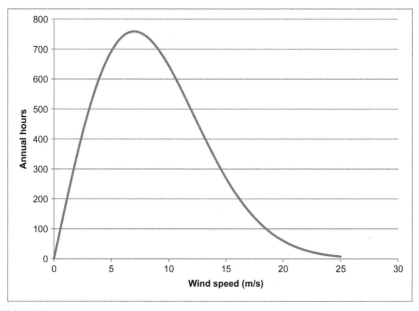

FIGURE 9.5 Typical Rayleigh Distribution of Wind Speed in a Particular Area ($v = 7$ m/s).

put all the wind speeds into bins that are 1 m/s wide and plotted the annual hours for each wind speed. The Rayleigh distribution does not model the zero wind speed particularly well because it assumes that the wind blows at zero speed for zero hours, which is obviously not true. The fact that this distribution is not valid for zero speed is of no consequence because no power is generated at zero speed anyway and, in the case of the S88, the turbine does not even start turning until a cut-in wind speed of 4 m/s is reached.

In the example given in Figure 9.5, the most common wind speed (the modal speed) is 7 m/s (15.6 mph). At 7 m/s, the Suzlon S88 turbine would only generate 511 kW, even though it has a capacity of 2.1 MW. The S88's minimum wind speed is 4 m/s where the turbine first starts generating power. At 4 m/s the turbine generates only 95 kW. This illustrates one of the problems with wind turbines. Most of the time they cannot generate their full capacity and, when the wind is not blowing, they cannot generate any power. Moreover, the generator is deliberately undersized so that it matches the more common wind speeds, not the high wind speeds. Doing this raises the so-called capacity factor for the turbine but not the power output. It also allows the turbine to produce constant power over a wider range of wind speeds. The capacity factor is the actual energy generated divided by the theoretical maximum energy that could be generated if the turbine was generating electricity at the maximum rate 100% of the time.

Theoretically, the Suzlon manufacturers could put any size generator on the S88 up to the full 23.3-MW capacities. Figure 9.6 shows what the power profile would look like with four different generators ranging from 1.1 MW up to 6 MW. As the size of the generator increases, the amount of energy that can be extracted from the wind increases, but not directly proportional to the size of the generator, and the capacity factor decreases (see Figure 9.7). The amount of energy that can be extracted from the wind is not directly proportional to the size of the generator because of the wind profile. For example, if a 6-MW generator was attached to the Suzlon S88, it would not reach full capacity until the wind speed reached 16 m/s (36 mph). If this turbine was placed in an area where the Rayleigh distribution of the wind speed looked like Figure 9.5, it would only be generating the full 6 MW when the wind speed was at or above 16 m/s, which would only be 8% of the time. If the wind frequency profile looked like Figure 9.5, the S88 could generate 14 GWh/yr with a 6 MW generator compared to 9 GWh/yr with the 2.1-MW generator normally supplied with the S88. The capacity factor, however, would decrease from 49% with the 2.1-MW generator to 27% with the 6 MW generator. These calculations do not allow for repair and maintenance downtime, which would decrease the estimates for the capacity factor and total energy generated. It also does not take into account the economics of using a larger generator. I have no data on the increased

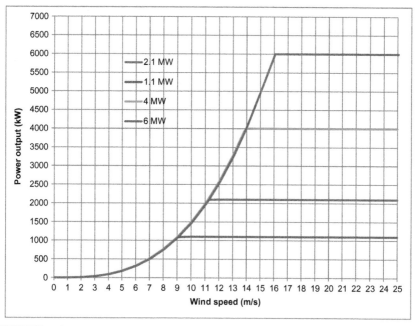

FIGURE 9.6   Power Output of the Suzlon S88 Wind Turbine as a Function of Wind Speed, Using Four Different Generators.

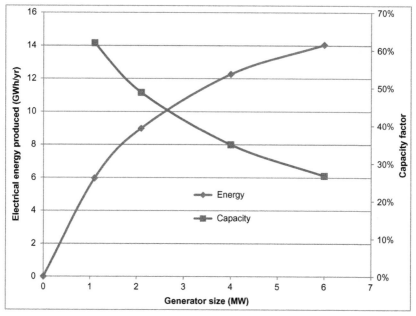

FIGURE 9.7   Annual Electrical Energy Output and Capacity Factor as a Function of Generator Size for the Suzlon S88 Wind Turbine Using the Wind Frequency Data of Figure 9.5.

capital and maintenance cost of the larger generator to determine if this would be worthwhile.

Figure 9.6 shows how both the energy generated and the capacity factor change when the generator size changes.

The technical specifications for other commercial wind turbines are shown in Table 9.1. Blade diameters range from 62 to 100 m with installed generator capacity in the range of 1-3 MW per turbine. Notice that the Vestas V90 comes with generators of two different sizes. As its name implies, the turbine blades have a diameter of 90 m, but they can be ordered with either a 1.8-MW generator or a 3.0-MW generator. The smaller

TABLE 9.1   Technical Specifications on Selected Commercial Wind Turbines

Model	Capacity (MW)	Blade Length (m)	Tower Height (m)	Total Height (m)	Blade Tip Speed (mph)	Rated Wind Speed (m/s)
GE 1.5 s	1.5	35.25	64.7	99.95	183	12
GE 1.5sle	1.5	38.5	80	118.5	183	14
Vestas V82	1.65	41	70	111	138	13
Vestas V90	1.8	45	80	125	157	11
Vestas V100	2.75	50	80	130	179	15
Vestas V90	3.0	45	80	125	200	15
Vestas V112	3.0	56	84	136	232	12
Gamesa G87	2.0	43.5	78	121.5	194	13.5
Siemens	2.3	46.5	80	126.5	169	14
Bonus (Siemens)	1.3	31	68	99	138	14
Bonus (Siemens)	2.0	38	60	98	151	15
Bonus (Siemens)	2.3	41.2	80	121.2	164	15
Suzlon 950	0.95	32	65	97	156	11
Suzlon S64	1.25	32	73	105	156	12
Suzlon S88	2.1	44	79	124	181	14
Repower MM92	2.0	46.25	100	146.25	163	11.2
Clipper Liberty	2.5	44.5	80	124.5	163	11.5
Mitsubishi MWT95	2.4	47.5	80	127.5	188	12.5

generator is designed to be used in areas with lighter winds so that the capacity factor can be maintained at a higher value in the less windy areas.

Most of the turbines sit on a tower around 80-m high. The wind speed increases at higher heights above the ground. At about 80 m, the wind speed levels off. This turns out to be an optimum height for the tower.

The top 10 wind turbine manufacturers in 2010, with their market share, are:

1. Vestas 14.8%
2. Sinovel 11.1%
3. GE Wind Energy 9.6%
4. Goldwind 9.5%
5. Enercon 7.2%
6. Suzlon Group 6.9%
7. Dongfang Electric 6.7%
8. Gamesa 6.6%
9. Siemens Wind Power 5.9%
10. United Power 4.2%

The largest is Vestas, a Danish company, and 4 of the top 10—Sinovel, Goldwind, Dongfang and United—are Chinese companies. This is not surprising given that wind energy has been growing the most rapidly in China. The only American company on the list is General Electric at number 3. As well as Vestas, Siemens also manufacturers their turbines in Denmark where wind turbine technology has been given a favored economic status.

CARBON DIOXIDE EMISSIONS AND POLLUTION

Wind power consumes no fuel and thus has no emissions directly related to electricity production. Wind power plants, however, do produce emissions during the initial manufacturing and construction. During the manufacture of the wind turbine, steel, concrete, copper wire, rubber, aluminum and other materials are manufactured and transported using processes that generate some emissions. The majority of the CO_2 emissions come from producing the concrete for the wind turbine foundations. The wind turbine manufacturer, Vestas, claims that the initial carbon dioxide emissions payback is within about nine months of operation. This is compared to a standard coal-fired power plant. Another 2006 study reported in *Wikipedia* found that CO_2 emissions of wind power range from 14 to 33 tons/GWh of energy produced.

A study by the Irish National Grid stated that "Producing electricity from wind reduces the consumption of fossil fuels and therefore leads to emissions savings." It found reductions in CO_2 emissions ranging from 330 to 590 tons of CO_2 per GWh.

LAND USE ISSUES

A wind farm requires roughly 0.1 km² (0.039 sq mi, 25 acres) of clear land per megawatt of turbine capacity. A 1-GW (1000-MW) wind farm, for example, might extend over an area of approximately 100 km² (25,000 acres). If they are sited closer together, the turbines start to interfere with each other's efficiency and they begin to harness less energy. 1 GW per 100 km² amounts to 10 W/m², and that is the maximum energy that can be harnessed if the wind blows at a constant rate all the time. If you put larger turbines in the same area, you have to put them farther apart so that they do not interfere with each other. The area required per turbine increases linearly as the turbine power increases. The maximum recoverable wind energy, however, remains constant at 10 W/m². Moreover, the average capacity factor for modern wind turbines is about 20%-30%, so this means that the turbines can harness about 2-3 W/m² (Figures 9.8 and 9.9).

In a typical wind prone area, the total available power from the wind is about 500 W/m². It would appear that wind turbines are capable of capturing less than 1% of this energy. It is not possible to harness more than 3 W/m² from even the most favorable windy areas. This may seem like a relatively large footprint for an energy harnessing installation, but the land can still be used for other activities, particularly crops and animal grazing. Wind energy experts contend that less than 1% of the land needs to be used for foundations and access roads; the other 99% can still be used for farming. Some clearing of trees around tower bases would be

FIGURE 9.8   Wind Farm at Arlington, Wyoming.

FIGURE 9.9   750-kW Turbines at Arlington Wind Farm, Wyoming.

necessary for the installation of tower sites in wooded areas, but the affected footprint would remain small.

At 2-3 W/m^2, the current installed wind-generating capacity of 46 GW occupies an area of 15-23 Gm^2 (3.7-5.7 million acres). If you wanted to increase production to 200 TWh/month—half of the U.S. electricity demand—you would need to install another 875 GW of wind-generating capacity for a total capacity of 920 GW. This would occupy 310-460 Gm^2 (76-115 million acres). This is approximately equal to the entire states of Colorado and Nebraska combined. While it is feasible to do this, the 875 GW would have to be backed up by other energy sources or energy storage capacity.

Figure. 9.10 shows the average wind speed in all areas of the United States. The best areas for wind power generation are in the central plains areas reaching from North Dakota and Montana in the north to Texas in the south, to Wyoming in the west, and as far east as Indiana. There are clearly large areas of the country that are suitable for wind generation. This map does not include offshore areas where additional potential exists.

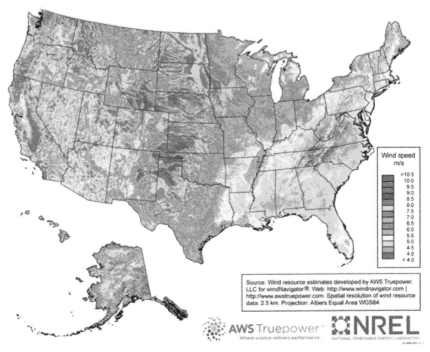

FIGURE 9.10   Average Wind Speeds in the United States. *Source: NREL, USA.*

In the United States, landowners typically receive $3,000 to $5,000 per year in rental income from each wind turbine, depending on the size. These turbines are usually deployed at a density of about one every thirty acres. A 1,000 acre tract of land, for example, could generate about $100,000 per year for the landowner in addition to the income gained from growing crops or grazing animals on the land.

Turbines are not generally installed in urban areas, except as smaller single units (see Figure 9.11). Buildings interfere with the wind flow, and turbines must be sited a safe distance from residences to guard against failure. The 20-MW Steel Winds project south of Buffalo, New York is an exception. It is situated in an urban location, but is separated from residences by siting the turbines on an uninhabited lake shore.

Wind turbines located in agricultural areas do interfere with crop-dusting operations. Operating rules prohibit the approach of aircrafts within a stated distance of the turbine towers. Even if turbine operators agree to shut down the turbines during crop-dusting operations, flying between turbines is still hazardous and limits the effectiveness of crop dusting.

In Ireland and Scotland, there has also been concern about the damage caused to peat bogs, with one Scottish politician campaigning for a moratorium on wind developments on peat areas claiming that, "Damaging the peat causes the release of more carbon dioxide than wind farms save."

THE PORTSMOUTH ABBEY WIND TURBINE

Portsmouth Abbey operates a Benedictine High School in Portsmouth, Rhode Island, and they have installed a single 660-kW turbine on a small hill on school grounds, set back sufficiently from the school buildings and neighboring houses (see Figures 9.11 and 9.12). This project is representative of other projects in the United States because there are similar small wind-generating projects in other urban areas around the world. Some details of the Portsmouth Abbey project are provided on the Portsmouth Abbey website and include the following information:

- Portsmouth Abbey and the Benedictine High School are one of the larger customers of national grid in this area and have been working diligently to combat rising energy costs through the implementation of energy efficiency and conservation programs.

FIGURE 9.11   The 660-kW Vestas Wind Turbine at Portsmouth Abbey. *Source: Br Joseph, Portsmouth Abbey.*

FIGURE 9.12   The 660-kW Vestas Wind Turbine at Portsmouth Abbey School. *Source: Br Joseph, Portsmouth Abbey.*

- In addition to energy conservation measures, the Abbey wished to promote renewable energy and had looked into a number of options, including solar, geothermal and wind power.
- As wind has been a resource on the site of the school since colonial times, it was determined that a wind turbine could possibly provide the Abbey and the school with the best option for a renewable energy source and the greatest savings in energy costs.
- With the assistance of the R.I. Renewable Energy Fund and the Roger Williams University, the school initiated a detailed study to quantify the local wind resource, including the installation of a special 50-m (164-ft.) tall meteorological monitoring pole.
- Additional analysis was made of the electrical usage patterns of the school in order to select a wind turbine that would produce the needed energy using the available winds. The desire was to select a turbine that could meet the school's needs and yet be sensitive to the setting of the

Abbey and the neighborhood. Studies found that a mid-sized wind turbine would suffice. In addition, the Abbey had enough space to place the turbine away from its boundaries. The turbine is approximately 750 ft. from the nearest neighbor.

- A Vestas 660-kW wind turbine generator was selected as the best choice for this project. This turbine has three 77-ft carbon fiber blades atop a 164-ft tapered tubular steel tower. The rotor turns at a constant 28.5 revolutions per minute. The structure stands a proud 240 ft from the ground to the tip of the highest blade. The turbine is secured by a concrete foundation that sits in a 30-ft hole, and its tower is firmly bolted to 80, one-inch diameter, 27-ft-long rods set firmly into the foundation.

- Over the last two decades, modern wind turbines generators have advanced in technology and are proven, reliable sources of electricity that are pollution-free and safe for the community. This wind energy project is providing a unique benefit to local schools. Portsmouth Abbey continues to invite local student groups for presentations and tours and hopes, in the future, to make available data that can be used for any number of academic disciplines, including meteorology, environmental or earth science, physics, engineering, etc.

- Modern wind turbines have special design features that have significantly reduced operating noise. The wind turbine produces less than 45 dB (decibels) at the edge of Cory's Lane. This is comparable to the ambient sounds of the wind blowing through the trees. The wind turbine is not a hazard to wildlife.

- Power from the wind turbine at the school reduces the load on the local utility electrical distribution feeder. This provides higher voltage and improved power quality to the entire neighborhood during peak power use periods.

- In December 2004, the Abbey applied to the State of Rhode Island Renewable Energy Fund for its support of this wind power project. After careful evaluation of the proposed plan, the Fund Board generously made available their advisory resources as well as a grant for more than one-third of the estimated project costs.

- On March 18, 2005, the Abbey applied to the Town of Portsmouth for the special-use, permit and variance needed to bring a wind turbine to fruition. With strong support from neighbors, the permit was unanimously granted.

- The foundation was completed at the beginning of January 2006. The State of Connecticut Police Department gave special permission for the lower part of the tower to travel through the state on a Friday, due to the truck's oversize load.

- The turbine arrived during the week of March 20 and was erected during the last week in March 2006. Brother Joseph watched the sun rise from the top of the turbine on March 31, 2006, the day the blades began to turn. The turbine began providing electricity for the grid at 10 a.m. that morning.

- As of March 31, 2007, one year to the day after the wind turbine began its operation, it has generated nearly 1.3 million kWh of clean electricity and has supplied 39.35% of the school's electrical energy use. This successful trend has continued.
- During its first year of operation, the highest documented wind gust was 67 mph. The turbine generates electricity at wind speeds up to the maximum of 55 mph, then pitches the blades to 90° angles and waits for the wind to subside to 45 mph before starting to turn again.
- Total wind turbine revenues during its first year of operation were $222,710, including $64,661 in renewable energy credits, $28,496 in wholesale electricity sold back to the grid, and $129,553 in retail electricity displaced.
- For its important contributions to conservation in Rhode Island, Portsmouth, the Abbey and the school received the 2007 Environmental Merit Award by the U.S. Environmental Protection Agency, the 2007 Conservation Award by the Garden Club of America, and the 2006 Senator John H. Chafee Award for Outstanding Conservation Project.

The Portsmouth turbine is monitored weekly for its energy output. The bar chart in Figure 9.13 and the graph in Figure 9.14 show the electrical energy output of the turbine's first five years, which began on March 31, 2006. The blue dots on Figure 9.14 show the weekly energy production and the red line shows the seven-week moving average of the weekly data. The moving average line shows that the energy production has an annual cycle of being higher in the winter months of November to February and lower in the summer months of July to September. This coincides with lower wind speeds in the summer and higher speeds in the winter.

The National Oceanic and Atmospheric Administration (NOAA) publish wind speed data for various locations around the United States. Figure 9.15 shows the average monthly wind speed for Block Island off Rhode Island. This data is reasonably representative of the Portsmouth area, where the Portsmouth Abbey Wind Turbine is located. This shows the lower average winds speeds in the summer. The average wind speed year-round is 4.75 m/s, and this can be represented with a Rayleigh distribution plot with a characteristic modal wind speed of 4.5 m/s, as shown in Figure 9.16. The Vestas turbine used here has a blade length of 77 ft, giving a blade diameter of 47 m. The energy output of this turbine, calculated from Equation 9.7, is shown in Figure 9.17. From this plot, it can be seen that the rated power output of 660 kW is reached at 11.6 m/s, well

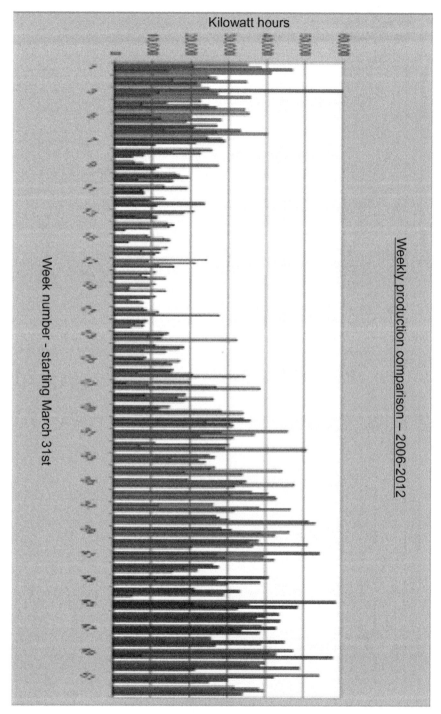

FIGURE 9.13 Weekly Electricity Production Statistics for Wind Turbine at Portsmouth Abbey. *Source: Br Joseph, Portsmouth Abbey.*

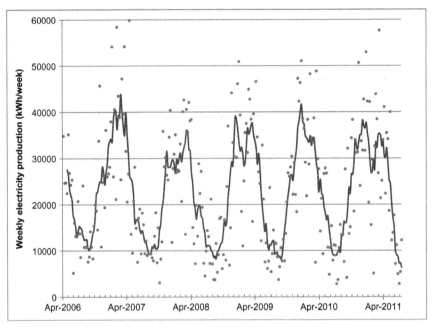

FIGURE 9.14   Weekly Electricity Production Statistics for Wind Turbine at Portsmouth Abbey (Blue Dots) Compared to 7-Week Moving Average (Red Line). *Data Source: Br Joseph, Portsmouth Abbey.*

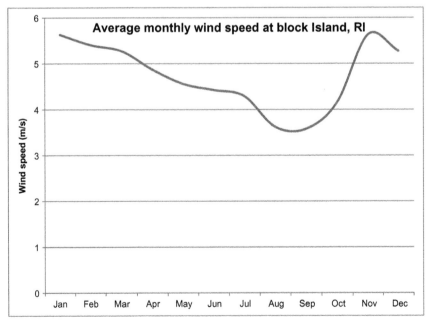

FIGURE 9.15   Average Monthly Wind Speed at Block Island, Rhode Island. *Data Source: NOAA, USA.*

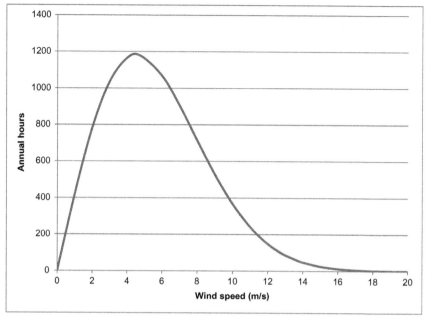

FIGURE 9.16   Rayleigh Distribution of Wind Speed in Rhode Island (V=4.5 m/s).

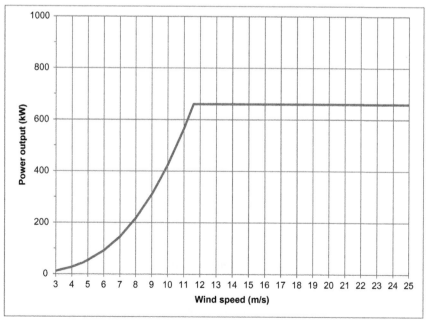

FIGURE 9.17   Calculated Power Output as a Function of Wind Speed for the 660-kW Vestas Wind Turbine at Portsmouth Abbey.

above the average wind speed. At the average wind speed of 4.75 m/s, only 45 kW can be generated by this turbine.

In order to analyze the optimum generator size, four generator sizes were investigated with regard to this turbine: 53 kW, 660 kW, 1.1 MW and 2.08 MW. The power output profiles for these four generators are shown in Figure 9.18. It is easy to see that the 53-kW turbine is close to the maximum power output for the average wind speed and it is tempting to argue that a bigger generator is not justified. Figure 9.19, however, shows the energy generated over one year and the resulting capacity factor for each generator size. The smallest generator (53 kW) would enjoy a high capacity factor of 65%, meaning that it would be operating most of the time and usually generating power at its maximum capacity; however, it would only generate 0.3 GWh per year. The 660-kW generator, by contrast, would be able to generate almost four times that amount of electricity at 1.12 GWh per year, even though its capacity factor drops down to only 19%. The 1.1-MW generator would only be able to generate a little more than this at 1.2 GWh per year and the 2.08-MW generator would only generate about 1.23 GWh per year. The bigger generators don't generate much more electricity because it takes much higher wind speeds for them to reach their maximum generating potential and in that area those

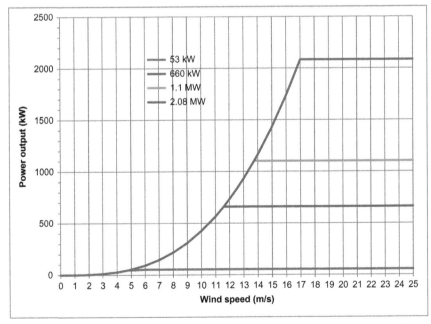

FIGURE 9.18   Calculated Power Output for the 47-m Vestas Wind Turbine Used at Portsmouth Abbey School Matched to Four Generator Sizes.

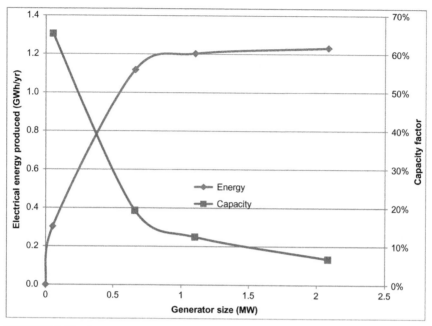

FIGURE 9.19 Energy That Could Be Captured by a 47 m Vestas Wind Turbine at Portsmouth Abbey as a Function of Generator Size.

wind speeds are rarely seen. The 1.1-MW generator needs wind at 13.75 m/s (31 mph) to reach its full 1.1 MW of power, while the 2.08-MW generator needs wind at 17 m/s (38 mph) to reach its full 2.08 MW of potential. Their capacity factors also drop down to 12.5% and 6.75%, respectively. Consequently, the 660-kW generator is the optimum size for this turbine and for the wind profile in this area.

Table 9.2 below shows the actual electricity generated during the first five years of operation. The average amount generated in the first five years was 1.214 GWh per year, which is very close to what was predicted for the wind profile with this turbine and is about as well as can be expected.

TABLE 9.2 Electricity Generated by Wind Turbine at Portsmouth Abbey in First 5 Years of Operation (kWh)

YEAR 1	1,293,406
YEAR 2	1,146,647
YEAR 3	1,130,917
YEAR 4	1,251,247
YEAR 5	1,247,634

IMPACT ON BIRDS, BATS AND OTHER FLYING WILDLIFE

There is a lingering perception that wind turbines kill a lot of birds and bats. This is not really true. In fact, wind farms cause about 0.4 bird fatalities per gigawatt-hour of electricity generated. This is relatively low and on par with any large manmade structure. It is also small compared to the number that die as a result of automobile traffic, hunting, electric power transmission lines, high-rise buildings and cats. In the United States, turbines kill about 150,000 birds per year, compared to 80,000 killed by aircraft, 80 million killed by cars, 100 million killed by collisions with plate glass, 130 million killed by power lines and 200 million killed by cats. The impact of wind turbines on birds is really negligible compared to other hazards.

In 2009, the Royal Society for the Protection of Birds (RSPB) in the UK warned that "numbers of several breeding birds of high conservation concern are reduced close to wind turbines," probably because "... birds may use areas close to the turbines less often than would be expected, potentially reducing the carrying capacity of an area." In the United States, the National Audubon Society also warns against siting wind farms in areas especially important to birds and other affected wildlife.

The U.S. Fish and Wildlife Service have also expressed concern and have issued guidelines for the siting of wind energy facilities in the United States. These guidelines recommend that wind farm operators do not place turbines in the following areas: areas documented as the location of any species protected under the Endangered Species Act; local bird migration pathways or areas where birds concentrate; near landscape features that attract raptors; in a configuration that is likely to cause bird mortality; and, areas where fragmentation of large contiguous tracts of wildlife habitat will occur as a result of turbine placement.

The Peñascal Wind Power Project in Texas and the Altamont Pass Wind Farm in California are two projects that have had particular issues with bird mortality. The Peñascal Wind Farm is located in the path of a bird migration route and it uses avian radar detectors to detect birds as far as 4 miles away. If the system determines that the birds are in danger of running into the rotating blades, it shuts down the turbines. The system automatically restarts the turbines when the birds have passed. At the Altamont Pass Wind Farm in California, continued efforts to reduce raptor and other bird mortality, including seasonal shutdowns, have proved ineffective. The issue appears to be with the size of the turbines because they currently believe that larger modern turbines would substantially reduce the number of raptors killed per megawatt of power produced at APWRA.

Bat migration can also be affected by wind turbine activity, but much less is known about these effects. One 2004 study in the eastern United States estimated that over 2,200 bats were killed by 63 onshore turbines in a six-week study. One method of mitigating this is to install microwave transmitters on the turbines. As bats use radar to migrate, they avoid radar-transmitting devices. The microwave rays warn the bats to stay away.

IMPACTS ON PEOPLE

Safety

Any large mechanical device such as a wind turbine has some hazards associated with its construction and operation. Hazards include workers falling off the towers or becoming caught in machinery. The blades can also fail and disintegrate, creating hazards for those nearby. Sometimes ice can form on the blades and be projected onto the ground. Aircraft and parachutists have also crashed into them. Malfunctioning turbines can potentially catch fire. The average death rate associated with wind-power generation is about 0.1 deaths per TWh generated. This is fairly low and is comparable to hydroelectric power.

Health and Esthetics

Some people object to the presence of wind turbines because they are large and visible. Many people do not like to live next to them because of noise, shadow flicker or because they are an eyesore. In agricultural areas, people still complain that they destroy the viewshed in otherwise pristine wilderness areas. Even when they are driving down the highway, some feel that the towers interfere with their view of the wilderness. These are subjective concerns, but some of them, such as noise, strobing and shadow flicker, can be measured. Noise is only an issue for those who live and work close to a turbine or wind farm; however, the noise generated is always less than traffic noise. The reported effects of noise include dizziness, anxiety, headaches and interruption to sleep patterns.

In Ontario, Canada, the Ministry of the Environment created noise guidelines to limit wind turbine noise levels at 30-m away from a dwelling or campsite to 40 decibels. Other countries have created minimum setbacks between turbines and dwellings. Two kilometers is typical.

A 2008 European Union–sponsored report from The Netherlands found that "...the sound of wind turbines causes relatively much annoyance. The sound is perceived at relatively low levels and is thought to be more annoying than equally loud air or road traffic." The report found that, "annoyance with wind turbine noise was associated with psychological distress, stress, difficulties to fall asleep and sleep interruption," which they described as a health effect. Similarly, a 2007 Swedish study report stated, "Annoyance is an adverse health effect."

A 2007 report by the US National Research Council noted, "... Low-frequency vibration and its effects on humans are not well understood.

Sensitivity to such vibration resulting from wind-turbine noise is highly variable among humans." Although there are opposing views on the subject, it has recently been stated (Pierpont 2006) that "...some people feel disturbing amounts of vibration or pulsation from wind turbines, and can count in their bodies, especially their chests, the beats of the blades passing the towers, even when they can't hear or see them." More needs to be understood regarding the effects of low-frequency noise on humans.

A 2009 report sponsored by wind-industry groups made the following counterclaims:

- *Wind Turbine Syndrome* symptoms are the same as those seen in the general population due to stresses of daily life. They include headaches, insomnia, anxiety and dizziness.
- Low frequency and very low–frequency "infra-sounds" produced by wind turbines are the same as those produced by vehicular traffic and home appliances, even by the beating of people's hearts. Such infra-sounds are not special and convey no risk factors.

People also complain about shadow flicker and strobing from wind turbines. Shadow flicker is caused by the blade casting moving shadows on the ground and through windows. Strobing is the effect produced when a turbine blade reflects the sun onto the ground and into people's eyes. This may only occur for a short period during the day, but people have reported numerous health effects from these, ranging from distraction while operating equipment to vertigo, headaches, nausea and seizures.

On April 4, 2004, Der Spiegel (Germany) was quoted by *The Telegraph* (England) saying, "They introduced the world to 'environmentally friendly' energy, but now some of Europe's 'greenest' countries are under pressure to backtrack on wind farms in the face of public anger over their impact on the countryside...Voters are outraged by the unsightly turbines, the loud, low-frequency humming noise that they create and the stroboscopic effects of blades rotating in sunshine...The dream of environmentally friendly energy has turned into highly subsidized destruction of the countryside."

This returns to the point of the Towler Principle—you cannot extract energy from the environment without having an impact on the environment. The more successful the energy source, the bigger is its environmental effects and the more these effects are noticed.

Energy Reliability

Despite the reported health effects and other objections that people have, the big problem with wind power is that the wind does not blow all the time and so the electricity production rises and falls as the wind speed rises and falls. This does not coincide with the demand for electricity. Depending on the site and the turbine properties, the wind turbine is usually generating electricity about 20-30% of the time. In the Portsmouth Abbey Wind Turbine project described above, they used a 660-kW turbine and generated 1.3 GWh in the first year and 6.07 GWh in five years, at an average of 1.214 GWh per year. This represents a capacity factor of 21.6%. The average in the United States is now approaching 30%.

If we relied only on wind energy to generate our electricity, the nominal output of the turbines must be 3.33 times the average electricity demand, and 70% of the electricity generated must be stored as some other form of energy for future use. Alternatively, if the wind turbine only supplies a small fraction of the electricity, the base load source must be turned down when the wind is blowing and generating electricity and turned up again when the wind is not blowing.

Whenever wind power is deployed, there are two choices you can make with regard to the reliability issue: the wind power must be supplemented with a large energy storage project or it must be combined with an alternative and more reliable energy source. If the first option is used, the energy storage cost increases the electricity cost substantially. If the second option is preferred, it is equivalent to providing a redundant system which also increases the cost substantially. For example, if you have to build a gas- or coal-fired power plant to back up the wind turbines when the wind is not blowing, you might as well not build the wind turbine at all. Just build the gas- or coal-fired power plant and use it all the time. Then you don't have to turn it down when the wind is blowing. Alternatively, it is possible to operate a dual purpose gas- or coal-fired power plant that generates electricity 70% of the time when the wind is not blowing, but during the 30% of the time when the wind is blowing the plant can switch to generating liquid fuels from the gas or coal. The economics of this have not been studied, but this scheme has been proposed for investigation.

You can argue that when the wind is blowing, the electricity being generated is cleaner. If clean technologies are the issue, you can build a gas- or coal-fired plant that has zero emissions and is just as clean as the wind turbine. This is a cheaper alternative to clean energy. Moreover, the wind turbine does have environmental impacts that cause concern to those affected. In fact, there are scientists that argue when all the environmental impacts are combined with the fact that the wind has to be backed up by other energy sources such as coal-fired plants, wind power is not very green at all.

> *Conveniently, it is near impossible to estimate the quantity of CO_2 and other gases emitted in order to back-up intermittent wind power: there are too many variables involved. So the wind industry has an easy job dismissing the critics. But the fact remains that CO_2 emissions are increased, possibly to the point where they fully compensate the savings realized by wind power. The example of Denmark seems to confirm this: producing 15 to 20% of its electricity from the wind, this most wind powerized country in the world has failed to reduce its greenhouse gas*

emissions. And California, Germany, Spain are building more fossil-fuel power plants in spite of their massive investments in wind farms. This, better than anything else, proves the point that wind farm opponents across the world are trying to make. -**Mark Duchamp, Windfarms/ Birds Research Manager, Proact International, quoted on Wikipedia.**

Energy Storage

If wind power is to be combined with energy storage, various storage methods must be considered. The best method currently available is pumped storage (see Chapter 10). In this method, the excess energy generated while the wind is blowing is used to pump water from a low-lying reservoir uphill to a higher reservoir. When the wind is not blowing, the water is allowed to flow from the higher reservoir to the lower reservoir, driving a water powered turbine to generate electricity. In some places, this is a reliable cost effective method of energy storage but the number of places where it can be done is limited. At the moment, the United States has the capacity to generate about 0.35 TWh/mo from pumped storage. This is clearly inadequate to back up wind power. The total hydroelectric generating capacity is about 25 TWh/mo (this depends on the level of rainfall that has filled the dams that serve the hydroelectric generators). If all of this was dedicated to backing up the wind generators, you would be restricted to only generating about 8 TWh/mo from wind. This level of electricity has already been exceeded from wind generators.

An alternative method of energy storage is the use of batteries. When the wind is blowing, the batteries are charged up, and when the wind is not blowing the batteries are used to feed electricity into the grid. This is a very expensive method of providing backup for wind power and greatly adds to the cost.

Wind power is expanding rapidly all around the world, but the larger it gets, the more reliability and backup become an issue. At the moment, the only option seems to be that the wind generators must be backed up by the fossil-fuel generators, coal and natural gas. They have to be turned up when the wind is not blowing and conversely turned down when the wind is blowing.

SUMMARY

Despite the concerns and objections, wind power is increasing rapidly in both the United States and the world. People like wind power because it is perceived as clean energy and it is much less expensive than solar power. Consequently, its use will continue to increase for some time. It may even double in the next 10 years. Eventually, however, concerns about its environmental impacts, reliability and cost will bring a halt to its expansion.

Hydroelectricity

Hydroelectricity is probably the cheapest, cleanest and most reliable of the renewable energy sources; however, it does have its detractors. Hydroelectricity is heavily dependent on the building of dams, which greatly alters the ecology of the river basin and limits fish migration and spawning regions. Worldwide, an installed capacity of 777 GW supplied 2998 TWh of hydroelectricity in 2006. This was approximately 20% of the world's electricity and accounted for about 88% of electricity from renewable sources. The United States has an installed capacity of 80 GW and in 2009 generated 250 TWh of electricity. This represents 36% of the installed hydroelectric capacity, which is typical of worldwide capacity factors. A hydroelectric power station never operates at full capacity because the water flow in the river fluctuates on an annual basis. The demand for electricity can also fluctuate, on both a daily and an annual basis.

China has the largest hydroelectric generating capacity in the world with more than 200 GW of installed capacity. China also continues to increase this capacity by building dams at a rapid rate. The largest hydroelectric generating station in the world is in China at the Three Gorges Dam on the Yangtze River, which has a generating capacity of 22.5 GW. The largest in the United States is the Grand Coulee Dam on the Columbia River, which has a generating capacity of 6.8 GW. The generating stations on the Columbia River together have a combined generating capacity of 37 GW (see more details on this in a later section).

The hydroelectric generating capacity is not fully developed either in the United States or in the world. China and Canada are currently increasing their hydroelectric capacity the fastest, while the United States lags behind. The United States has about 400 GW of additional hydroelectric potential, which could generate 1250 TWh of electricity per year, assuming a similar 36% capacity factor. This would increase the hydroelectric share of U.S. electricity to about 40% of the total. Given that hydroelectric power is clean and reliable and has high turn-down ratios, increasing the level of hydroelectric power is entirely feasible. The disadvantage of this is the environmental impact. Dams change the ecology of the river, particularly altering

The Future of Energy
http://dx.doi.org/10.1016/B978-0-12-801027-3.00010-5

fish migration patterns. It should be noted that there are advantages too beyond mere energy gains. Flood control, additional water availability and recreational activities always benefit from the building of dams.

HYDROELECTRIC BASICS

In a hydroelectric power station, the potential energy of the water due to its height above the turbine is first converted to kinetic energy and then to electrical energy. The formula for calculating the electric power production at a hydroelectric plant is

$$P = \eta \rho V h g. \tag{10.1}$$

Where, using SI units,
P = power (W),
ρ = the density of water (1000 kg/m^3),
h = the difference in height between the water level behind the dam and the water level in the river below the dam (m),
V = water volumetric flow-rate (m^3/s),
ρV = the mass flow-rate of the water through the turbine (kg/s)
g = acceleration due to gravity of 9.81 m/s^2,
η = coefficient of efficiency due to losses.

A reasonable number for η in modern hydroelectric turbines is 0.85-0.9. This means that 10-15% of the potential energy of the water is lost due to friction of the water flowing in the penstock and in the turbine blades and gears. This "lost" energy is transmitted to the water and surrounding atmosphere as heat (see Chapter 3).

As an example, a particular turbine with an energy efficiency of 0.9 (90%) is designed to allow 1200 m^3/s through its turbine blades when there is a height (h) of 94.4 m between the top of the water in the dam and the river below. To calculate the installed power capacity of this turbine:

$$P = 0.9 \text{*} 1000 \text{*} 1200 \text{*} 94.4 \text{*} 9.81 = 1.00 \times 10^9 \text{ W} = 1 \text{ GW}$$

Figure 10.1 shows the elements of a hydroelectric generating station. The water from the reservoir behind the dam enters the intake near the bottom of the reservoir. It then flows through the penstock, where the potential energy of the water in the reservoir is converted to kinetic energy. The flowing water turns the turbine blades converting the kinetic energy of the water to the electric energy in the generator. The generator is then connected to the electrical grid for power use. The details of the turbine are shown in Figure 10.2. The water flowing through the penstock reaches the wicket gate which directs the water through the turbine blades causing them to turn the shaft. This turns the rotor through the stator,

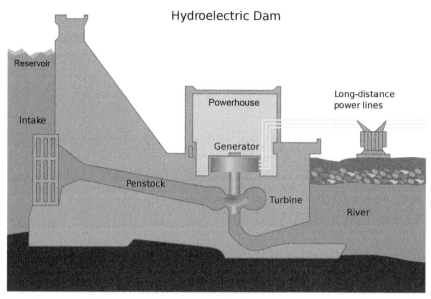

FIGURE 10.1 Diagram of a Hydroelectric Generating Station. *Source: Tomia, Wikimedia Commons.*

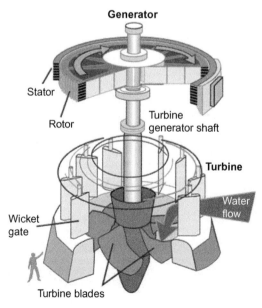

FIGURE 10.2 Hydroelectric Generating Turbine. *Source: Army Corps of Engineers.*

which causes the electrical current to flow in the stator wires. This electrical energy is then fed into the power grid. These turbines are often very large devices. Figure 10.2 compares the turbine to the size of a human being standing beside it.

In the next section, the Columbia River hydroelectric scheme is discussed in more detail as an example of river development.

THE COLUMBIA RIVER HYDROELECTRIC SYSTEM

The Columbia River system in the northwest United States, shown in Figure 10.3, is an excellent example of an integrated hydroelectric development. The Columbia and its tributaries drain a 260,000-square-mile

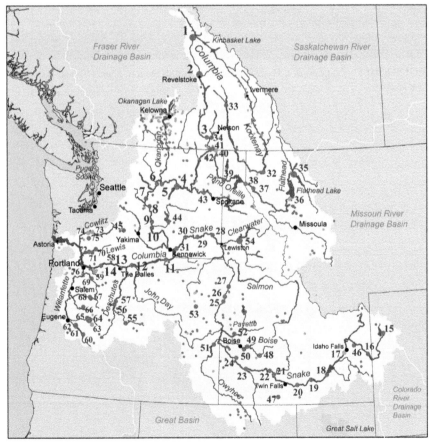

FIGURE 10.3 The Columbia-Snake River Basin and Hydroelectric Complex. *Source: K. Musser, Wikimedia Commons.*

basin that spans seven states (Oregon, Washington, Idaho, Montana, Nevada, Wyoming, and Utah) and the Canadian province of British Columbia. It flows for 1243 miles from where it originates in the Columbia Lake in southeastern British Columbia. It first flows north for about 200 miles, then turns south for another 300 miles before it crosses into the U.S. state of Washington. It then flows south through Washington and turns east to form the border between Washington and Oregon. Its average volumetric discharge rate is 265,000 cubic feet of water per second (192 million acre-feet per year), with the highest volumes in the wet summer months between April and September and the lowest in the drier winter months from November to February when precipitation is lower and water is stored in the catchment basin as snow. Its source is at 2690 ft above sea level and from there it drops an average of 2.1 ft per mile, but in some sections it falls more than 5 ft per mile.

Its major tributary, the Snake River, has a source in Yellowstone Park in Wyoming at an elevation of 8930 ft above sea level. This contributes a great deal of potential energy to the system. The Snake contributes an average flow-rate of 57,000 ft^3/s to the Columbia's total of 265,000 ft^3/s. The Columbia has 14 major tributaries, shown in Table 10.1 with the average discharge contribution of each tributary. The most important tributary,

TABLE 10.1   Average Discharge Rates of the Columbia River Tributaries

Tributary	Average Discharge (ft^3/s)
Snake River	56,900
Willamette River	37,400
Kootenay River	30,650
Pend Oreille River	26,430
Cowlitz River	9140
Spokane River	7900
Lewis River	6125
Deschutes River	5845
Yakima River	3542
Wenatchee River	3079
Okanogan River	3039
Kettle River	2925
Sandy River	2257
John Day River	2060

the Snake, has 13 tributaries of its own. It is 1078 miles long, which is longer than the 918 miles that the Columbia flows from its source to the confluence with the Snake. The Deschutes, John Day and Willamette tributaries drain basins south of the Columbia, while the Yakima, Lewis and Cowlitz tributaries drain areas on the north side of the river.

The Columbia-Snake River Basin is arguably the most hydroelectrically developed river system in the world. More than 400 dams and many other structures on tributaries restrict the river flows and tap a large portion of the Columbia's hydroelectric generating capacity, more than 34.3 GW. Figure 10.3 shows the 76 dams that generate hydroelectricity. The key to the dam numbers shown in Figure 10.3 is shown in Table 10.2. Of the 34.3 GW of generating capacity, 20 GW is on the 11 main dams on the U.S. portion of the Columbia, 4.7 GW is on the Snake, 1.6 GW is on the Kootenay, 4 GW is on the Canadian portion of the Columbia, and the other 3.7 GW is on other tributaries on the U.S. side. The 24-GW hydroelectric

TABLE 10.2 Map Key for Figure 10.3, The 76 Dams on the Columbia River Hydroelectric System

Map Number	Name	River	Height (m)	Date
1	Mica Dam	Columbia River	243	1973
2	Revelstoke Dam	Columbia River	152	1984
3	Keenleyside Dam	Columbia River	59	1968
4	Grand Coulee Dam	Columbia River	168	1942
5	Chief Joseph Dam	Columbia River	70	1955
6	Wells Dam	Columbia River	49	1968
7	Rocky Reach Dam	Columbia River	36	1962
8	Rock Island Dam	Columbia River	22	1932
9	Wanapum Dam	Columbia River	31	1963
10	Priest Rapids Dam	Columbia River	29	1959
11	McNary Dam	Columbia River	67	1954
12	John Day Dam	Columbia River	70	1968
13	The Dalles Dam	Columbia River	61	1957
14	Bonneville Dam	Columbia River	60	1937
15	Jackson Lake Dam	Snake River	21	1911
16	Palisades Dam	Snake River	82	1957
17	Gem State Dam	Snake River	12	1988

TABLE 10.2   Map Key for Figure 10.3, The 76 Dams on the Columbia River Hydroelectric System—cont'd

Map Number	Name	River	Height (m)	Date
18	American Falls Dam	Snake River	32	1978
19	Minidoka Dam	Snake River	26	1906
20	Milner Dam	Snake River	24	1905
21	Lower Salmon Dam	Snake River	16	1949
22	Bliss Dam	Snake River	43	1950
23	C.J. Strike Dam	Snake River	35	1952
24	Swan Falls Dam	Snake River	25	1901
25	Brownlee Dam	Snake River	128	1958
26	Oxbow Dam	Snake River	64	1961
27	Hells Canyon Dam	Snake River	98	1967
28	Lower Granite	Snake River	69	1975
29	Little Goose Lock and Dam	Snake River	69	1970
30	Lower Monumental	Snake River	69	1969
31	Ice Harbor Lock and Dam	Snake River	63	1962
32	Libby Dam	Kootenay River	129	1973
33	Brilliant Dam	Kootenay River	42	1944
34	Duncan Dam	Duncan River	40	1967
35	Hungry Horse Dam	Flathead River	172	1953
36	Kerr Dam	Flathead River	60	1953
37	Noxon Rapids Dam	Clark Fork	52	1960
38	Cabinet Gorge Dam	Clark Fork	52	1952
39	Albeni Falls Dam	Pend Oreille River	55	1955
40	Boundary Dam	Pend Oreille River	104	1967
41	Seven Mile Dam	Pend Oreille River	67	1979
42	Waneta Dam	Pend Oreille River	67	1954
43	Long Lake Dam	Spokane River	65	1915
44	O'Sullivan Dam	Crab Creek	61	1949
45	Tieton Dam	Tieton River	97	1925
46	Ririe Dam	Willow Creek	77	1977

Continued

TABLE 10.2 Map Key for Figure 10.3, The 76 Dams on the Columbia River
Hydroelectric System—cont'd

Map Number	Name	River	Height (m)	Date
47	Salmon Falls Dam	Salmon Falls Creek	66	1912
48	Anderson Ranch Dam	Boise River	139	1947
49	Arrowrock Dam	Boise River	107	1915
50	Lucky Peak Dam	Boise River	104	1955
51	Owyhee Dam	Owyhee River	127	1932
52	Black Canyon Dam	Payette River	56	1924
53	Mason Dam	Powder River	53	1968
54	Dworshak Dam	Clearwater River	219	1973
55	Arthur R. Bowman Dam	Crooked River	75	1961
56	Round Butte Dam	Deschutes River	134	1964
57	Pelton Dam	Deschutes River	62	1957
58	Condit Dam	White Salmon River	38	1913
59	Portland No. 1 Dam	Bull Run River	56	1929
60	Hills Creek Dam	Willamette River	104	1962
61	Lookout Point Dam	Willamette River	84	1953
62	Fall Creek Dam	Willamette River	62	1965
63	Smith Dam	McKenzie River	66	1962
64	Cougar Dam	McKenzie River	158	1964
65	Blue River Dam	Blue River	95	1968
66	Green Peter Dam	Santiam River	115	1968
67	Detroit Dam	Santiam River	141	1953
68	Big Cliff Dam	Santiam River	52	1953
69	North Fork Dam	Clackamas River	63	1953
70	Swift No. 1 Dam	Lewis River	126	1958
71	Yale Dam	Lewis River	98	1953
72	Merwin Dam	Lewis River	95	1931
73	Mossyrock Dam	Cowlitz River	112	1968
74	Mayfield Dam	Cowlitz River	61	1963
75	Sediment Retention	Toutle River	73	1988
76	Willamette Falls Locks	Willamette River	15	1873

TABLE 10.3   Hydroelectric Generating Capacity on the Columbia River

Dam	Capacity (MW)	Rank	Location	Date	Lake
Mica Dam	1805	5	British Columbia	1973	Kinbasket Lake
Revelstoke Dam	1980	4	British Columbia	1984	Revelstoke Lake
Keenleyside Dam	185	14	British Columbia	1968	Raised Arrow Lakes
Grand Coulee	6809	1	Washington	1942	Franklin D. Roosevelt
Chief Joseph Dam	2620	2	Washington	1955	Rufus Woods Lake
Wells Dam	840	12	Washington	1967	Lake Pateros
Rocky Reach Dam	1287	7	Washington	1961	Lake Entiat
Rock Island Dam	660	13	Washington	1933	Rock Island Pool
Wanapum Dam	1038	9	Washington	1963	Lake Wanapum
Priest Rapids Dam	955	11	Washington	1961	Priest Rapids Lake
McNary Dam	980	10	Washington; Oregon	1954	Lake Wallula
John Day Dam	2160	3	Washington; Oregon	1971	Lake Umatilla
The Dalles Dam	1780	6	Washington; Oregon	1960	Lake Celilo
Bonneville Dam	1050	8	Washington; Oregon	1937	Lake Bonneville

generating capacity in the 14 dams on the main Columbia River section is shown in Table 10.3.

Rock Island Dam on the middle river was the first major hydroelectric power project on the Columbia. Completed in 1932, Rock Island Dam is small compared to the much larger dams that followed it. The Bonneville and Grand Coulee were built by the federal government and completed, respectively, in 1937 and 1942. The last dam built on the U.S. portion of the Columbia, the John Day Dam, came on line in 1971. In 1973 and 1984, Canada completed two large dams—the Mica and the Revelstoke—on the upper portion of the river on the Canadian side. These two dams had a generating capacity of almost 4 GW. They were built after a treaty between Canada and the United States was negotiated and signed in 1964.

Six of the dams on the Columbia create large storage reservoirs that are not only used to generate hydroelectricity, but also serve a very important flood control function and provide water for irrigation systems in agricultural projects on the Columbia Plateau in Washington. With the completion of four dams on the lower Snake River from 1962 to 1972, a series of lakes also allowed barges to navigate more than 465 miles from the Pacific Ocean to the inland port of Lewiston, Idaho (see Figure 10.3).

THE GRAND COULEE DAM

The largest dam on the river is the Grand Coulee dam and its associated Franklin D. Roosevelt Lake (Figure 10.4). This dam not only generates electricity but also plays a major role in the supply of Columbia River water to the irrigation projects on the Columbia Plateau. To support the irrigation projects, the John W. Keys III Pump-Generating Plant pumps water uphill 280 ft from the Franklin D. Roosevelt Lake to the Banks Lake. This water is then used to irrigate approximately 670,000 acres of farmland in the Columbia Basin Project. More than 60 crops are grown in the basin and distributed across the nation.

Congress authorized the Grand Coulee Dam in 1935, with its primary purpose to provide water for irrigation. When the United States entered

FIGURE 10.4   Grand Coulee Dam and Hydroelectric Generators, Columbia River. *Source: Army Corps of Engineers.*

World War II in 1941, the focus of the dam changed from irrigation to power production. In 1943, Congress authorized the Columbia Basin Project to deliver water to the farmers of the Columbia Plateau in central Washington State. Construction of the irrigation facilities began in 1948. Components of the project include the pump-generating plant, the feeder canal and the storage reservoir, which is now named Banks Lake.

Banks Lake was formed by damming the northern 27 miles of the Grand Coulee canyon and has a storage capacity of 715,000 acre-feet of water. The Banks Lake stores water for irrigation and also is used for recreational boating. The pumping plant began operation in 1951. From 1951 to 1953, six pumping units—each rated at 65,000 horsepower and with a capacity to pump 1600 ft^3/s—were installed in the plant. In the early 1960s, the potential for pumped electricity storage was investigated and planned. Reversible pumps were installed to allow water from Banks Lake to flow back through the units to generate power during periods of peak demand. The first three generating pumps came online in 1973. Two more generating pumps were installed in 1983, with the final generating pump being installed in January 1984. The total generating capacity of the plant is now 314 MW. Further details on its use for pumped storage are discussed in the section on pumped storage.

It is interesting to read the stated purposes and benefits of the Grand Coulee dam, which according to the U.S. government, "... include flood control and river regulation, water storage and delivery (including irrigation), power generation, recreation, and fish and wildlife." The Grand Coulee is the largest hydroelectric dam in the United States with a generating capacity of 6809 MW. Other details of the dam are shown in Table 10.4 in both English and SI units.

The U.S. government reports the daily height of the water in the dam as well as the water inflow and outflow estimates. The height of the water in the dam is directly related to the net inflow minus the outflow. The data for the period from January 2009 to July 2011 is shown in Figure 10.5. There is a lot of fluctuation in the daily inflow and outflow data, so Figure 10.5 shows a 30-day moving average of the net water inflow minus the outflow to and from the lake. This is calculated by taking the gross volume of water that has flowed into the lake from the river, minus the gross volume of water that has flowed out of the lake through the generating turbines to the river below. This net volume is averaged over the previous 30 days to smooth out the daily variations. Notice that in Figure 10.5 when the net water flow increases, so too does the height of the water in the dam, and vice versa.

The electricity generated by the dam can be calculated from the water height (shown in Figure 10.5) relative to the river below as well as the discharge volumes, using Equation 10.1. To calculate the power output, you

TABLE 10.4 Specifications of the Grand Coulee Dam

Grand Coulee Dam	English Units	SI Units
Total length of dam wall (axis)	5223 ft	1592 m
Length of main dam wall	3867 ft	1178 m
Length of Forebay Dam	1170 ft	356 m
Length of Wing Dam	186 ft	56 m
Height above Bedrock	550 ft	167 m
Height above original Streambed	401 ft	122 m
Height above current river level	345 ft	105 m
Lake area	82,300 acres	33,306 ha
Lake length	151 miles	243 km
Lake Shoreline	600 miles	965 km
Active and joint use capacity	5,185,000 acre-feet	6,395,603,340 m^3
Maximum elevation above sea level	1290 ft	393 m
Total generating capacity	6809 MW	6809 MW

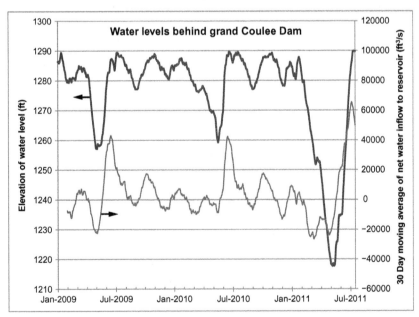

FIGURE 10.5 Water Levels (Gray Line) and Net Inflow (Light Gray Line) to Franklin D. Roosevelt Lake Behind Grand Coulee Dam from January 2009 to July 2011. *Data Source: U.S. Bureau of Reclamation.*

need to first estimate the turbine efficiencies, which turn out to be 87%. This number can be obtained by matching the maximum turbine capacity [reported to be 6809 MW and calculated using the maximum head height of 105.2 m (345 ft)] and the maximum reported volumetric discharge of 7592 m^3/s. The maximum water height is assumed to be 345 ft above the river below, as shown in Table 10.4. This corresponds to a water elevation of 1290 ft above sea level for the water in the dam. From these values, you can then calculate the electricity generated on any day at Grand Coulee dam. As an example, on July 17, 2011, the water level was at the maximum 1290 ft above sea level corresponding to a water head height (h) of 345 ft above the discharge water level. The average volumetric flow-rate of water discharged on that day was 174,000 ft^3/s. Converting these numbers to SI units gives $h = 105.2$ m and $V = 4927$ m^3/s. The power output that day was

$$P = \eta \rho V h g$$

$$P = 0.87 * 1000 * 4927 * 105.2 * 9.81 = 4.421 \times 10^9 \, \text{W} = 4.421 \, \text{GW}.$$

The net power generated each day since January 1, 2009 is shown plotted in Figure 10.6. From January 1, 2009 to July 17, 2011, 53.97 TWh of electricity was generated. This represents about 36% of the total capacity and is typical for a hydroelectric operation.

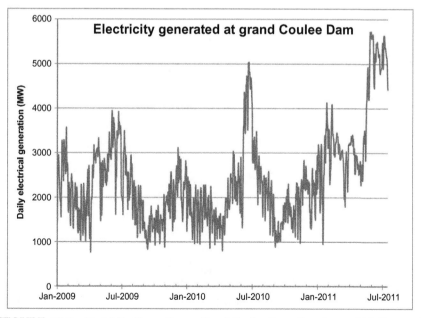

FIGURE 10.6   Electricity Generation at Grand Coulee Dam. *Data Source: U.S. Bureau of Reclamation.*

THE OTHER COLUMBIA RIVER DAMS

The 11 hydroelectric generating dams in the U.S. portion of the river are shown in Figure 10.8. Five of these dams (and another three on the Snake River) are "run-of-the-river" dams, so called because they have only a small capacity reservoir. Consequently, the water coming from the upstream must be used for generation at the moment it is received or must be allowed to bypass the dam. These dams usually generate electricity constantly. The other dams can be turned on and off as the electricity demand changes. The biggest of these run-of-the-river dams is the Chief Joseph Dam (see Figures 10.7 and 10.8), which has a generating capacity of 2620 MW and is located immediately below the Grand Coulee Dam. The others are the four lowest level dams: the McNary, the John Day, the Dalles, and the Bonneville.

The Columbia River dams have altered the migration of the previously strong anadromous fish populations (fish that migrate between freshwater rivers and saltwater oceans) in the river. This has led to significant declines in the numbers of these fish. Prior to the construction of the dams, commercial fishing operations had harvested millions of pounds of fish each year, especially salmon. The largest portion of the harvest came from the lower river and the estuary by the ocean. Since then, the combined effects of dams, increased ocean fishing, deterioration of the river habitats

FIGURE 10.7 Chief Joseph Dam and Hydroelectric Generators, Columbia River, an Example of a "Run of the River" Dam. *Source: Army Corps of Engineers.*

FIGURE 10.8 Eleven U.S. Dams on the Columbia River. *Source: U.S. Bureau of Reclamation.*

and changing river conditions have made the Columbia less and less habitable for the anadromous fish.

Ever since the early 1970s, the fish catch has continuously declined, with hatchery-raised species now making up more than 80% of commercially caught salmon in the river. Fish hatcheries began operation in the basin in 1877 and helped mitigate the dam-caused salmon declines. Nevertheless, some species of salmon have now been listed as endangered. In order for fish to migrate up the river, the dams on the lower river have fish ladders that allow the migrating fish to bypass the dams and continue up the river to their spawning grounds. These fish ladders are chutes that the fish can swim up to reach the water above the dam. The Chief Joseph Dam (see Figure 10.7)—which is located on the middle reaches of the river just below the Grand Coulee Dam, as well as the Grand Coulee Dam—have no fish ladders, which blocks fish migration to the upper half of the Columbia River system.

Despite the barriers that the dams ostensibly cause, navigation of the Columbia as a shipping channel has actually been enhanced by the changes to the river. Previously, large commercial ships could not travel very far up-river. Because of the dredging of a deep channel in the lower river—which was deepened to 40 ft in 1976—and the deep man-made lakes on the middle river, ocean-going freighters can navigate up the Columbia to Portland, Oregon, and Vancouver, Washington (not to be confused with the Canadian city of Vancouver, British Columbia). Barges can also transport goods from there to the interior. Towboats pull the barges up through locks on the Bonneville, the Dalles, John Day, McNary and on four

Snake River dams carrying fuel and other commodities upriver and returning with wood chips, lumber and agricultural products downriver.

In the nineteenth century, before any dams were built, agricultural projects began to use water from the Columbia River. By the 1920s, major irrigation projects along the Columbia and its tributaries such as the Yakima, Wenatchee and Umatilla rivers operated with the benefit of federal assistance. During the 1930s and 1940s, however, the construction of the big dams, especially the Grand Coulee Dam on the upper river and the McNary Dam on the middle river, greatly increased irrigated agriculture on the Columbia Plateau. In 1948, the Columbia Basin Project began transporting Columbia River water by canal to more than 670 thousand acres on farms in central Washington. This project required large pumping stations, a network of canals and large sprinkler systems. There were more than 60 major irrigated crops including alfalfa, potatoes, mint, beets, beans, orchard fruit and grapes.

Recreation on the Columbia, which includes sailing, swimming, water skiing, canoeing and other water sports have become commonplace on the river since World War II. This was made more feasible by the creation of the lakes that resulted from the building of the dams. More recently, sailboarding and windsurfing have also become popular. In 1986, Congress passed the Columbia River Gorge National Scenic Area Act, which mandates environmental protection of the Columbia River Gorge through cooperation among federal, state, municipal and county governments in the Gorge.

PUMPED HYDROELECTRIC STORAGE

Pumped storage is a variation on hydroelectric power generation, which can be used to store energy, especially electricity. The procedure is to store energy by pumping water from a lower elevation reservoir to a higher elevation reservoir. When the energy is needed, the higher elevation reservoir water can be run through a hydroelectric turbine to generate electricity. The turbine is designed to operate as the pump during the water storage operation. The method so far has mainly been used to store or buy electricity at lower off-peak rates and then generate electricity for use or sale at higher peak rates. Some energy (10-15%) is lost in the process to friction, so pumped storage actually consumes energy. Overall, it makes money and allows power utilities to balance the electrical demand. It also has the potential to be expanded to include wind and solar energy projects to balance the load when the wind is not blowing or the sun is not shining.

In 2008, the world pumped storage-generating capacity was 104 GW. The European Union (EU) had 38.3 GW of pumped storage capacity (36.8% of world capacity) out of a total of 140 GW of hydropower, which represented 5% of the total net electrical capacity in the EU. In 2009, the

United States had 21.5 GW of pumped storage-generating capacity (20.6% of world capacity) accounting for 2.5% of the net electrical generating capacity.

One example of a pumped storage plant in the United States is the *Raccoon Mountain Pumped-Storage Plant* in Marion County, Tennessee, 6 miles west of Chattanooga. It is owned and operated by the Tennessee Valley Authority (TVA) and has been in operation since 1978. It cost $300 million to build. Water is pumped from the Nickajack Reservoir on the Tennessee River to a storage reservoir at the top of Raccoon Mountain (see Figures 10.9, 10.10, and 10.11). The storage reservoir

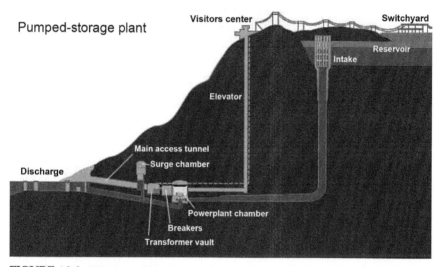

FIGURE 10.9   Diagram of Raccoon Mountain Pumped-Storage Plant, Tennessee. *Source: Tennessee Valley Authority.*

FIGURE 10.10   Raccoon Mountain Storage Reservoir. *Source: Tennessee Valley Authority.*

FIGURE 10.11   Concrete Penstock Inside Raccoon Mountain. *Source: Tennessee Valley Authority.*

on top of the mountain covers an area of 528 acres and was created by a dam wall that is 230 ft high and 8500 ft long and curves around the contours of the mountain to contain the water. It takes 28 h to fill the upper reservoir and 22 h to drain it. During periods of high-electric demand, water is released from the reservoir through a tunnel drilled through the mountain, driving four 383 MW electric generators (a total of 1532 MW; see Figure 10.12) in the same tunnel inside the mountain. It is used almost every day to balance the electrical load in the TVA system.

The largest-capacity pumped-storage power station in the world (and in the United States) is the *Bath County Pumped Storage Station* in Virginia, with a generating capacity of 2772 MW. The pump-turbine system is composed of six 462-MW units manufactured by the Allis-Chalmers Manufacturing Company. The two reservoirs are separated by 1263 ft (385 m) in elevation. It cost $1.6 billion to build and went into operation in 1985. The upper reservoir is 265 acres (110 ha) in area, and the lower reservoir is 555 acres (220 ha). The upper reservoir can discharge water to the hydroelectric generators at the maximum rate of 852 m^3/s. During operations, the water level can fluctuate by 105 ft (32 m) in the upper reservoir and 60 ft (18 m) in the lower reservoir. The generating efficiency in this system has not been reported but, using the methods shown previously in the Grand Coulee Dam section, it can be calculated as 86.1%. Other pertinent details of the plant are

Owners: Dominion (60%) and Allegheny Power (40%)

FIGURE 10.12 Hydroelectric Pump-Turbine Hybrids at Raccoon Mountain Pumped-Storage Plant. *Source: Tennessee Valley Authority.*

Lower Reservoir Dam

- 135 ft high (41 m)
- 2400 ft long (732 m)
- Contains 4 million cubic yards (3.1 million cubic meters) of earth and rock fill

Lower Reservoir

- 555 surface acres (2.25 km^2)
- Water level fluctuates 60 ft (18 m) during operations

Upper Reservoir Dam

- 460 ft high (140 m)
- 2200 ft long (671 m)
- Contains 18 million cubic yards (13.8 million cubic meters) of earth and rock fill

Upper Reservoir

- 265 surface acres (1.07 km^2)
- Water level fluctuates 105 ft (32 m) during operations

Water flow—Pumping: 12.7 million gallons/minute (801 m^3/s)
Water flow—Generating: 13.5 million gallons/minute (852 m^3/s)
Turbine generators: Six Francis-type 462-MW units manufactured by Allis-Chalmers
Maximum pumping power (per unit): 642,800 horsepower (479,300 kW)
Using these numbers, the generating capacity of the system can be verified using Equation 10.1.

$$P = \eta \rho Vhg$$

$$P = 0.861^*1000^*852^*385^*9.81 = 2.77 \times 10^9 \, W = 2.77 \, GW$$

At the end of the generating cycle, when the water level in the upper reservoir has been lowered by 32 m and the water level in the lower reservoir has been raised by 18 m, the height difference between the two reservoirs has been reduced to 335 m. The water flow-rate has also been reduced to 740 m^3/s. The generating capacity becomes

$$P = 0.861^*1000^*740^*335^*9.81 = 2.094 \times 10^9 \, W = 2.094 \, GW.$$

There is a smaller pumped storage facility on the Columbia River associated with the Grand Coulee Dam. The *John W. Keys III Pump-Generating Plant* pumps water 280 ft uphill from the Franklin D. Roosevelt Lake to the Banks Lake. This water is used to irrigate farmland in the Columbia Basin Project. Construction of the irrigation facilities began in 1948. Components of the project include the pump-generating plant, feeder canal and the Banks Lake equalizing reservoir.

Banks Lake was formed by damming the northern 27 miles of the Grand Coulee canyon and has an active storage capacity of 715,000 acre-feet of water. The plant was originally designed just as a pumping operation and began to pump water for irrigation in 1951. From 1951 to 1953, six pumping units were installed in the plant to pump the water to the upper reservoir. Each pump was rated at 65,000 horsepower and with a capacity to pump 1600 ft^3/s. In the early 1960s, it was decided that this arrangement had the potential for pumped power storage. Reversible Francis-type pump-generators were installed to allow water from Banks Lake to flow back through the units to generate power during periods of peak demand. The total generating capacity of the pumped storage plant is now 314 MW.

SUMMARY

Hydroelectric power generates a clean, renewable and reliable energy supply. According to a March 2007 report released by the Electric Power Research Institute (EPRI), there is 90 GW of untapped hydroelectric power generation potential in the United States. According to EPRI, this could produce enough energy to serve the needs of 22 cities the size of Washington, D.C. and equates to over 250 million tons of potential reductions in greenhouse gas emissions per year that is currently unrealized. EPRI reports that, by the year 2025, the United States also has the potential to develop 10 GW (or 11% of the total above) from new, small hydro capacity gains at existing hydro sites and new generating facilities at existing dams.

The U.S. Hydropower Association website states that the industry could install 60 GW of new capacity by 2025, which is only 15% of the total untapped hydropower resource potential in the United States, meaning that the Hydropower Association has identified 400 GW of additional hydropower potential. At 36% capacity, this 400 GW of additional capacity could generate an additional 1250 TWh of clean renewable electricity per year, 40% of the U.S. demand. EPRI only considers developing 90 GW of this potential because it is better to leave some rivers wild to maintain some diversity in the ecosystems.

Developing this hydroelectric potential has the following added benefits as stated by the U.S. Department of the Interior for the Grand Coulee Dam: "... flood control and river regulation, water storage and delivery (including irrigation), power generation, recreation, and fish and wildlife development." Flood control is an important benefit of hydroelectric dams. In the summer of 2011, many communities along many rivers in the United States (including the Mississippi, Missouri, and Red Rivers) were flooded, creating significant dislocation. Prior to the building of the three Gorges Dam on the Yangtze River in China, millions of people lost their lives when they drowned in floods that occurred periodically on the river. The Three Gorges Dam certainly had an impact in the areas behind the dam and the whole ecology of the river was changed, but the lives that it saves every year and the electricity that it generates makes it worthwhile. There is still plenty of further potential for this in the United States and around the world.

Geothermal Energy

Geothermal energy is a rather interesting source of energy. It has a number of similarities to standard fuel-based energy sources such as coal, oil, and natural gas. All of these burn fuel to generate steam, which is then used to drive a steam turbine to generate electricity. In certain locations, there are hot formations close to the surface that already contain hot water or steam under pressure and that can be used to drive the same steam turbines without burning any fuel. This type of energy source is not cheap to develop and operate because wells have to be drilled and produced and steam turbines have to be built and run. Nevertheless, under certain circumstances, it is cost effective to develop geothermal energy.

The word geothermal comes from the Greek words *geo* (Earth) and *therme* (heat). Geothermal energy is heat from within the Earth. Most energy sources used on Earth, including fossil fuels, originated in the sun. Geothermal energy is different; it originates in processes occurring within the Earth. This energy can be recovered as steam or hot water and can then be used to heat buildings or to generate electricity. Geothermal energy can be classified as a renewable energy source because the energy is continuously being produced inside the Earth due to the radioactive decay of elements such as uranium, radium, thorium, and potassium. This produces very high temperatures, hot enough to keep much of the interior core of the Earth in a molten state.

Earth actually has a number of different layers. The central core has two layers: a solid iron core and an outer core made of magma, which is melted rock. The next layer is the mantle, which surrounds the core, and is about 1800 miles thick. It is made up of magma and solid rock. The crust is the outermost layer of the Earth, which is 3-5 miles thick under the oceans and 15-35 miles thick on the continents. The Earth's crust is broken into continental plates, which are constantly in motion. Magma can come close to the Earth's surface near the edges of these plates. This is where volcanoes and earthquakes occur. The lava that erupts from volcanoes is partly magma. Deep underground, the rocks and water absorb heat energy from this magma. As the water rises to the surface, natural hot springs and geysers occur, such as Old Faithful at Yellowstone National Park (see Figure 11.5). The water in these systems ranges from 400 to 700 °F.

Generally, geothermal energy is recovered by drilling wells into the hot water and steam reservoirs underground and pumping the heated water or steam to the surface. This is then used to heat buildings, such as homes and offices, and to generate electricity. The most active geothermal resources are found along major plate boundaries where earthquakes and volcanoes are concentrated. Much of the geothermal activity in the world occurs along the Ring of Fire, which encircles the Pacific Ocean and runs down the west coast of North and South America, through the coast of Alaska, around the Aleutian Island chain, down the Kamchatka peninsula of Russia, through Japan and the Philippines and down to New Zealand. Iceland, which sits at the conjunction of the North Atlantic Ridge and two continental plates, is also very geologically active. This activity allows a significant portion of its energy to be derived from geothermal and hydroelectric resources.

In Iceland, hot water and steam are recovered from shallow hot spots and this is then used to heat buildings and generate electricity. A city utility, commonly known as a district heating system, produces the hot water and pipes it into buildings to keep people warm through the long cold winters. In the capital city of Reykjavik (population 120,000), hot water is piped from a spring 25 km away, and residents use it not only for heating their houses but also for hot tap water. It is doubtful that the Icelandic community would thrive the way it has without access to this energy source.

Figure 11.1 shows where the potential geothermal resources of the United States are located. Most of the geothermal reservoirs are located

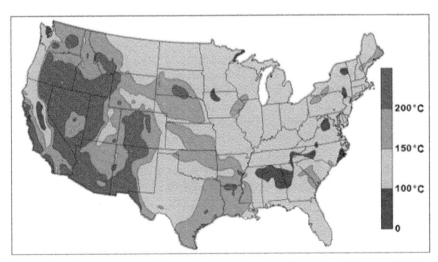

FIGURE 11.1   U.S. Geothermal Resource Map. *Source: U.S. Department of Energy and NREL.*

in the western states and Hawaii. California generates the most electricity from geothermal energy. The Geysers, a dry steam reservoir north of San Francisco in Northern California, is the largest known dry steam field in the world. It has been producing electricity since 1921 and has been greatly expanded since 1990. Figure 11.1 gives the impression that geothermal resources are widespread and abundant; however, not all of the areas shown in this figure are economic to develop yet because they are much too deep.

The three main categories of utilizing geothermal energy are

(a) Direct use and district heating systems, which use hot water from springs or reservoirs near the surface
(b) Electricity generation power plants, which require water or steam at very high temperatures (300-700 °F) close to the surface. Geothermal power plants are generally built where geothermal reservoirs are located within 6000 ft of the surface
(c) Geothermal heat pumps, which use stable temperatures near the Earth's surface to control building temperatures above ground. Heat pumps can be used to heat or cool buildings very efficiently, but they generally use water within 1000 ft of the surface.

There have been direct uses of geothermal hot water as an energy source throughout history. Many cultures have used hot mineral springs for bathing, cooking and heating. Many people firmly believe that the hot, mineral-rich waters have beneficial healing powers. After bathing, the most common direct use of geothermal energy is for heating buildings through district heating systems. Hot water near the Earth's surface can be piped directly into buildings and industries for heat. In addition to building heat and electricity generation, the drying of vegetable and fruit products for storage and later use is another small industrial use of geothermal energy.

Table 11.1 shows the world's installed geothermal electrical generating capacity, with the United States as the world leader. The world total installed capacity now exceeds 9000 MW (9 GW), which is a very small part of the world's electricity supply. In 2009, U.S. geothermal power plants produced 15 billion kilowatt-hours (kWh) from the total 3102 MW capacity. This represented a 55% utilization or capacity factor and about 0.4% of the total U.S. electricity supply. In 2009, 5 states had geothermal power plants: California, which had 35 geothermal power plants and produced 85% of U.S. geothermal electricity; Nevada, which had 18 geothermal power plants and produced 11% of U.S. geothermal electricity; and, Hawaii, Idaho, and Utah, which each had one geothermal plant and account for the other 4%.

Twenty countries including the United States had geothermal power plants in 2008, which generated a total of about 60.4 billion kWh.

TABLE 11.1   Installed Geothermal Capacity (MW)

1	USA	3102
2	Philippines	1966
3	Indonesia	1189
4	Mexico	958
5	Italy	863
6	New Zealand	769
7	Iceland	575
8	Japan	502
9	El Salvador	204
10	Kenya	167
11	Costa Rica	166
12	Nicaragua	88
13	Russia (Kamchatka)	82
14	Turkey	82
15	Papua New Guinea	56
16	Guatemala	52
17	Portugal (The Azores)	29
18	China	24
19	France (Guadeloupe)	16
20	Ethiopia	7.3

The Philippines was the second largest geothermal power producer after the United States at 9.8 billion kWh, which equaled about 17% of the country's total power generation. Geothermal power plants in El Salvador and Iceland produced about 1.4 and 3.8 billion kWh, respectively, which was equal to about 25% of the total power generated in those countries. The other 75% of Iceland's electricity is generated from hydroelectricity.

There are three basic types of geothermal power plants: dry steam plants, flash steam plants, and binary cycle power plants. A dry steam plant uses steam piped directly from a geothermal reservoir to turn the generator turbines (see Figure 11.2). After the steam passes through the turbines, it is condensed into water and normally reinjected into the reservoir from which it came. This step is usually necessary to maintain the steam supply underground. The largest power plant of this type is The Geysers in California. For many years, The Geysers did not reinject their

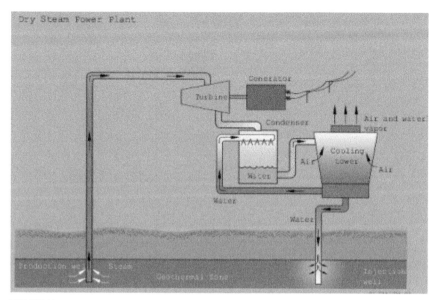

FIGURE 11.2 Dry Steam Geothermal Power Plant. *Source: DOE, INL.*

used water, as is implied in Figure 11.2, because of the process used. This process is discussed in more detail later in the chapter. Now, they are beginning to inject make-up water obtained from other sources.

A flash steam plant uses high-pressure hot water from the Earth and runs it through a flash drum where the pressure is allowed to decrease, which allows some of the hot water to flash into steam at a lower pressure. This steam is then used to drive the steam generator turbines. When the steam cools, it condenses to water and is injected back into the ground to be heated again. The water from the bottom of the flash drum is also added to this steam to be reinjected. The majority of geothermal power plants are flash steam plants, which are illustrated in Figure 11.3.

Binary cycle power plants transfer the heat from geothermal hot water to another liquid such as butane. The heat causes the second liquid to turn to vapor, which is used to drive a generator turbine, as shown in Figure 11.4. This process is used when the hot water does not have a high enough temperature and pressure to use the dry steam or flash steam process.

To analyze the power output of any steam turbine, both the first and second laws of thermodynamics must be applied. To illustrate a sample calculation, assume that a geothermal plant is producing hot water at a pressure of 2 MPa and is utilizing a flash steam process as illustrated in Figure 11.3. The steam exits the flash drum as a saturated vapor at the pressure of 2 MPa. It travels through the turbine and exits into the

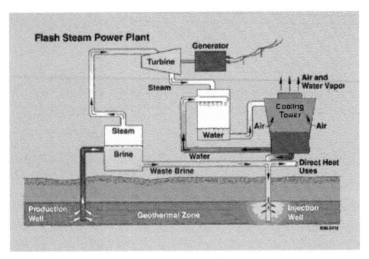

FIGURE 11.3   Flash Steam Geothermal Power Plant. *Source: DOE, INL.*

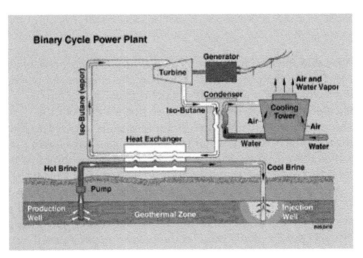

FIGURE 11.4   Binary Cycle Geothermal Power Plant. *Source: DOE, INL.*

condenser at a pressure of 10 kPa as partially condensed steam and liquid water droplets. The second law of thermodynamics will calculate what the maximum properties of the steam can be at the exit of the turbine. Assume that this calculation gives us a steam quality at the turbine exit of 76%. These numbers allow us to calculate the energy of the steam entering and exiting the turbine. The relevant energy parameter of the steam at each point is called the enthalpy of the steam. Enthalpy includes the internal energy as well as the flow energy of the steam, both of which are a function of its temperature and pressure. Saturated steam at the pressure of 2 MPa has an enthalpy of 2800 kJ/kg and the 76% quality steam at 10 kPa has an enthalpy of 2008 kJ/kg. Consequently, if the entire plant was flashing 100 kg/s of steam from the flash drum, using the first law of thermodynamics the turbine could generate:

$$Power = 100^*(2800 - 2008) = 79{,}200\,kW = 79.2\,MW$$

Note the low efficiency of this process. Only the flashed steam is being used to generate power in the turbine. Most of the brine (about 75%) being produced by the production well is bypassing the turbine and is being reinjected. Of the steam that is being sent through the turbine to generate electricity, only 28% of its energy is being captured by the turbine. The rest of the energy remains with the low pressure steam, which will also be reinjected back into the ground. The Geysers power plant is more efficient than this because it produces pure steam directly from the ground.

THE GEYSERS DRY STEAM GEOTHERMAL POWER PLANT

The largest geothermal system in the world currently in operation is a steam-driven plant in an area called The Geysers, north of San Francisco, California. Despite the name, there are no actual geysers in the area. The energy source is dry steam, not hot water. The area was known for its hot springs (not geysers) at least as far back as 1847 when it was discovered by Bill Elliot who called it "The Geysers." Soon after this, it was developed as a resort spa. Hot springs like this have been popular throughout history. Bathing in the hot water is a pleasant experience and gives people the impression that it is beneficial to their health and well-being. The resort became a popular holiday destination and thrived for many years.

The first well and power plant at the site were completed in 1921, which generated electricity at the rate of 250 kW. This was a very small power plant and was barely large enough to power the resort, which had been operating continuously since 1852. Deeper wells were drilled in the 1950s, but real development did not occur until the 1970s and 1980s. By 1990, 26 power plants had been built for a capacity of more than 2000 MW.

Because of the rapid development of the area in the 1980s and the technology used, the steam resource has been declining since 1988. Today, owned primarily by Californian utility Calpine and with a total operating capacity of 2000 MW, The Geysers' facilities still meet nearly 60% of the average electrical demand for California's North Coast region (from the

Golden Gate Bridge north to the Oregon border). The plants at The Geysers use an evaporative water-cooling process to create a vacuum that pulls the steam through the turbine, producing power more efficiently. This process, however, loses 60-80% of the steam to the air without reinjecting it underground. This led to the decline of the volume of steam coming from the underground reservoir and consequently the power that it was able to generate. To recharge the reservoir and to increase the declining steam volume and pressure, various interested partners created the Santa Rosa Geysers Recharge Project. This project involves transporting 11 million gallons per day of treated wastewater from neighboring communities through a 40-mile pipeline and injecting it into the ground to provide make-up water to be converted to steam by the hot rocks below. This part of the project came online in 2003. The city of Santa Rosa plans to further expand this program by increasing the amount of wastewater sent to the Geysers to 20 million gallons per day.

OTHER GEOTHERMAL APPLICATIONS

Geothermal springs can also be used directly for many varied heating purposes. Hot spring water is used to heat greenhouses, to make dried fish and jerky, for enhancing oil recovery and to heat fish farms and spas. Klamath Falls, Oregon; Boise, Idaho; and Warm Springs, Virginia are a few of the cities in the United States where geothermal spring water has been used to heat homes and buildings for more than a century. In Iceland, virtually every building in the entire country is heated with hot spring water.

Another method of utilizing geothermal energy is via a heat pump. The temperatures just below the Earth's surface hold nearly constant between 35 and 65 °F, depending on the location. For most areas, this means that soil temperatures are usually warmer than the air in winter and cooler than the air in summer. Geothermal heat pumps use the Earth's constant temperatures to heat and cool buildings. They transfer heat from the groundwater into buildings in winter and reverse the process in the summer. A heat pump is a kind of reversible refrigerator. A refrigerator uses electrical energy to pump heat from a cold box inside the refrigerator into the warm room outside the refrigerator. A heat pump acts the same way: it pumps heat from the Earth in the wintertime into the house, and in the summer it pumps heat from the house back into the Earth. The U.S. Department of Energy and the EPA have partnered with industry to promote the use of geothermal heat pumps because they can be more efficient than direct heating.

In regions with temperature extremes, ground-source heat pumps are the most energy-efficient heating and cooling systems available. Far more efficient than electric heating and cooling, these systems can pump four times the energy they use in the process. That statement might seem like it violates the first law of thermodynamics, but it does not. This is because a heat pump moves energy from one location to another. For example, it can move the energy required to heat a house from the Earth to the house, but it typically uses only one quarter of the energy gained. The U.S. Department of Energy believes that heat pumps can save a typical home $350 a year in energy costs, with the system often paying for itself in 8-12 years. Tax credits and other incentives can reduce the payback period to 5 years or less.

More than 750,000 ground-source heat pumps supply climate control in U.S. homes and other buildings, with new installations occurring at a rate of about 50,000 per year. Ground-source heat pumps are especially popular in rural areas without access to natural gas pipelines, where homes must use propane or electricity for heating and cooling. Recent government policy initiatives are offering strong incentives for homeowners to install these systems. The 2008 economic stimulus bill— the Emergency Economic Stabilization Act of 2008—includes an 8-year extension (through 2016) of the 30% investment tax credit, with no upper limit in home installations, for Energy Star certified geothermal heat pumps.

ENHANCED GEOTHERMAL SYSTEMS

There is a vast amount of heat energy available from dry rock formations very deep below the surface (4-10 km). Using a set of emerging technologies known as Enhanced Geothermal Systems (EGSs), it may be possible to capture this heat for electricity production on a much larger scale than conventional technologies allow. Geothermal heat occurs everywhere under the surface of the Earth, but the conditions that make water circulate to the surface are found only in a small percentage of the Earth's land mass. EGS is an approach to capturing the heat in "hot dry rock," but it is still not commercial, although it does show promise. The hot rock reservoirs, typically at greater depths below the Earth's surface than conventional sources, are first fracture-stimulated by pumping high-pressure water through them. This opens up pathways for water to flow through the rocks from one well to another. The plants then inject water through the fractured hot rocks, where it heats up, returns to the surface as high pressure steam and powers the steam turbines to generate electricity. After it exits the turbine as cooler low pressure steam, it is condensed into hot water. This water is then reinjected into the reservoir through the injection wells to complete the circulation loop in the same manner as shown in Figures 11.2-11.4. The Department of Energy is collaborating with universities and the geothermal industry in the United States on research to demonstrate the potential of EGS in hot dry rock. Australia, France, Germany, and Japan also have similar research and development programs.

Oil and gas fields already under production represent another large potential source of geothermal energy. In many existing oil and gas reservoirs, a significant amount of high-temperature water is coproduced with the oil and gas. This could be used for the production of electricity whether in a flash system (Figure 11.3) or in a binary system (Figure 11.4). The Rocky Mountain Oilfield Testing Center, located at the Teapot Dome Oilfield in the Powder River Basin in Wyoming, has been investigating the technology to do this.

An MIT study, commissioned by the U.S. Department of Energy in 2006, estimated that the United States has the potential to develop 44 GW of geothermal capacity by 2050 by coproducing electricity, oil and natural gas at oil and gas fields primarily in the southeast and southern Plains states. Further, it would be possible to develop 100 GW of EGS geothermal capacity by 2056. This study projects that such advanced geothermal systems could supply 10% of the U.S. base-load electricity by that year.

These new developments in geothermal energy will be supported by increased levels of federal research and development funding. Under the American Recovery and Investment Act of 2009, $400 million of new funding was allocated to the DOE's Geothermal Technologies Program. Of this, $90 million is expected to go toward a series of demonstration projects to prove the feasibility of EGS technology. Another $50 million will fund demonstration projects for other new technologies, including coproduction of hot water with oil and gas wells. The remaining funds will go to exploration technologies to expand the deployment of geothermal heat pumps and other uses.

If these resources can be tapped, they offer greatly increased potential for geothermal electricity production capacity. In a recent assessment, the U.S. Geological Survey (USGS) estimated that conventional geothermal sources on private and accessible public lands across 13 western states have the potential capacity to produce 8-73 GW, with a mean estimate of 33 GW. State and federal policies are likely to encourage developers to tap some of this potential in the next few years. The Geothermal Energy Association estimates that 132 projects now under development around the country could provide up to 6 GW of new capacity. As EGS technologies improve and become competitive, even more of the largely untapped geothermal resources could be developed. The USGS study found that hot dry rock resources could provide another 345-727 GW of capacity, with a mean estimate of 517 GW. At a 75% capacity factor, 517 GW could generate 3400 TWh of electricity, which compares to the total 4326 TWh of electricity that was generated in the United States in 2010. If further improvements can be made in EGS technology, this resource would be capable of supplying the majority of U.S. electricity needs.

Not only do geothermal resources in the United States offer great potential, they can also provide continuous base-load electricity. It is not an intermittent power source like solar and wind power. According to the U.S. National Renewable Energy Laboratory, the capacity factors of geothermal plants (the ratio of the actual electricity generated compared to what would be produced if the plant was running 100% of the time) are comparable with those of coal and nuclear power. The capacity factor of the Geysers Power Plant has averaged 55% recently, but geothermal plants are capable of running at 75-80% capacity. With the combination of both the size of the resource base and its reliability, geothermal energy can play a significant role in the supply of clean renewable electricity; however, that day has not yet come.

The MIT study commissioned by the U.S. Department of Energy in 2006, referred to above, made some specific recommendations for successful EGS projects. In particular, they pointed to the following improvements that would make EGS technology cost competitive:

1. *"High flow rates with long path lengths are needed.* Natural hydrothermal systems require each production well to produce about 5 MW per well, which requires flow rates ranging from 30 to 100 kg/s (depending on the fluid temperature). At the same time, we need a large heat-exchange area or residence time for water to reheat to production temperatures; this could imply large-pressure drops. Better understanding of successful natural systems (in comparable geological settings) should lead to improved methods of generating artificially EGSs. For instance, the residence time of water injected at Dixie Valley is 3-6 months, and the production wells show little or no cooling due to the aggressive injection program. At Steamboat, though, the residence time for the water is closer to 2 weeks and there is fairly significant cooling.

    The well spacing between injectors and producers at Dixie Valley is about 800 m, and there are probably at least two fractures with a somewhat complex connection between the injectors and producers resulting in a long fluid-path length. At the Steamboat hydrothermal site, the distance between producers and injectors is more than 1000 m; however, because there are so many fractures, the transmissivity is so high that there is low residence time for injected fluids. At the East Mesa hydrothermal site, the reservoir is in fractured sandstone and the residence time varies from one part of the field to another. Some injectors perform well in the center of the field, while other injectors are in areas with either high matrix permeability in some zones or fractures that cause cold water to break through faster. The large volume of hot water stored in the porous matrix at East Mesa made it possible to operate the field for a long time before problems with cooling developed.

2. *Stimulation is through shearing of pre-existing fractures.* In strong crystalline rock, hydraulic properties are determined by the natural fracture system and the stresses on that system. The expectation of scientists planning the early experiments in enhancing geothermal reservoirs was that fracturing would be tensile. While it may be possible to create tensile fractures, it appears to be much more effective to stimulate pre-existing natural fractures and cause them to fail in shear. Understanding the orientation of the stress field is crucial to designing a successful stimulation. Fortunately, in even the most unpromising tectonic settings, many fractures seem to be oriented for shear failure. At Cooper Basin, which is in compression, stimulation of two nearly horizontal pre-existing fracture systems appears to have been successful in creating a connected reservoir of large size. The shearing of natural fractures increases hydraulic apertures and this improvement remains after pressures are reduced. Fortunately, stress fields in strong rocks are anisotropic so the critically aligned natural joints and fractures shear at relatively low overpressures (2-10 MPa).

3. *Fractures that are stimulated are those that will take fluid during prestimulation injection.* The fractures that are found to be open and capable of receiving fluid during evaluation of the well before stimulation are almost always those that are stimulated and form large-scale connections over a large reservoir volume. This may be because these fractures are connected anyway or because the fractures that are open are those oriented with the current stress state. It is important, therefore, to target areas that will have some pre-existing fractures due to their stress history and the degree of current differential stress. Even in areas with high compressional stresses, such as Cooper Basin in Australia, there are natural, open fractures.

With present technology, connected fractures cannot be created where none exist. It may be possible to initiate new fractures, but it is not known whether these will form large-scale flow paths and connect over large volumes of the reservoir. This means that the fracture spacing in the final reservoir is governed by the initial, natural fracture spacing. The number of fractures in a wellbore that will take fluid is important to assess in each well. The total heat that can be recovered is governed by the fracture spacing because the temperature drops rapidly away from the fracture face that is in contact with the injected cool fluid.

4. *There is currently no reliable open-hole packer to isolate some zones for stimulation.* This is routine in the oil and gas industry; but in the geothermal industry, high-temperature packers for the open hole are not reliable so we stimulate the entire open interval. Logging shows that the first set of open fractures is the one most improved. If you want to stimulate some zones more than others, or if you want to create new fractures, you will need a good, reliable, high-temperature open-hole packer. Although earlier testing at Soultz (France) using a cement inflatable aluminum packer has been encouraging, more development work remains to be done to improve reliability and increase temperature capability.

5. *Hydraulic stimulation is most effective in the near-wellbore region.* The near-wellbore region experiences the highest pressure drop so stimulation of this region is important. Connectivity in the far field away from the wells is also required to maintain circulation and accomplish heat mining. A variety of techniques, both from the oil and gas industry and from geothermal experience, can be effectively used to improve near-wellbore permeability. Hydraulic stimulation through pumping large volumes of cool fluid over long time periods and acidizing with large volumes of cool fluid and acid (of low concentration) have been most effective. Use of high-viscosity fluids, proppants, and high-rate high-pressure stimulation has been tried with mixed success and may still have potential in some settings, particularly in sedimentary reservoirs with high temperatures. There are, however, limits to the temperature that packers, proppants, and fracturing fluids can withstand. Some of these techniques are impossible or very costly in a geothermal setting.

   In crystalline rocks with pre-existing fractures, oil and gas stimulation techniques have failed to result in connection to other fractures and may form short circuits that damage the reservoir. Current efforts to stimulate geothermal wells and EGS wells, in particular, are limited to pumping large volumes of cold water from the wellhead. This means that the fractures that take fluid most readily anyway are stimulated the most. Only a small portion of natural fractures in the wellbore support flow. Because these more open fractures may also be the ones that connect our producers to our injectors, this may not be a disadvantage. There may be a large number of fractures observed in the wellbore; however, an ability to identify and target the best ones for stimulation is limited because of a lack of research.

6. *The first well needs to be drilled and stimulated in order to design the entire system.* Early efforts to create reservoirs through stimulation relied on drilling two wells that were oriented in such a way that there appeared to be a good chance of connecting them, given the stress

fields observed in the wellbore and the regional stress patterns. At Fenton Hill, Rosemanowes, Hijiori, and Ogachi, this method did not yield a connected reservoir. Stress orientation changes with depth, or with the crossing of structural boundaries. The presence of natural fractures already connected (and at least somewhat permeable) makes evaluating the stimulated volume difficult. It seems much easier to drill the first well, then stimulate it to create as large a volume as possible of fractured rock, then drill into what we think is the most likely place, and stimulate again. Because of this, we can design wells as either producers or injectors, whereas it would be better if we could design wells for both production and injection. This emphasis on the first well demands that it be properly sited with respect to the local stress conditions. Careful scientific exploration is needed to characterize the region as to the stress field, pre-existing fractures, rock lithology, etc.

7. *Monitoring acoustic emissions is the best tool for understanding the system.* Mapping of acoustic events is one of the most important tools for understanding the reservoir. In hydrothermal systems, well tests and tracer tests demonstrate that water is circulating and in contact with large areas of rock. Stimulated fractures can be assessed in the same way, once there are two or more wells in hydraulic connection to allow for circulation tests. The location of acoustic emissions generated during stimulation and circulation can also be mapped with accuracies of around ± 10-30 m. While no one is completely sure what the presence or absence of acoustic emissions means in terms of fluid flow paths or reservoir connectivity, knowledge of the location and intensity of these events is certainly important. This information helps define targets for future wells.

If we drill into a zone that has already been stimulated and shows a large number of acoustic emissions events, it is commonly assumed that the well is connected to the active reservoir. This fact, however, does not always result in a good system for heat extraction. For example, at Soultz, GPK4 was drilled into an area that was within the volume of mapped acoustic emissions. Even after repeated stimulations, it did not produce a connected fracture system between the production and injection wells.

Mapping of acoustic emissions has improved so that we can locate acoustic emissions and determine the focal mechanism for these events more accurately than in the past. As a result, there is a better understanding of the stress field away from the wellbore and how stimulation affects it. Methods for mapping fractures in the borehole have been developed and the upper limit for temperatures at which they can operate is being extended. Ultrasonic borehole televiewers, microresistivity fracture imaging, and wellbore stress tests have all proved very useful in understanding the stress state, nature of existing fractures, and the fluid flow paths (before and after stimulation). Correlating the image logs with high-resolution temperature surveys and with lithology from core and cuttings allows a better determination of which fractures might be productive.

8. *Rock-fluid interactions may have a long-term effect on reservoir operation.* While studies of the interaction of the reservoir rock with the injected fluid have been made at most of the sites where EGS has been tested, there is still a good deal to learn about how the injected fluid will interact with the rock over the long term. The most conductive fractures often show evidence of fluid flow in earlier geologic time such as hydrothermal alteration and mineral deposition. This is encouraging in that it suggests that the most connected pathways will already have experienced some reaction between water and the rock fracture surface.

Fresh rock surfaces will not be protected by a layer of deposited minerals or alteration products. Currently, there is no way to know how much surface water (which cannot be in equilibrium with the reservoir rock) will need to be added to the system over the long term. The longest field tests that have been conducted have seen some evidence for the dissolution of rock and this has led to the development of preferred pathways and short circuits.

Regardless, the produced fluids will need to be cooled through the surface equipment, possibly resulting in precipitation of scale or corrosion (Vuatarez, 2000). Although not analyzed in this study, the use of carbon dioxide (CO_2) as the circulating heat transfer fluid in an EGS reservoir has been proposed (Pruess, 2006). Brown (2000) has developed a conceptual model for such a system, based on the Fenton Hill Hot Dry Rock reservoir. The argument is made that supercritical CO_2 holds certain thermodynamic advantages over water in EGS applications and could be used to sequester this important greenhouse gas.

9. *Pumping the production well to get the high-pressure drops needed for high flow rates without increasing overall reservoir pressure seems to reduce the risk of short circuiting while producing at high rates.* High pressures on the injection well during long-term circulation can result in short circuits. Circulating the fluid by injecting at high pressures was found to consume energy while, at the same time, tending to develop shorter pathways through the system from the injector to the producer. High-pressure injection during circulation may also cause the reservoir to continue to extend and grow, which may be useful for the portion of time the field is operating. It may not, however, create fractures that are in active heat exchange, given the system of wells that are in place.

High-pressure injection can also result in fluid losses to those parts of the reservoir that are not accessed by the production wells. By pumping the production wells in conjunction with moderate pressurization of the injection well, however, the circulating fluid is drawn to the producers from throughout the stimulated volume of fractured rock, minimizing fluid loss to the far field.

10. *The wells needed to access the stimulated volume can be targeted and drilled into the fractures.* While drilling deep wells in hard, crystalline rock may still be fairly expensive, the cost technology has improved dramatically since the first EGS wells were drilled at Fenton Hill. Drill bits have much longer life and better performance, typically lasting as long as 50 h even in deep, high-temperature environments. The rate of penetration achievable in hard, crystalline rock and in high temperature environments is continually increasing, partly due to technology developments with funding from the U.S. government.

As the oil and gas industry drills deeper, and into areas that previously could not be drilled economically, they will encounter higher temperatures and more difficult drilling environments. This will increase the petroleum industry's demand for geothermal-type drilling. Most geothermal wells need to have fairly large diameters to reduce pressure drop when flow rates are high. Directional control is now done with mud motors, reducing casing wear and allowing better control. Although high temperatures are a challenge for the use of measurement-while-drilling tools for controlling well direction, they did not exist when the first EGS well was drilled. Furthermore, the temperature range of these tools has been extended since they first became available. Mud motors are now being developed that can function not only at high temperatures, but also with aerated fluids.

11. *Circulation for extended time periods without temperature drop is possible.* Although early stimulated reservoirs were small and long-term circulation tests showed measurable temperature drop, later reservoirs were large enough that no temperature drop could be measured during the extended circulation tests. It is difficult to predict how long the large reservoirs will last because there is not enough measurable temperature change with time to validate the numerical models. Tracer test data can be used for model verification, but in cases where extremely large reservoirs have been created, tracer data may not be adequate for determining the important parameters of heat-exchange area and swept volume.

12. *Models are available for characterizing fractures and for managing the reservoir.* Numerical simulation can model fluid flow in discrete fractures. It can also model flow with heat exchange in simple to complex fractures, in porous media and in fractured, porous media. Changes in permeability, temperature changes, and pressure changes

in fractures can be fit to data to provide predictive methods. Because long-term tests have not been carried out in the larger, commercial-sized reservoirs, it is not yet known whether the models will adequately predict the behavior of such reservoirs. Rock-fluid interactions in porous media or fractured, porous media can also be modeled, but their long-term effects are equally uncertain. Commercial fracture design codes do not take thermal effects into consideration in determining the fracturing outcome. Geothermal codes for fracture stimulation design purposes that do consider thermal or hydraulic effects in fracture growth have not yet been developed.

13. *There are a number of induced seismicity concerns.* In EGS tests at the Soultz site, microseismic events generated in the reservoir during stimulation and circulation were large enough to be felt on the surface. Efforts to understand how microearthquakes are produced by stimulation are ongoing, and new practices for controlling the generation of detectable microseismic events are developing. A predictive model that connects reservoir properties and operating parameters such as flow rate, volume injected, and pressure which might affect the generation of detectable microearthquakes is important to realizing the potential of EGS. Such a model has not been quantitatively established." MIT Report to DOE, "The Future of Geothermal Energy," 2006.

ENVIRONMENTAL ISSUES

Geothermal energy is not without environmental impact according to the Towler Principle (Chapter 1). So what are those impacts? On the one hand, geothermal power plants do not burn fuel to generate their electricity so their air emissions are very low. They release less than 1% of the carbon dioxide emissions of a fossil fuel plant. Geothermal plants, however, usually emit hydrogen sulfide (also known as "rotten egg gas") that is often found naturally in steam and hot water underground. This must be captured and converted to elemental sulfur or sulfuric acid or reinjected underground. Some plants also emit trace amounts of arsenic and other harmful metals. Some geothermal plants, such as the one at the Salton Sea reservoir in Southern California, also produce a significant amount of salt that builds up in the pipes and other plant equipment. Initially, the Salton power plant recovered this salt and put it into a landfill, which caused contamination of the groundwater supply. Now, they dissolve the salt in the wastewater and reinject it back into the same formation that produces the hot water via the reinjection process.

If the condensed steam or cooled water is not reinjected, the water in the reservoir underground becomes depleted. This can affect the water table and the groundwater supplies of other users and also results in a depleting geothermal resource. Consider, for example, if there was a proposal to build a geothermal power plant in the Yellowstone National Park adjacent to the Old Faithful geyser (Figure 11.5). This would be a natural place to

FIGURE 11.5   Old Faithful Geyser at Yellowstone National Park, Wyoming. *Source: public domain photo by Jon Sullivan.*

put a geothermal plant because there is clearly hot water and steam close to the surface that could easily be captured and utilized in a commercial power plant. The result of this, however, would likely be that the Old Faithful geyser would cease to erupt once every hour, as it currently does. When people visit Yellowstone National Park, they come to see Old Faithful. If it ceased to erupt, many people would be upset and some people would not come to visit the National Park at all, despite the many other natural wonders there are to see. Clearly, the environmental impact of a geothermal power plant in Yellowstone National Park would not be acceptable.

The large Geysers project in Northern California suffered from the problem of a depleting water table and declining power output until they began injecting the make-up water to remedy this. Apart from the impacts on other users of the water table, if the geothermal plant does not reinject water, the result is a depleting nonrenewable energy source.

With closed-loop systems, there are minimal emissions and minimal environmental impacts because everything brought to the surface is returned underground. The energy extracted is renewed by ongoing radioactivity in the Earth's interior. In general, the surface is disturbed by the drilling of wells and deployment of turbines and associated equipment that have a visual impact and other disturbances to the environment. There could be other environmental impacts that are not immediately apparent but, if developed in the proper manner, the overall environmental impact of a geothermal project should be small; but it is not always acceptable. To illustrate this point, the Glass Mountain project in Northern California was an EGS test case which targeted low-permeability, high-temperature rocks just outside the defined hydrothermal boundary of The Geysers geothermal project. The operator, Calpine, was unable to obtain an environmental permit to initiate the project, which led to its cancelation. The Glass Mountain project would have been relatively benign and was located adjacent to an area that was already developed for geothermal activity, but it was still deemed to be unacceptable by the permitting authorities. Calpine then moved the project to its operating plant in an area inside The Geysers boundary where there is low permeability rock and high hydrogen sulfide content and a resulting acidic steam. The new project targeted the stimulation of low-permeability rock on the fringe of their production area to improve the steam quality and to recharge the reservoir while recovering heat energy.

THE FUTURE OF GEOTHERMAL ENERGY

With current technology, geothermal energy is likely to remain a small player in the United States and the world energy supply. There are only limited areas where geothermal energy can be extracted economically. It does have the potential to play a larger role if EGSs technology can be made cost effective. If this happens, the payoff could result in a cleaner sustainable energy system that is cost competitive. Like hydroelectricity, it is one of the few renewable energy technologies that can supply continuous and reliable base-load power and can be turned up and down as the load changes. The costs for electricity from geothermal facilities are highly variable and generally higher than fossil fuels. The costs, however, are declining and The Geysers geothermal plants have realized a 50% reduction in the price of electricity since 1980. The aim is to open up a large swathe of potential geothermal resources that might be able to produce electricity for about 8 cent/kWh (including a production tax credit), a cost that would be competitive with new conventional fossil fuel-fired power plants. There is also increased potential for the direct use of geothermal resources as a heating source for homes and businesses in many locations. In order to tap into the full potential of geothermal energy, the emerging technologies require further development and cost improvements.

12

Ethanol, Biodiesel, and Biomass

In many respects, biomass has had a bright past. Prior to 1900, it was the chief energy source used in the world, and it continues to maintain a strong presence today. In Chapter 1, Figure 1.1 shows that biomass was overtaken by coal as the number one energy source in the world around 1910 (depending on the data source). The main biomass energy source represented in Figure 1.1 is wood. People have cut down trees or gathered fallen wood and burned them for fuel since prehistoric times. Wood and other biomass was burned to heat houses and other buildings and to provide fuel for cooking food. In fact, it continues to be used that way today. When I was a boy growing up in the small town of Chinchilla (in Queensland, Australia), our house had a wood-burning stove in the kitchen. Not only was it used for cooking, but also it heated the hot water for the house and was the only source of space heating. We regularly bought a load of eucalyptus wood from a supplier, and one of the chores required of my brothers and me was to split the wood with an ax into pieces small enough to be burned in the stove. We kept that stove burning almost continuously. Apart from this wood, the only other energy source used in our house was electricity that powered the lights, a washing machine (no dryer) and a single radio. We even solar dried our clothes on a clothesline. Today, homes are typically heated with natural gas, propane or heating oil. Clothes are dried in a dryer powered by gas or electricity. While wood is a less common energy source today, particularly in the Western world, many people would still like to see its use expanded in the future.

Biomass is viewed as a clean and renewable energy source because it can be renewed and replaced by growing new plants and trees. When new trees are grown, they capture carbon dioxide from the atmosphere that is created when they are burned. If every tree that is burned is replaced by a new tree, biomass is truly renewable and carbon neutral. This, however, has not always been true. In the middle ages in Europe, most of the European forests were cut down and burned, but were not replaced. This was done not only to provide an energy source, but also to provide potash for glass and soap manufacturing. This practice was not carbon neutral or renewable because new trees were not grown to replace the ones chopped and burned.

Various forms of biomass, including wood, can be also converted into ethanol and other fuels to provide transportation fuels. In the United States, ethanol is generated from corn and has been strongly promoted as a renewable and clean energy source. Careful analysis, however, shows that it is not carbon neutral because significant pollution is generated from the farming machinery used to grow the corn and from the fuel used for distillation of the alcohol. Moreover, diverting the corn to energy use has driven up the price of corn-based foods. The idea that a vital food resource is being used to make fuel to drive our automobiles is distasteful to some people. Due to this, ethanol is rapidly falling out of favor again in the United States. In Brazil, a large ethanol fuel program is based on sugarcane which has many advantages over corn-based fuel. It is cheaper and produces bagasse (the fibrous residue that remains after sugarcane is crushed to extract the juice)—a by-product that can be used as an energy source to distill the ethanol.

Biomass has three relevant components: cellulose, starches, and sugars. All of these can be burned directly to release energy. Starches and sugars can also be readily converted into ethanol via the biological fermentation process; however, this process is only capable of making ethanol up to about 12% purity. After that, the microorganisms begin to die because the ethanol they are producing as a waste product becomes poisonous to them. Consequently, after fermentation, the ethanol produced must be concentrated by a distillation process. This takes energy to achieve. Cellulose can also be converted into liquid fuels (such as ethanol, gasoline, or diesel) or burned to provide heat.

ETHANOL PRODUCTION

The largest producers of fuel ethanol in the world are shown in Table 12.1. The top two are the United States and Brazil that dominate world production. The largest producer, the United States, makes all of its ethanol from corn stocks. The second largest producer, Brazil, uses sugarcane.

The monthly fuel ethanol and biodiesel production (in barrels per day) in the United States, for the past 30 years, is shown in Figure 12.1. The United States is now producing over 900,000 barrels per day of fuel ethanol, around 5% of the country's liquid fuel supply. Beginning in 2002, the fuel ethanol production was ramped up in response to subsidies instituted by the federal government. The main argument for these subsidies was that ethanol is a clean-burning renewable fuel with a high octane number.

TABLE 12.1 Fuel Ethanol Production in the World (Thousands of Barrels of Ethanol per Day)

Rank	Country	2005	2006	2007	2008	2009
1	USA	254.7	318.6	425.4	605.6	713.5
2	Brazil	276.4	306.1	388.7	466.3	449.8
3	China	20.7	24.1	28.7	34.3	37.0
4	France	2.5	5.0	9.3	17.0	21.5
5	Canada	4.4	4.4	13.8	16.3	18.7
6	Germany	2.8	7.4	6.8	10.0	13.0
7	Spain	5.2	6.9	6.2	5.9	8.0
8	Jamaica	2.2	5.2	4.9	6.4	6.9
9	Thailand	1.2	2.2	3.0	5.7	6.9
10	India	3.7	4.1	4.5	4.6	5.8
11	Colombia	0.5	4.6	4.7	4.4	5.2
12	Australia	0.4	1.3	1.4	2.5	3.5

Data Source: EIA.

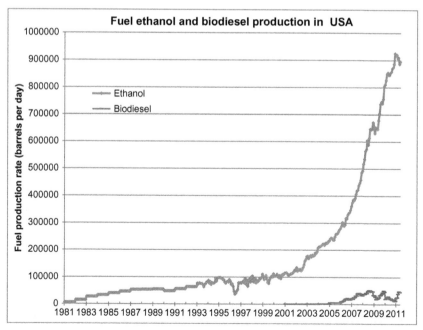

FIGURE 12.1 Monthly fuel ethanol and biodiesel production in the United States *Data Source: EIA.*

Its octane value using the $(R+M)/2$ formula is 99, which is significantly higher than regular gasoline (which is around 85-87). As explained in the next section, however, ethanol also has a low-energy density that results in a lower fuel economy than gasoline.

The second supporting argument for ethanol subsidies is that it is produced from indigenous sources in the United States and displaces imported oil from the fuel supply. The arguments against this are that ethanol is primarily produced from corn. This leads to an increase in the price of corn which is an important element of the U.S. food supply. Moreover, if imported oil is used to plow the fields and harvest the corn and to distill the fermented ethanol, the amount of displaced imported oil is not nearly as high as the ethanol production numbers indicate. This factor is a function of the fuel production efficiency measured by the energy balance in producing ethanol.

Most of the ethanol produced in the world is generated from corn or sugarcane. Whether corn or sugarcane is used, it needs to be grown, collected, crushed, processed into pure sugar, fermented and distilled into pure ethanol. All of these steps require energy use. The total amount of energy input into the process compared to the energy released by burning the resulting ethanol fuel is known as the energy balance. Energy balance estimates from different sources can be contradictory and can inspire passionate intellectual debates. Several credible researchers (e.g., Pimentel, Keeney and DeLuca, Ho) have concluded that the energy balance is actually negative. In other words, it takes more energy to produce a gallon of ethanol than is recovered when it burns as fuel. A comprehensive study by Shapouri, Duffield and Wang of the USDA, completed in 2002, concluded that the ratio of energy out to energy in (the energy balance's defining number) is 1.34. This means that 34% more energy is obtained from the ethanol fuel than is required to produce it. To get to that number, the authors include energy credits for co-products produced along with the ethanol. These include distiller's dried grains, corn oil, corn gluten meal, and corn gluten feed from wet milling.

Opponents often overstate this number, assuming that the one unit of energy used to produce the 1.34 energy units of ethanol is all imported oil. In actual fact, a lot of the energy used to distill the ethanol is domestic coal or natural gas. In 2007, *National Geographic Magazine* estimated that the energy balance number is 1.3 (i.e., one barrel equivalent of fossil-fuel energy is required to create 1.3 barrels of oil equivalent energy from the resulting corn ethanol). The energy balance for sugarcane ethanol produced in Brazil is more favorable: one unit of fossil-fuel energy is required to create eight energy units from the sugarcane-derived ethanol. Sugarcane is easier to grow than corn and the sugarcane bagasse (the dried cane cellulosic residue) is usually the fuel burned to distill the fermented ethanol.

Most of the corn ethanol produced in the United States is blended at a ratio of 10% ethanol to 90% gasoline and thus is known as E10. Pure ethanol (E100) and an 85% ethanol blend (E85) are also available. E85 is the most common blend available in Brazil.

FUEL ECONOMY OF ETHANOL

Pure ethanol has a high octane number (99), but an energy density that is 32.5% lower than gasoline (see Table 2.4). This means that adding ethanol in any proportion to regular gasoline will increase the octane number, but will lower the fuel economy. This might seem contradictory because most people believe that high octane fuels will give them a higher fuel economy (measured as miles per gallon). Octane rating, however, is not a measure of the energy content of the fuel, but instead is a measure of its ability to burn at high compression ratios without auto-igniting.

In theory, all vehicles have a fuel economy that is directly proportional to the fuel's energy density; however, there are many variables that affect the performance of a particular fuel in a particular engine. The compression ratio of the engine has a significant effect. Using the data in Table 2.4, ethanol contains approximately 32.5% less energy per unit volume than gasoline. In theory, burning pure ethanol in a vehicle will result in a 32.5% reduction in miles per gallon compared to burning pure gasoline. Given that ethanol has a higher octane rating, the engine can be made more efficient by raising its compression ratio. This can make the fuel economy of ethanol almost constant for any blend. This is not easy to do. So, ethanol blends in practice give a lower fuel economy. For E10 (10% ethanol and 90% gasoline), the effect is small (3.3%) when compared to conventional gasoline and even smaller when compared to boutique and reformulated gasoline blends. For E85 (85% ethanol), the effect becomes more significant. E85 will produce about 27.5% lower mileage than gasoline and hence will require more frequent refueling. Flex fuel vehicles are more optimized towards ethanol blends. Based on EPA tests for 2006 E85 models, the average fuel economy for E85 vehicles resulted in 25.6% lower mileage than regular unleaded gasoline.

BIODIESEL PRODUCTION

Figure 12.1 also shows the production of biodiesel in the United States. Biodiesel production now has a capacity of about 45,000 barrels a day, significantly less than ethanol production. It is also somewhat cyclical in nature, although Figure 12.1 only shows the commercial production of biodiesel. There is also a lot of biodiesel produced in small batch systems for personal use. The EIA data plotted in Figure 12.1 does not reflect this noncommercial production.

The name biodiesel implies that the same diesel fuel made from crude oil can be manufactured from biological materials, but this is not true. Diesel fuel is a mixture of long-chain alkane molecules, such as decane (the C10 alkane), dodecane (the C12 alkane), and eicosane (the C20 alkane). What has come to be known as biodiesel is actually a mixture of methyl esters that are manufactured from oils and fats. All oils and fats, whether from animal or vegetable sources, are made up of fatty acids called triglycerides. There are 14 principal triglycerides or fatty acids found in oils and fats. These are shown in Table 12.2. The percentage of the 14 triglycerides found in common fat and oil sources is shown in Table 12.3. All of these triglycerides can be reacted with methanol to make methyl esters. Methyl esters can be burned in diesel engines as diesel fuel. These esters have a

TABLE 12.2 The Fourteen Principal Fatty Acids Found in Fats and Oils

Fatty Acid	No. of Carbons and Double Bonds	Chemical Structure	Melting Point (°C)	Boiling Point (°C)
Caprylic	C8	CH3(CH2)6COOH	16.5	239
Capric	C10	CH3(CH2)8COOH	31.3	269
Lauric	C12	CH3(CH2)10COOH	43.6	304
Myristic	C14	CH3(CH2)12COOH	58	332
Palmitic	C16:0	CH3(CH2)14COOH	62.9	349
Palmitoleic	C16:1	CH3(CH2)5CH=CH(CH2)7COOH	33	–
Stearic	C18:0	CH3(CH2)16COOH	69.9	371
Oleic	C18:1	CH3(CH2)7CH=CH(CH2)7COOH	16.3	–
Linoleic	C18:2	CH3(CH2)4CH=CHCH2CH=CH(CH2)7COOH	−5	–
Linolenic	C18:3	CH3(CH2)2CH=CHCH2CH=CHCH2CH=CH(CH2)7COOH	−11	–
Arachidic	C20:0	CH3(CH2)18COOH	75.2	–
Eicosenoic	C20:1	CH3(CH2)7CH=CH(CH2)9COOH	23	–
Behenic	C22:0	CH3(CH2)20COOH	80	–
Eurcic	C22:1	CH3(CH2)7CH=CH(CH2)11COOH	34	–

TABLE 12.3 Percentage of Fatty Acid Types in Various Oils and Fats

Fatty Acid Fat or Oil	C8:0	C10:0	C12:0	C14:0	C16:0	C16:1	C18:0	C18:1	C18:2	C18:3	C20:0 C22:0	C20:1 C22:1	Other
Yellow Grease	–	–	–	1	23	1	10	50	15	–	–	–	–
Tallow	–	–	0.2	2-3	25-30	2-3	21-26	39-42	2	–	0.4-1	0.3	0.5
Lard	–	–	–	1	25-30	2-5	12-16	41-51	4-22	–	,	2-3	0.2
Butter	1-2	2-3	1-4	8-13	25-32	2-5	25-32	22-29	3	–	0.4-2	2-1.5	1-2
Coconut	5-9	4-10	44-51	13-18	7-10	–	1-4	5-8	1-3	–	–	–	–
Palm Kernal	2-4	3-7	45-52	14-19	6-9	0-1	1-3	10-18	1-2	–	1-2	–	–
Palm	–	–	–	1-6	32-47	–	1-6	40-52	2-11	–	–	–	–
Safflower	–	–	–	–	5.2	–	2.2	76.3	16.2	–	–	–	–
Peanut	–	–	–	0.5	6-11	1-2	3-6	39-66	17-38	–	5-10	–	–
Cottonseed	–	–	–	0-3	17-23	–	1-3	23-41	34-55	–	–	2-3	–
Corn	–	–	–	0-2	8-10	1-2	1-4	30-50	34-56	–	–	0-2	–
Sunflower	–	–	–	–	6	–	4.2	18.7	69.3	0.3	1.4	–	–
Soybean	–	–	–	0.3	7-11	0-1	3-6	22-34	50-60	2-10	5-10	–	–
Rapeseed	–	–	–	–	2-5	0.2	1-2	10-15	10-20	5-10	0.9	50-60	–
Linseed	–	–	–	0.2	5-9	–	0-1	9-29	8-29	45-67	–	–	–
Tung	–	–	–	–	–	–	–	4-13	8-15	72-88	–	–	–

high cetane number. This makes them a good blending stock to combine with regular crude oil-based diesel fuel, but they can also be burned directly in diesel engines. They do, however, have one major disadvantage: they have a relatively high pour point, which means they tend to gel up (solidify) in cold weather. This disadvantage can be overcome when they are blended with regular crude oil-based diesel fuel at low proportions. They can only be used directly as diesel fuel (known as B100) in the summer or in warm climates.

Biodiesel is typically made by reacting Straight Vegetable Oil (SVO), Waste Vegetable Oil (WVO), or animal tallow with methanol, which is known as the trans-esterification reaction. WVO, also known as yellow grease, is mostly composed of C18 fatty acids with one double bond and C16s with no double bonds. The volume of biodiesel produced is about equal to 90% of the volume of WVO that is supplied to the reaction process; the other 10% is primarily glycerol. In general, the reaction is

$$\text{Triglyceride} + \text{Methanol} \rightarrow \text{Methyl Ester} + \text{Glycerol}.$$

The catalyst used is either sodium hydroxide (NaOH) or potassium hydroxide (KOH). The glycerol is a by-product that has many uses. While most uses require small volumes, it can also be made into methanol in a process described later.

A lot of biodiesel is made from WVO in small home-processing kits similar to the one shown in Figure 12.2 from Reliance Energy Resources.

The WVO is fed from a transportation container to one of the large tanks where the biodiesel process starts. The small central tank holds a mixture of methanol and catalyst (sodium or potassium hydroxide) to react with the vegetable oil. These chemicals are mixed in the correct proportions and fed into one or both of the large tanks containing the WVO and then the components react for a set period of time. After the reaction has finished and the glycerol has separated from the biodiesel by gravity, the glycerol is drained off and a water wash occurs. Mist washing is a critical part of making high-quality biodiesel. The mist heads are metered at a precise flow rate to achieve thorough washing. The water wash occurs in the same tank where the reaction occurred.

After the biodiesel is water washed, it is allowed to settle so that all of the water comes out of the biodiesel. The settling of the glycerol and water from the biodiesel is assisted by using a sonic agitator that reduces the duration of the settling period from about a day to a few hours. Once the settling is complete, the biodiesel can be separated from the water and pure biodiesel remains. The purpose of washing the methyl ester layer is to get rid of impurities such as dissolving glycerols and the remaining traces of catalyst. The methyl ester is usually tested to make sure that the quality meets the minimum standards for diesel fuel.

1
2
3
4
5
6
7
8
9
10
11
12
13
14

Three-Tank System

15
16
17
18
19
20
21
22
23
24
25

FIGURE 12.2 Personal home biodiesel kit (three-tank system) model 6016-1 from Reliance Energy Resources *Source: Reliance Energy Resources.*

The main waste by-product from this process, glycerol, is currently not very valuable due to the high production that has occurred in the United States by home and commercial biodiesel producers. There is, however, a solution to this problem as discussed below.

The trans-esterification of oil involves a reaction between the triglycerides and methanol in the presence of the catalyst, yielding a mixture of fatty acid methyl esters, which is a combination of the biodiesel and glycerol. One problem associated with the use of WVO is the high content of free fatty acids (FFAs), which can neutralize the caustic (KOH or NaOH) catalyst and form fatty acid salts as an unwanted product. Free fatty acids are components of the triglyceride molecule that break off during the cooking process. When there are too many FFAs in the cooking oil mixture, the oil is no longer useful for cooking and must be discarded or converted into biodiesel. The triglyceride-methanol reaction is catalyzed (sped up) by the hydroxides (NaOH or KOH). These hydroxides are not consumed by the triglyceride-methanol reaction, but the FFAs do react with the hydroxides resulting in loss of catalyst. Elimination of FFAs can be done through esterification using methanol and a sulfuric acid catalyst before proceeding to the trans-esterification step. After the esterification pretreatment, WVO is heated to about 120 °F in order to speed up the trans-esterification reaction. The trans-esterification product obtained is allowed to separate, and the aqueous layer containing some traces of glycerol and catalyst is drained out.

Another important part of the system shown in Figure 12.2 is an Ultra-Sonic Frequency Generator (USFG). The USFG improves the performance of the system in two ways: it causes the reaction to proceed at a faster rate and in higher yields, and it aids in the settling process between the organic and aqueous layers. It is thought that the USFG does this by improving agitation and mixing throughout the system. The improvement in mixing is due to induced cavitation of the liquid. This is effected by creating alternating low and high pressure bubbles, which cause certain areas to collapse rapidly and increases the shear force gradients. The ultra-sonication also aids in the separation process. Typically, a biodiesel batch will separate in 10-24 h. With the addition of an USFG, this time can be reduced to less than one hour. An USFG can also improve catalyst efficiency. The amount of catalyst needed for the average batch can be reduced by 50%. Other benefits from ultra-sonication include higher purity in the glycerol by-product and less necessity for excess methanol in the reaction.

Glycerol is the only major by-product in this process. Its main characteristics include

- It is a neutral, sweet tasting, colorless, thick liquid that can be frozen to a thick paste and has a high boiling point
- It is not poisonous or toxic
- It can easily dissolve in water or alcohol, but not in oils. It is also a good solvent. There are many substances which will dissolve in glycerol that will not dissolve in water or alcohol
- It is highly hygroscopic, which means that it absorbs water from the air. For example, if you leave a beaker of glycerol exposed to the air, it would quickly become a miscible mixture that is 80% glycerol and 20% water
- It is widely used to manufacture capsules, suppositories, ear infection remedies, anesthetics, cough remedies, lozenges, and gargles
- It is also used to moisten, sweeten or preserve many foods and drinks such as soft drinks, candies, cakes, meat, and cheese
- It is used in textiles to soften the yarn and to lubricate fibers of different kinds

- It is used as a lubricant in food processing to manufacture resin coatings, to add flexibility to rubber and plastic and as a building block to manufacture flexible foams
- It is used to manufacture dynamite and to create components that are used in radios and neon lights.

Despite all of these uses, the biodiesel industry has created an oversupply of glycerol, which has caused its price to plummet.

It is possible to transform glycerol back into methanol, which is the preferred method of disposal. In fact, there are catalysts available on the market which can perform this conversion at low temperatures and pressures. Edman Tsang, a researcher in the Department of Chemistry at the University of Oxford (England), has patented a technology that uses a metal catalyst to convert glycerol into methanol by reacting it with hydrogen. This process works at a temperature of 100 °C (212 °F) and at 20 bars of pressure. The process produces only methanol and no by-products according to the reaction

$$Glycerol + Hydrogen =\rightarrow Methanol$$

$$CH_2OH - CH_2OH - CH_2OH + 1.5H_2 \rightarrow 3CH_3OH.$$

Once produced, the methanol can then be reused in the next triglyceride batch. The net overall biodiesel reaction then becomes

$$Triglyceride + Hydrogen \rightarrow Methyl Ester.$$

Adoption of this solution will solve the glycerol oversupply problem.

DIRECT CONVERSION OF WOODY BIOMASS INTO GASOLINE, DIESEL AND OTHER LIQUID FUELS

The cellulose in woody lignin-based biomass (such as trees) and the bagasse from sugarcane and switch-grass cannot be converted into liquid fuels as easily as sugars and starches. This is the subject of much research. Throughout history, the energy available in woody biomass has been traditionally released by direct burning. The resulting energy can be captured as heat to maintain the temperature of a living space, which has been its primary use. It can also be converted into electricity by using the heat energy to generate steam to drive a steam turbine. Because so much energy is consumed as liquid fuels, a process for converting the woody biomass into liquid fuels is very desirable. One way to do this is via the syngas pathway. To do this, wood lignins are first converted into syngas (hydrogen and carbon monoxide) in a gasifier by reacting it with steam and oxygen. The syngas is then used to make diesel or gasoline typically using the Fischer-Tropsch process or the methanol pathway. The same methods are used to make liquid fuels from natural gas or coal. These processes are covered in Chapter 6 for natural gas and in Chapter 13 for coal.

One of the more innovative biomass-to-liquid fuel conversion processes belongs to a newly formed company called KiOR, which was founded in 2007 by Khosla Ventures (KV) and a group of catalyst scientists who had an idea for making renewable fuels from cellulosic biomass through a one-step catalytic process. The KiOR process for catalytic conversion of biomass to a hydrocarbon fuel has significant implications because it is a simple and inexpensive process. It uses a relatively abundant low-cost feedstock, trees and, in KiOR's particular case, the Southern Yellow Pine.

KV made a $10 million investment in KiOR to fund the Company's R&D program and pilot plant. The pilot plant was an opportunity for KiOR's proof of concept: that cellulosic feedstock could be converted into renewable fuel through a scalable, relatively inexpensive catalytic process. Encouraged by technical success at the pilot scale, KiOR accelerated the time scale after just a few months of pilot testing and began work on a demonstration unit, with a capacity to process 10 dry tons of wood and produce up to 15 barrels of product per day. This is a small production volume, but a commercial unit would be 1000 barrels per day. KiOR then arranged meetings with three southeastern state governors to discuss the facility's location. Within one week, KiOR had agreed to build a small commercial scale plant in Mississippi due to Governor Haley Barbour's enthusiasm for the project and the benefits it could offer his state. The MS legislature approved a $75 million interest-free loan for the plant in September 2010. By the first quarter of 2011, construction was under way for an initial-scale commercial production facility in Columbus, Mississippi, which is shown in Figure 12.3.

The company's technology platform combines its new catalyst systems with a process based on existing FCC technology, a standard process used in oil refining. The efficiency of KiOR's process, called biomass fluid catalytic cracking, and the proven nature of catalytic cracking technologies allow for significant cost advantages including lower capital and operating costs versus traditional biofuel producers. KiOR processes its renewable oil product in a conventional hydrotreater, which is a standard process unit used in oil refineries, into gasoline and diesel blend-stocks that can be combined with existing fossil-based fuels and used in vehicles on the road today.

KiOR began construction of its first commercial scale facility, located in Columbus, Mississippi, in early 2011. The facility will produce gasoline and diesel from Southern Yellow Pine. The gasoline and diesel product is intended as a blending-stock with regular fossil fuels. KiOR has already

FIGURE 12.3   KiOR's biomass fuel processing plant under construction in Columbus, MS
Source: KiOR.

established purchase agreements for the site's entire fuel output. The company chose the Columbus location due to the site's proximity to abundant, locally grown tress and its access to a shipping infrastructure. The facility is designed as an initial scale commercial facility, processing 500 dry tons of wood per day. Once fully operational, it is expected to produce over 11 million gallons of gasoline, diesel and fuel oil blend-stocks annually. This is equivalent to 720 barrels of oil per day, a very small commercial size production plant, but economically significant.

KiOR currently has sales agreements in place with Hunt Refining, Catchlight Energy (a joint venture between Chevron Corporation and Weyerhaeuser Company) and FedEx Corporate Services. Catchlight Energy also has an agreement with KiOR to supply the biomass feedstock for the facility. The facility will cost approximately $190 million to build, and production is scheduled to commence in the second half of 2012.

In the third quarter of 2012, KiOR planned to begin construction of a larger production facility in Newton, Mississippi. KiOR is designing this plant as its standard commercial production plant, which will process approximately 1500 tons of biomass per day, three times the size of its Columbus facility at 2150 barrels per day. By expanding its standard commercial facilities in discrete, independent projects that are each viable on their own, KiOR believes it will have the flexibility to coordinate its growth in response to capital availability and market conditions.

ENVIRONMENTAL ISSUES

The biomass energy source is not without environmental impact. The burning of wood may be renewable, but the environmental impacts can be so severe that the burning of wood in open fireplaces is highly regulated in most communities. These effects, however, can be mitigated. In general, the more efficient the stove, the less pollution it produces. In most places, there are regulations covering how efficient a wood-burning stove must be. According to the U.S. Environmental Protection Agency, the use of wood for residential heating contributes up to 50% of the organic air pollutants being released into the atmosphere, some of which are carcinogenic. During winter months, in areas where wood is the principal heating fuel, wood stoves produce as much as 80% of these types of pollutants. Robert McCrillis of the EPA says, "... In the field, it is the installation and how the stove is operated that has the largest effect on how it performs. Public education on the correct use of a word-burning stove is part of the current regulations...." To meet federal clean air standards, some areas are regulating the use of wood stoves and banning open fireplaces in new constructions. In order to curb pollution, some communities only allow the installation of EPA-approved Phase II stoves, which combust the wood more completely and are more energy efficient.

In the United States, the manufacturing of ethanol from corn generates significant amounts of pollutants in the planting and harvesting of the corn and the distillation of the fermented alcohol. Moreover, the low energy balance ratio (1.34) for corn-based ethanol means that it generates more pollution than just carbon dioxide. If a policy was adopted in the United States which stated that the corn used for ethanol production had to be cultivated and harvested with equipment powered by pure ethanol and that distillation and processing had to be fueled by ethanol or waste corn stalks and husks, the exact net production of ethanol would be demonstrated. This may result in a lot less ethanol being produced in the country.

THE FUTURE OF BIOMASS

There is really only a small and limited future for biomass in the energy supply. It does not make sense to be converting part of the nation's food supply into fuel. The production of ethanol from corn in the United States should be reduced and the corn redirected back into the food supply. Some additional trees could be grown and burned for fuel or processed into liquid fuels via the gasification and Fisher-Tropsch routes or using the KiOR catalysis process. The arable land available to cultivate trees for this purpose is, however, limited. Most of the available land is already being devoted to growing crops and wood for paper. While it is possible to grow some trees, this will not have a substantial impact on future energy supply. Similarly, growing food crops such as rapeseed, soy and corn to be processed directly from SVO into biodiesel also does not make sense. Again, these crops are needed for the food supply. It does make sense to process waste material, such as WVO and tallow into biodiesel; however, the supply of these is so limited that it will always be a small niche market. Generally, this means that energy from biomass cannot and should not increase much.

13

Coal and Clean Coal Technologies

Coal was once the largest source of energy in the world. While it no longer holds the number one position, it still maintains second place behind oil and just ahead of gas. Moreover, in terms of the energy reserves tied up in coal, it is far and away the number one source of energy reserves in the world. In fact, the U.S. state of Wyoming alone has more energy in their proved coal reserves than Saudi Arabia has in oil. Consequently, coal will likely play a major role in future energy supply.

Coal is a reliable source of energy that can be cranked up or turned down as the demand changes. Coal produces electricity at a lower cost compared to all of the other fuel sources, except natural gas. For example, 91% of Wyoming's electricity is produced from coal, resulting in the cheapest retail electricity prices in the nation at 6.1 cents per kilowatt-hour (¢/kWh) (2009 prices). In comparison, New York gets just 10% of their electricity from coal and their average price is 15.5 ¢/kWh. On a national level, the average price of electricity is 9.8 ¢/kWh, with the residential sector paying the highest price at 11.5 ¢/kWh. Coal is a cheap and abundant source of energy; however, if it is burned without controls, it can have the most severe impact on the environment.

The environmental impact of coal is perhaps its greatest weakness or deficit. Coal produces twice as much carbon dioxide as natural gas per unit of energy generated. It can also produce a lot of other pollutants ranging from fly ash, mercury, sulfur dioxide to nitrous oxide, as well as several others. The solution to this problem is the use of clean coal technologies. These technologies capture the undesirable waste products and sequester them underground or convert them to useful products so that they do not pollute the atmosphere and lead to global warming. To ensure a positive outlook for coal, efforts need to be made to reduce its environmental impact to acceptable levels.

WORLD COAL PRODUCTION

The proved reserves, annual production, exports and reserve life of the countries with the 15 largest coal reserves in the world are shown in Table 13.1. The United States has the largest reserves at 237 billion tons (237 Gt). The total proved coal reserves in the world are 861 Gt. To put this into perspective, each ton of coal has the same energy content as three to four barrels of oil (depending on the rank and energy content of the coal and the specific gravity and energy content of the oil). 237 billion tons of coal is the energy equivalent of about 800 billion barrels of oil. Contrast that with the Saudi Arabian (who has the world's largest oil reserves) proved oil reserves of 263 billion barrels and it becomes clear that the United States and the world have a lot of coal.

The coal reserves shown in Table 13.1 represent only the proved reserves. There is a lot more coal in the United States and the world that remains unproven because the owners have no need to invest resources in proving up more coal. They already have plenty of coal resources to satisfy

TABLE 13.1   Largest Coal Reserves and Producers (2010)

Rank	Country	Reserves (Gt)	Production (Mt/year)	Reserve Life (years)	Exports (Mt/year)
1	USA	237.30	984.6	241.0	54.8
2	Russia	157.01	316.9	495.5	118.7
3	China	114.50	3240.0	35.3	34.8
4	Australia	76.40	423.9	180.2	261.7
5	India	60.60	569.9	106.3	2.3
6	Germany	40.70	182.3	223.3	0.5
7	Ukraine	33.87	73.3	462.1	6.3
8	Kazakhstan	33.60	110.8	303.2	23.3
9	South Africa	30.16	253.8	118.8	66.9
10	Colombia	6.75	74.4	90.7	68.7
11	Canada	6.58	67.9	97.0	29.0
12	Poland	5.71	133.2	42.9	13.2
13	Indonesia	5.53	305.9	18.1	237.1
14	Brazil	4.56	5.5	832.7	0.0
15	Greece	3.02	68.5	44.1	0.0

Data Source: BP

their current needs. For example, on the western half of the north slope of Alaska, there are an estimated 4 trillion tons of coal, equivalent to at least 12 trillion barrels of oil. None of that coal is ever counted in the U.S. coal reserve picture because there is no infrastructure in place to mine and transport that coal to market. The full extent of the coal has not been properly delineated with exploration probes.

The largest producer of coal in the world is China, which produces more than three times the amount of coal as the second largest producer, the United States. The coal production in China is mostly used domestically to power its rapidly expanding economy and its massive domestic demand, which in some respects is also a product of the world's largest population. The largest exporters of coal are Australia and Indonesia, which are only the fourth and sixth largest producers in the world. The majority of coal produced in both Australia and Indonesia is exported to Asia, particularly Japan, South Korea, and China. China also exports some of its own coal production as well (but it only exports 1% of its domestic production while importing 7% of its domestic demand). The total world production of coal in 2010 was 7273 Mt/year, but only 13% (943 Mt) was exported to other countries. This means that 87% of the world's coal production is used indigenously. This fact is likely to change in the future as a stronger world import/export trade develops.

The reader should also look at the reserve life column in Table 13.1. Most coal producers have large reserve lives, indicating there is significant potential for increased coal production around the world. The only exception to this is China and Indonesia, who have ramped up their production for different reasons. It is likely that each of them can increase their reserves with a program of exploration and delineation.

Figure 13.1 shows the history of coal production in the world since 1800. In Chapter 1, the reader learned that coal was the energy source that fueled the industrial revolution in the nineteenth century and that by 1910 it had overtaken biomass (particularly wood) as the prime energy source in the world. The data source used to plot Figure 13.1 was compiled by Professor David Rutledge of the California Institute of Technology. The total cumulative production of coal from 1800 until the present time has been 316 billion tons of coal (or 316 Gt) with current production more than 7 Gt/year. The growth rate has been steady since 1800 with two periods of reduced growth from 1920 to 1945 and more recently from 1990 to 2000. Since the turn of the twenty-first century, coal production has resumed its strong growth, primarily due to rapid expansion in China and Indonesia.

Current proved reserves in the world are 861 Gt, and combined with the 316 Gt already produced, this represents an ultimate coal recovery of 1177 Gt. It should be noted that this represents the minimum ultimate

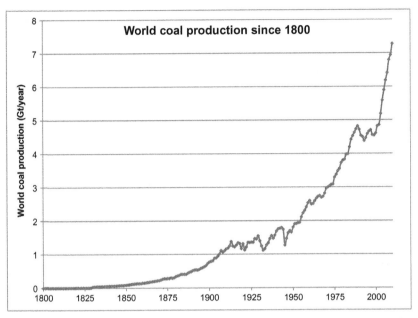

FIGURE 13.1   World coal production since 1800. *Data Source: David Rutledge, Caltech.*

recovery because there is probably a lot more coal that can be proved up if the need arises.

With so much energy tied up in coal reserves, you would expect that coal has many years of production of the current reserves before it peaks and begins to decline. Professor Rutledge (whose data set is plotted in Figure 13.1) published an analysis of his data in 2010 using probit and logit transforms to determine that the total expected cumulative coal production will only be 680 Gt, leaving only 364 Gt of reserves still to be produced. He did not calculate the date for the peak production, but he did estimate that 90% of the world's coal reserves would be depleted by 2070, leaving it with a less important future. Essentially, Professor Rutledge is saying that he does not believe that the reserves published by the world's current coal producers will ever be produced. He believes that the current estimate of future production of at least 861 Gt (remember this is only proved reserves) is overestimated.

To back up his claim, Rutledge points to an estimate of future British coal production made by geologist Edward Hull in 1861 who calculated "an available quantity of 79,843 million long tons of coal (81 Gt), which if divided by 72 million long tons (the quantity produced by the United Kingdom in the year 1859), would last for no less than 1100 years." For his assessment, Hull included seams with a thickness of 2 ft. or more, down to a depth of 4000 ft. Hull's estimates were later increased by the

UK Royal Commission on Coal Supplies, which in 1871 counted seams down to a thickness of one foot and arrived at 149 Gt of minable coal. Production in the United Kingdom actually peaked in 1913 at 292 Mt/year. British coal mines have mostly been depleted and so far have produced 27 Gt of coal, much less than Hull's 81 Gt or the Royal Commission's 149 Gt. In retrospect, both Hull's numbers and the Royal Commission's numbers were not good estimates of future UK coal production because it is not economic to mine one-foot or two-foot coal seams down to a depth of 4000 ft.

A separate analysis of future world coal production was made by Patzek and Croft in 2010. These researchers concluded that,

> The global peak of coal production from existing coalfields is predicted to occur close to the year 2011. The peak coal production rate is 160 EJ/year, and the peak carbon emissions from coal burning are 4.0 Gt C (15 Gt CO_2) per year. After 2011, the production rates of coal and CO_2 decline, reaching 1990 levels by the year 2037, and reaching 50% of the peak value in the year 2047. It is unlikely that future mines will reverse the trend predicted in this business-as-usual scenario. (Patzek and Croft, 2010)

These two forecasts are quite surprising given the large reserves of energy tied up in coal. Consequently, it is necessary in this chapter as well as Appendix B to examine the world coal production data of Rutledge using the Hubbert theory (shown in Appendix B) to see what it forecasts.

In Figure 13.2, the coal production rate (shown in Figure 13.1) is divided by the cumulative coal production and the result is plotted versus the cumulative coal production (for the mathematically minded, the plot is q_t/Q_t V Q_t). This plot shows that since 1931 the growth rate in coal production has been fairly constant at 2.2%. This plot is also a key plot used in Hubbert's model for predicting future production. The fact that the growth rate has been constant for many years suggests that the coal production rate is not about to peak and is likely to continue with this growth rate for many more years. This is discussed in more detail in Appendix B.

In Appendix B, the nonlinear regression method is also used to fit Rutledge's data to Hubbert's model and obtain the forecast shown in Figure 13.3. The red line in this chart is a plot of the Hubbert model and the blue points are the Rutledge data points. The parameters obtained from nonlinear regression and used to plot Hubbert's model are:

$$Q_{max} = 3529 \, Gt, \quad a = 1042.15 \quad \text{and} \quad b = 0.02203 \, \text{per year}.$$

The values for the parameters a and b are not highly important, but are merely shown here for completeness. The value for Q_{max}, is the ultimate cumulative coal production. Its value, 3529 Gt, is well in excess of the 1177 Gt estimated above for the current proved reserves plus the past cumulative production. It is also much more than Rutledge's estimate of 680 Gt.

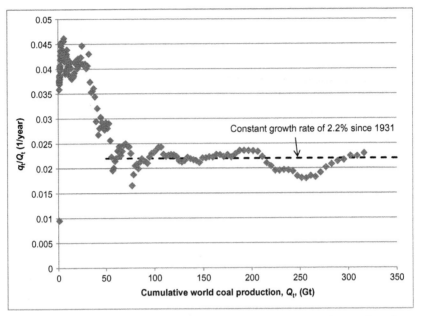

FIGURE 13.2   Plot of Equation B.3 of Appendix B for world coal production. *Data Source: David Rutledge, Caltech.*

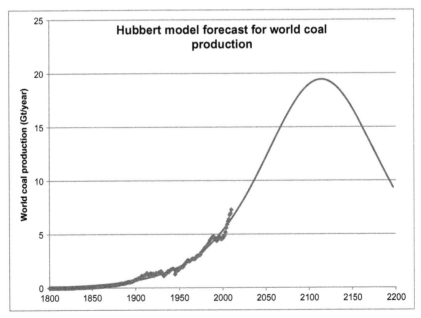

FIGURE 13.3   Hubbert model fit of world coal production. *Data Source: David Rutledge, Caltech.*

It is, however, not an unreasonable number given that there has been no real effort to establish just how much coal exists in the world and what can be economically produced. It was previously mentioned that 4000 Gt of coal is known to exist on the north slope of Alaska. Russia also has the Tungus and the Lena coal deposits, which are similarly large unproven and undeveloped coal deposits in central Siberia. There could also be other similarly large coal deposits located elsewhere in the world.

The timing for the peak coal production is calculated in Appendix B to occur in the year 2115. This estimate has an accuracy limited by the model fit of the data, which is about 15%.The peak could realistically occur anywhere from 2070 to 2160.

In Chapter 5, it was explained that when making forecasts for world production of any resource, such as coal, too much faith should not be placed in the backside of the curve. Once you hit the peak, or as you approach the peak, the resulting shortages will drive up the price of coal and change the dynamics of the forecast.

If a more conservative approach is taken, you could easily use the current proved reserves number (861 Gt) and the recent sharp increase in world coal production to make a forecast based on those numbers. This results in a peak world production of 17 Gt/year with the peak occurring in 2044. This is probably not as realistic a forecast as the one in Figure 13.3.

U.S. COAL PRODUCTION

The United States is the second largest coal producer in the world and also has the largest coal reserves at 237.5 billion tons (Gt); however, it was not always that way. From 1880 until 1918, the United States experienced rapid expansion of the coal mining industry. From 1918 until 1960, the coal industry was in the doldrums, with periods of decline and recovery followed by more decline. Most of the production came from underground mines in the Appalachian regions of Kentucky and West Virginia. Life in these underground mines was tough and many miners contracted health problems such as black lung and other respiratory diseases, as well as lung cancer.

Modern underground mines have been made much safer by better ventilation and very stringent Occupational Safety and Health Administration (OSHA) and Mine Safety and Health Administration (MSHA) regulations. Around 1970, the Clean Air Act introduced much tougher air emission standards which led to the development of surface mines in the western United States, particularly in the Powder River Basin of Wyoming. The Powder River Basin coal is a lower rank (sub-bituminous) coal that has higher moisture content, but also much lower sulfur content. The lower rank means that it contains less energy than anthracite or bituminous coals, but it also contains much less sulfur. Consequently, it is

TABLE 13.2   Top Coal Producing States in the United States

1	Wyoming	391,027
2	West Virginia	124,378
3	Kentucky	97,359
4	Pennsylvania	52,589
5	Montana	35,815
6	Indiana	32,340
7	Texas	31,830
8	Illinois	30,610
9	North Dakota	27,161
10	Colorado	25,639

Based on 2009 Production in Mega-Tons (Mt).

much easier to comply with the Clean Air Act with Powder River Basin coal than with the eastern U.S. coals. Today, most of the U.S. production comes from the Powder River Basin, where massive open cut mines easily extract huge amounts of coal. The top 10 coal-producing states in the United States are shown in Table 13.2. The largest coal-producing state is Wyoming, which produces more coal than the next 6 states combined.

The historical U.S. production is shown in Figure 13.4. Since 1997, it has been relatively constant at approximately 1 billion tons per year (1 Gt/year).

To forecast future coal production for the United States, the familiar Hubbert model is used and shown in Figure 13.5. The only way to fit the data was to use the same nonlinear regression technique that was applied to the world coal production data; however, Q_{max} is fixed at 308 Gt. This number was obtained by adding the current proved reserves of 237.5 Gt plus the total cumulative production to date of 70.5 Gt. The regression technique determined the other two parameters of the Hubbert model to be $a = 651.4$ and $b = 0.02387$ per year. Figure 13.5 is a forecast of the future production of the current proved reserves (237.5 Gt), but this production could be increased if the proved reserves are increased. The forecast peak rate is 1.84 Gt/year, which occurs in 2070.

AUSTRALIAN COAL PRODUCTION

Australia is the world's largest exporter of coal, and it is useful to look at their production growth. Figure 13.6 shows that growth was relatively slow until about 1950 when they were producing 27 Mt/year. Production

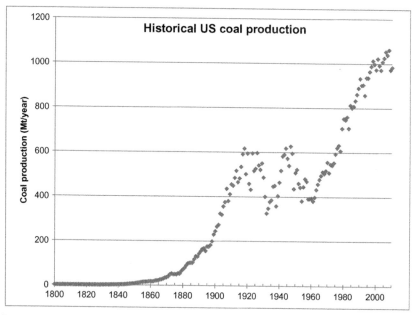

FIGURE 13.4 Coal Production History in the United States. *Data Source: David Rutledge, Caltech.*

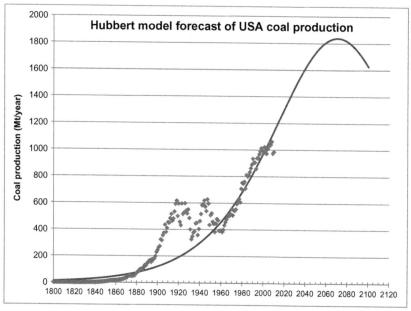

FIGURE 13.5 Hubbert model fit of U.S. coal production. *Data Source: David Rutledge, Caltech.*

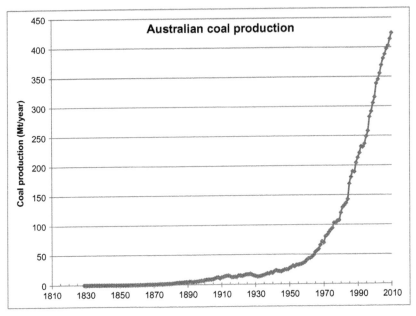

FIGURE 13.6   Coal production history in Australia. *Data Source: David Rutledge, Caltech.*

has grown rapidly since then, and in 2010, they produced 424 Mt of coal. The majority of this coal production comes from the Bowen Basin in Queensland. This coal is mostly shipped to Asia, particularly Japan, Taiwan, and South Korea, whose economies are powered by Australian coal. 75% of Australian coal production is exported and the other 25% is used to generate electricity and to make steel. 85% of Australia's electricity is generated from their indigenous coal production.

The Hubbert model is employed to forecast Australia's future production in the same manner as was done for the world and U.S. coal production. There is one exception: there is first an attempt to forecast their ultimate production, Q_{max}, using the plot of Equation B.3 of Appendix B. This plot is shown in Figure 13.7, which shows that the recent production data can be extrapolated to an ultimate cumulative of 44.4 Gt of coal. This cannot be correct because their current proved reserves plus their historical production add up to 87.3 Gt of coal. The extrapolation in Figure 13.7 is probably unreliable because the Hubbert model trend has not been properly established yet. Accepting the 87.3 Gt value for Q_{max}, and applying the same nonlinear regression technique that was applied to the world coal production data, a forecast for Australian coal production was obtained and is shown in Figure 13.8. The regression technique determined the other two parameters to be $a = 57991$ and $b = 0.05268$ per year. Figure 13.8 is essentially a

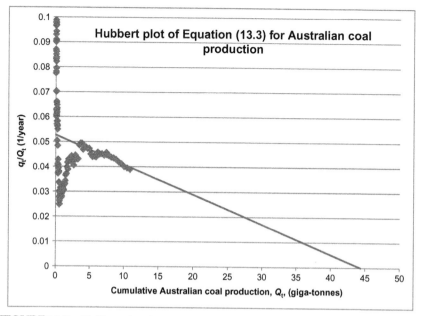

FIGURE 13.7 Hubbert plot of Equation B.3 of Appendix B for Australian coal production shown in Figure 13.6. *Data Source: David Rutledge, Caltech.*

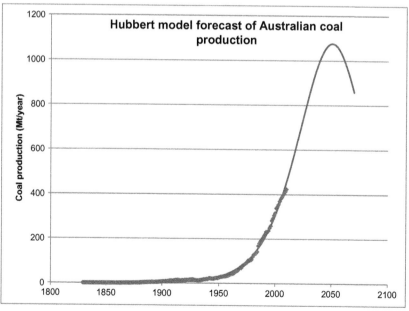

FIGURE 13.8 Hubbert model fit of Australian coal production. *Data Source: David Rutledge, Caltech.*

forecast of the future production of their current proved reserves (76.4 Gt). The forecast peak rate is 1.07 Gt/year, which occurs in 2050.

CHINESE COAL PRODUCTION

China overtook the United States as the world's largest coal producer in 1985, as China was beginning a vast economic expansion. Coal was its fuel of choice to power this economic miracle. It now produces more than three times as much coal as the United States and the expansion continues unabated. It has, however, less than half the proved reserves of the United States. If its coal mining is to continue its rapid expansion, new reserves will have to be found quickly. Its coal production history is shown in Figure 13.9, and the Hubbert model forecast of its future production is shown in Figure 13.10. The parameters used in the forecast were:

$$Q_{max} = 168.68\,Gt, \quad a = 33{,}587{,}978 \text{ and } b = 0.13806 \text{ per year}$$

168.68 Gt represents its ultimate cumulative production, made up of its cumulative production to date (54.18 Gt) plus its published proved reserves (114.5 Gt). The other two parameters (a and b) were determined by fitting its recent rapid growth to the Hubbert model. The parameter b shows that its production growth has been 13.8% per year, which is also a

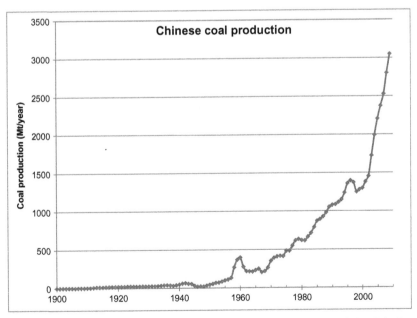

FIGURE 13.9   Coal production history in China. *Data Source: David Rutledge, Caltech.*

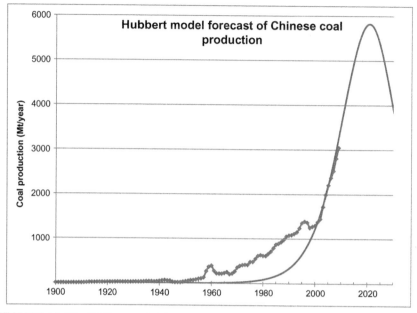

FIGURE 13.10 Hubbert model forecast of Chinese coal production. *Data Source: David Rutledge, Caltech.*

growth rate of 6% of its cumulative production. If this rapid production growth continues, it will peak at 5.8 Gt/year in 2021. From there, it will begin a rapid decline and will have to switch to other fuel sources or ramp up its coal imports to similarly high levels. If there are people reading this book who would like to become very rich, mining coal, then exporting it to China could be one way to do it.

China is already the world's largest coal importer; however, after 2021, its local production will have peaked and, therefore, it will be importing even larger quantities of coal. If you expect that rising coal prices will generate new reserves for China, remember that Australian coal production will not peak until 2050, American coal production will not peak until 2070 and world coal production will not peak until 2115. It will be much cheaper for China to begin rapidly increasing its coal imports soon to make up for the looming shortfall in its production.

Although China does export 1% of its coal production to its neighbors, it is already importing about 7% of its coal consumption (200 Mt/year) primarily from Australia and Indonesia. This level of dependency on coal imports is rising rapidly. These two largest coal exporters, Australia and Indonesia, had supply disruptions in 2010 that led to blackouts throughout China and record imports from nontraditional suppliers. Queensland, the main coal-producing state in Australia, experienced massive floods that largely halted Australian coal exports for months, and Indonesia had heavy rains, limiting its production as well. Following these weather disruptions, Northeast China experienced a bitterly cold winter, demanding historically high levels of coal to heat millions of homes and power hundreds of thousands of factories. As a result, there was high demand for coal on the international market. Since 2010, the highest prices paid internationally for coal have been in Asia, particularly China. Even though China is the world's largest coal producer and it is only importing 7% of its domestic coal demand, in 2010 it overtook Japan as the world's largest coal importer. China will be the primary driver of an expanding coal import/export market.

The United States is currently exporting 1–2 Mt/year to China, and there is a proposal to build two further ports in Washington aimed at exporting an additional 100 Mt/year of Powder River Basin coal to China. This will likely become a common occurrence as the United States and other countries compete for the expanding Chinese market.

HEALTH, SAFETY, AND ENVIRONMENTAL ISSUES IN COAL USE

Coal mining in underground mines has long been recognized as being arduous work, liable to injury and disease. Consequently, stringent regulations have been implemented that require dust suppression, ventilation, and respiratory protective equipment for modern underground mines. Nevertheless, even today there are still increased health and safety hazards. Coal dust has been a serious problem in mining, causing coal workers' pneumoconiosis, or "black lung disease," and chronic obstructive pulmonary disease. While this has been largely eradicated, diesel fuel particulate exposures still occur in underground mines because of diesel-powered mobile equipment that is used primarily for drilling and hauling coal. Diesel particulates pose a risk of lung cancer. Coal miners are also exposed to safety hazards that include explosions and cave-ins, as well as the usual dangers associated with the heavy machinery that is used today.

Environmentally, the burning of coal has been responsible for many environmental hazards. If burned in an uncontrolled environment, coal puts out many air pollutants ranging from sulfur dioxide to nitrous oxide, mercury, arsenic, and coal ash. These emissions are now strongly regulated to prevent them from being emitted to the atmosphere. Coal also emits twice as much carbon dioxide per unit of energy produced than does natural gas and 25% more than oil. This emission is now recognized as a potential cause of global warming, and consequently regulations on carbon dioxide emissions are likely to be implemented in the future. As of 2011, this has not yet been done but USA has recently restricted the carbon dioxide emissions from new coal fired power plants, which caused the planned construction of several new coal fired power plants to be cancelled.

CLEAN COAL TECHNOLOGIES

When coal is burned, the carbon in the coal becomes carbon dioxide; the sulfur in the coal becomes sulfur dioxide; and the nitrogen in the coal, and a little of the nitrogen in the air used to burn the coal, become various oxides of nitrogen. Normally, these chemicals are all released into the atmosphere. The solution to this pollution is to capture these undesirable emissions and sequester them. The collection of technologies that do this is called clean coal technologies. Up until now, the approach to solving these problems has been to capture the emissions after they have been generated. Electrostatic precipitators capture the coal ash and soot particles, while scrubbers are used to capture the sulfur and nitrogen compounds. A better approach is to gasify the coal first to generate syngas (a mixture of hydrogen and carbon monoxide) and then separate out the gaseous components. They can then be used to not only generate electricity, but also to make liquid fuels, petrochemicals and fertilizers, among other things.

All clean coal technologies begin with the coal gasification process. In a coal gasifier, the coal is reacted with steam and oxygen to produce syngas products (hydrogen and carbon monoxide) as well as small amounts of methane and carbon dioxide. There are six principal reactions that take place at relatively high temperatures (2000-3000 °F):

$$\text{Full Oxidation} \quad C + O_2 \rightarrow CO_2 \quad -394 \, MJ/kmol$$

$$\text{Partial Oxidation} \quad C + \tfrac{1}{2}O_2 \rightarrow CO \quad -111 \, MJ/kmol$$

$$\text{Methanation} \quad C + 2H_2 \rightarrow 2CH_4 \quad -75 \, MJ/kmol$$

$$\text{Gasification} \quad C + H_2O \rightarrow CO + H_2 \quad +131 \, MJ/kmol$$

$$\text{Water} - \text{Gas Shift} \quad CO + H_2O \rightarrow CO_2 + H_2 \quad +151 \, MJ/kmol$$

$$\text{Boudouard Reaction} \quad C + CO_2 \rightarrow 2CO \quad +172 \, MJ/kmol$$

The main gasification reaction between coal and steam is an endothermic reaction. This means that for every kilo-mole of reactant or product, 131 MJ/kmol of energy must be input to the reacting system. This energy comes from the oxidation reactions, which is why it is necessary to input oxygen into the reacting chamber (the gasifier). The oxidation reactions are exothermic; when the coal reacts with the oxygen, energy is released.

There are many commercial gasifiers available, and Figure 13.11 is one example that is produced and marketed by ConocoPhillips. In this gasifier, coal is fed into two left-hand ports as coal slurry, a mixture of finely

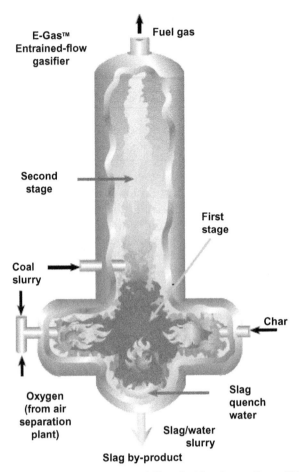

FIGURE 13.11   ConocoPhillips Entrained Flow Gasifier. *Source: ConocoPhillips.*

ground coal and water. Oxygen is mixed in with the coal slurry before being pumped into the gasifier. This gasifier operates at 2500 °F. The reaction chamber is kept at that high temperature by the exothermic oxidation reactions. If the temperature falls below 2500 °F, the oxygen feed rate is increased, and vice versa. The water content in the feed has the opposite effect. The higher the water content, the lower the chamber temperature, and vice versa. The ash in the coal does not react, but the temperature is high enough to melt the ash and it flows to the bottom of the reacting chamber. It is then resolidified with quenching water and it falls out of the bottom of the chamber as slag. Any unreacted coal is separated from the product gas and returned to the right-hand port of the gasifier in the form of char.

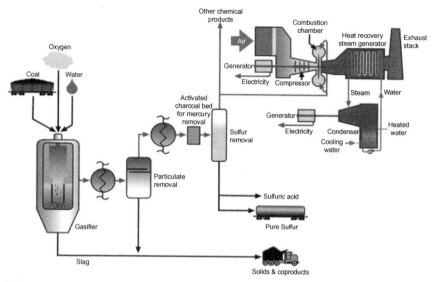

FIGURE 13.12   IGCC electrical power generation. *Source: Idaho National Labs.*

The most efficient way to make electricity from syngas is by using an integrated gasification combined cycle (IGCC) plant, as shown in Figure 13.12. After the syngas exits the gasifier, it undergoes some cleaning processes, after which it is sent to a gas turbine where it is burned to generate electricity. The hot exhaust gas from the gas turbine is also used to heat water into steam, which is also used to generate more electricity in a steam turbine. It is called a combined cycle process because electricity is generated in both the gas turbine cycle and the steam turbine cycle. It is also possible to capture all the carbon dioxide from the exhaust gas using an amine capture plant (see Figure 4.12).

An alternative for capturing the carbon dioxide is to burn the syngas in a pure oxygen environment in the gas turbine, which will then produce an exhaust gas that is mainly carbon dioxide and water vapor. The water vapor can be condensed out of the carbon dioxide by cooling the exhaust gas to produce a carbon dioxide stream that is ready for use or sequestration. Using this type of technology, it is possible to construct a coal-fired power plant that has zero emissions. All of the potential pollutants are captured and either used in other ways or sequestered underground. Figure 13.13 shows a schematic of a zero-emissions IGCC plant that was devised at the University of Wyoming by me and my colleagues (John Ackerman, David Bell, and Morris Argyle).

The zero-emissions IGCC plant shown in Figure 13.13 is a particularly simple design that is much cheaper to build. In this process, the mercury in the coal is removed at the beginning of the process using a simple heating

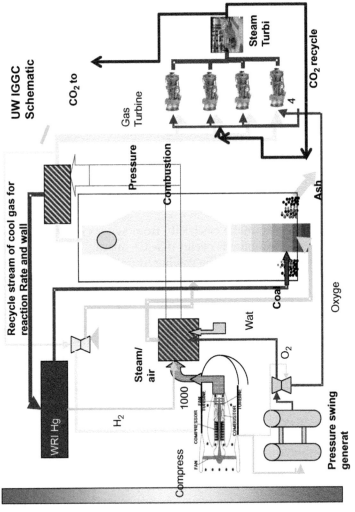

FIGURE 13.13   The University of Wyoming IGCC electrical power schematic. *Source: Ackerman.*

method that was developed at the Western Research Institute. This design also avoids expensive syngas treatments because no separation of syngas compounds is required between the gasifier and the turbine. No cooling or treatment, other than particulates removal, is required. The exhaust is primarily CO_2 and water. The water is removed by cooling and condensation. A compressor is used to produce compressed air for the process flow streams: the air cycle machines that are used to boost O_2/H_2O pressure to the gasifier and the pressure swing O_2 generators. The gas turbines not only drive the electric generation units, but the hot exhaust is used to produce heat for steam generation that is used in the steam turbine. The CO_2 from the exhaust is recycled as carrier gas in the gas turbines, which improves the thermodynamic efficiency of the process.

The U.S. Government had proposed to investigate an upgraded IGCC design called FutureGen, whose process flowsheet is shown in Figure 13.14. In the FutureGen project, the carbon monoxide in the syngas would be converted into carbon dioxide using the water gas shift reaction. This reaction produces more hydrogen; the result is that extra hydrogen is generated, leaving the product stream as hydrogen and carbon dioxide. The carbon dioxide is then separated from the hydrogen to produce both a pure hydrogen stream and a pure carbon dioxide stream. The hydrogen is burned in the gas turbine to produce electricity (Figure 13.15). The advantage of this system is that the pure hydrogen stream could be used for other hydrogen-based applications, as shown in Figure 13.16. Chapter 15 will discuss hydrogen applications in more detail. The carbon dioxide is sent for sequestration or to be used for enhanced oil and gas recovery applications. This is also depicted in Figure 13.16.

After the FutureGen project depicted in Figure 13.14 was abandoned, a FutureGen 2.0 was redesigned in 2011 as a purely carbon capture and storage project. In this project, a $1.3 billion retrofit of an existing power plant is proposed, adding oxy-combustion to the plant to produce a mostly pure CO_2 exhaust stream which is sequestered underground. The modified process schematic diagram for FutureGen 2.0 is shown in Figure 13.15. This project had not been approved at the time of writing.

COAL TO LIQUIDS TECHNOLOGIES

Coal can be made into liquid fuels, such as gasoline diesel and jet fuel, quite readily. This has been done many times in the past and is likely to occur again in the future. Technology has come a long way since the first coal-derived liquid fuel (a synthetic crude oil) was produced through direct liquefaction in the early 1900s.

In 1925, Franz Fischer and Hans Tropsch used an indirect liquefaction process, which still bears their name, to produce suitable transportation

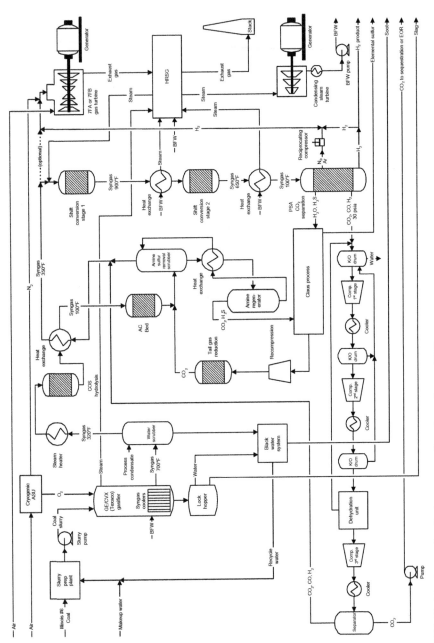

FIGURE 13.14   FutureGen Project process flowsheet. *Source: DOE.*

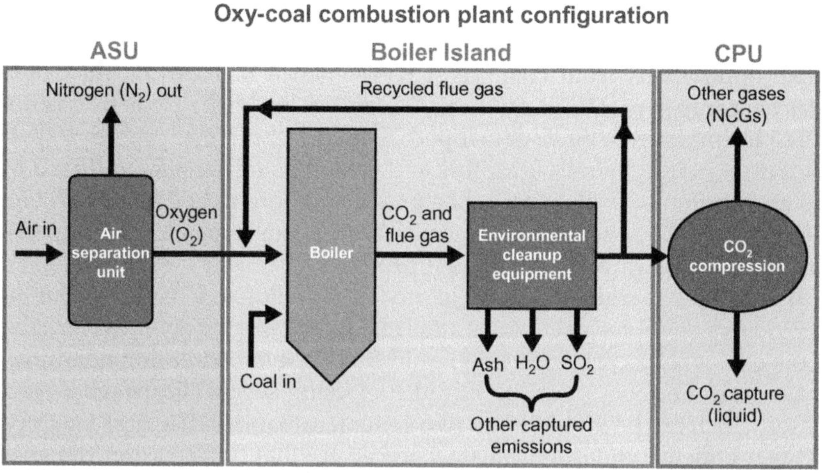

Oxy-coal combustion plant configuration

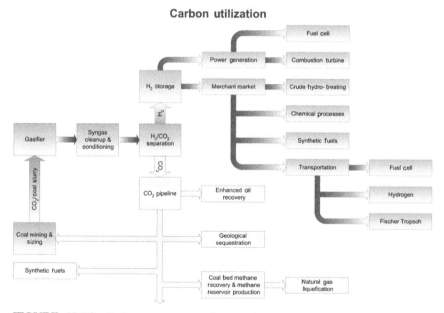

FIGURE 13.15   FutureGen 2.0 Process. *Source: FutureGen Alliance.*

Carbon utilization

FIGURE 13.16   Carbon management from coal gasification processes. *Source: Idaho National Labs.*

fuels such as diesel and gasoline. Germany had 25 liquefaction plants that, at their peak in 1944, produced 125,000 barrels of fuel per day and provided 90% of the nation's fuel needs. During World War II when Germany was cut off from its oil supplies, it ramped up the production of liquid fuels from coal to provide fuel for the war effort. When South Africa operated under its apartheid policy from 1948 to 1994, it became isolated from the world economy and no one would export oil to it. It also began to make liquid fuels from coal. As a consequence of that era, the South African company, SASOL, is now the world leader in Coal to Liquids (CTL) technology. SASOL developed its own commercial process to produce transportation fuels (gasoline and diesel) using syngas produced by the gasification of coal. SASOL has produced more than 750 million barrels of synthetic fuels from coal since 1980. It continues to do so today, although it now imports oil and oil products.

It is cheaper to convert natural gas to liquid fuels (see Chapter 6), but the world has a lot more coal than natural gas. As the world begins to run short of oil, coal will be converted into liquid fuels in ever increasing quantities.

Coal can be converted into liquid fuel using several liquefaction processes which can be divided into two general categories. The first category, indirect liquefaction, is a multistep procedure that first requires the gasification of coal to produce syngas. This process is shown schematically in Figure 13.17. The syngas from the gasifier is converted to liquid fuel via two methods: the Fischer-Tropsch process or the Mobil process. In the Fischer-Tropsch process, which is much more common, the syngas is first cleansed of impurities and subjected to further chemical refinement to produce a sulfur-free diesel or gasoline. The initial syngas can be derived from coal alone or from a coal/biomass mixture. The process is the same when biomass is included, but the amount of CO_2 emitted during the process decreases as the proportion of biomass increases. In the less-common Mobil process, the syngas can be converted to methanol, which is

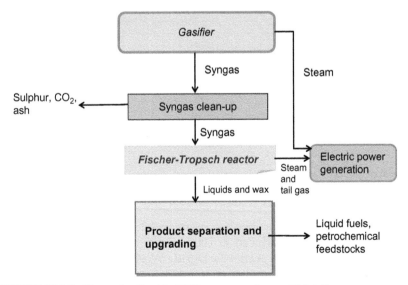

FIGURE 13.17   The coal to liquids (CTL) processes. *Source: FT Solutions.*

subsequently converted to gasoline via a dehydration sequence. Indirect liquefaction of coal using the Fischer-Tropsch process produces a significant amount of CO_2 that is removed from the fuel as a necessary step during the final stages of the process. This CO_2 must then be sequestered.

The second category, direct liquefaction, requires creating a chemical reaction of the coal at high temperatures with hydrogen gas using a catalyst to produce a liquid fuel. Direct liquefaction usually produces lower-quality liquid fuel that is more expensive to upgrade to make it comply with U.S. fuel standards. Although the process was used in Germany and is being used in China, it is not a viable option for meeting the U.S. liquid fuel requirements.

In terms of economics, coal-based liquid fuel becomes viable when the price of oil exceeds $50 per barrel, which is the case today. To fund a CTL plant, this price needs some guarantee or the risk becomes too high. This is because of high front-end expenditures—a 10,000 barrel-a-day plant could cost $750 million or more to construct.

Not included in the above estimate is the cost of sequestering the captured CO_2, which would increase the price of the end product by a projected $5 per barrel. The imposition of a strict carbon cap and trade regime would also raise the cost of fuel produced with CTL technology because of the CO_2 emissions associated with it. While there is significant uncertainty, a recent study by Rand Corporation estimated that CTL production plus carbon storage could produce fuel at a cost of anywhere from $1.40 to $2.20 per gallon or more by 2025. With current oil prices exceeding $80 per barrel, CTL is currently an economically attractive alternative, and there is a likelihood of CTL plants being built in the near future.

Proponents of using coal-based liquid fuel for the transportation industry say that it would help the United States achieve energy independence. Currently, the United States consumes nearly 20 million barrels of liquid fuel for transportation each day and imports almost 40% of its feed stocks. The 237 billion tons of proved recoverable coal that the country has in its reserves could be used to replace some of this imported oil. If the United States doubled its current coal production to 5 million tons per day and diverted the extra 2.5 million tons per day to make transportation fuels, this would displace about 4.5 million barrels of imported oil per day. Other benefits of CTL fuel cited by proponents are that it can be used to make gasoline, diesel, and jet fuel. Plants designed for indirect CTL processes can also be easily converted to hydrogen fuel cell production plants if fuel cell technology becomes more viable (see Chapter 15).

Opponents of CTL technology point out that the synthetic fuel derived from coal produces additional CO_2 as a byproduct. If the CO_2 removed from the fuel during the refinement process is not sequestered, the quantity of CO_2 released by extracting, refining and burning coal-based liquid fuel is 25% more than the amount emitted by conventional oil. Even with carbon sequestration, the emissions benefits over conventional fuel appear negligible at best, though there is significant debate over the true impact. While some studies have found CTL produces greater emissions than conventional fuel, even with sequestration, the recent Rand study found no difference between CTL and petroleum emissions. A Department of Energy study found that CTL emissions with sequestration were actually 5-12% lower.

UNDERGROUND COAL GASIFICATION

The coal gasification process can be conducted underground, and this process has an interesting and varied history. In the gasification process, the coal is reacted with steam and oxygen to produce syngas. If this is done with the coal remaining in-situ, it has many cost advantages. First, the coal does not have to be mined. Second, the gasification equipment does not have to be built. Finally, it can be applied to deep coal seams that are uneconomic to mine. Consequently, underground coal gasification (UCG) is cheaper than gasifying mined coal.

Although UCG was first proposed by a British engineer named William Siemens in 1868, much of the process development occurred in the Soviet Union in the 1930s. This was partly due to Russian chemist Dmitri Mendeleev who had heard Siemens' proposal in London in 1868 and further developed the idea over the next 30 years. Mendeleev was a very famous chemist who, among other things, invented the periodic table of the elements. Others were also working on developing Siemens' idea and in 1910 patents on UCG were granted to an American engineer named Anson Betts for "a method of using unmined coal." The first experimental work on UCG was planned to start in 1912, in Durham, England, under the leadership of 1904 Nobel Chemistry Prize winner William Ramsay. His project was to inject air and steam into an injection well and produce syngas out of a production well. He was unable to commence the project before the beginning of World War I and the project was abandoned after that.

Vladimir Lenin became familiar with Mendeleev's ideas, and during Russia's communist revolution in 1917, he promised the coal workers that he would develop the UCG process so that they would not have to venture underground to mine coal again. This promise was never totally fulfilled, but a considerable amount of research was conducted by the Soviet Union to make the process feasible and several projects were implemented there in the 1930s. Modern health and safety regulations in many nations, however, had greatly improved working conditions in underground mines, and the Soviet Union continued to operate these underground mines despite Lenin's earlier promises.

The UCG process is illustrated in Figure 13.18. An injection well is drilled into the coal seam and preferably horizontally along the coal seam.

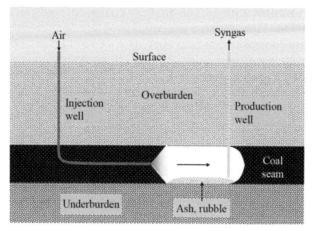

FIGURE 13.18 Underground coal gasification schematic. *Source: DTI.*

This intersects a vertical production well some distance away. Oxygen and steam are injected into the first well and initiate the coal gasification process. The resulting syngas is produced in the second well.

Besides the Soviet Union, this process has been tested in China at Ulanchap, in Australia at Chinchilla and Kingaroy, and in Wyoming at Hoe Creek. Tests at Hoe Creek, in northeastern Wyoming, showed that groundwater contamination is a potential problem. In an above-ground gasifier, the production of coal tars can be avoided by operating at sufficiently high temperatures. In UCG, regardless of the cavity temperature, a temperature gradient will form at the wall and coal will pyrolyze to form coal tar, a complex mixture of hydrocarbons that is potentially carcinogenic. Part of the problem at Hoe Creek was that the cavity pressure was too high, which forced some of the gas and tar into the surrounding formation. The test was conducted in a shallow seam, and when the cavity roof collapsed, water from a shallow freshwater aquifer mixed with the tar-contaminated coal and rock. Water contamination issues can be reduced by gasifying at a pressure less than the hydrostatic pressure. Water will tend to flow into the gasification cavity, flushing coal tars into the gasification zone and towards the production well. This strategy has been successfully demonstrated at Linc Energy's Chinchilla test in Australia. Groundwater contamination, however, appears to have killed Cougar Energy's UCG pilot project at Kingaroy, Australia. The state of Queensland's environmental authority shut it down in July 2010 after the project's gas-production well ruptured and Cougar reported trace levels of benzene and toluene in groundwater at the site. Cougar contested the shutdown and issued a statement in July 2011 asserting that "there have been no concerns with water quality in local water bores."

Another problem with UCG is that the coal seam roof can collapse as coal is removed. This is also a potential problem in underground mining, but some control techniques such as roof bolting, are not available when there are no miners underground. If uncontrolled, roof collapse could lead to a loss of gas seal and provide a path for the mixing of contaminated and fresh water. Roof collapse can be controlled, in part, by choosing coal seams with sufficient overburden strength and thickness. In the conventional room and pillar mining technique, pillars of unmined coal are left to help support the roof. In UCG, the equivalent technique is to limit the cavity cross-sectional area to limit the unsupported roof span and to space the cavities so that walls remaining between the cavities support the roof. If the UCG seam is much deeper than an overlying freshwater aquifer, there is less likelihood that a roof collapse will lead to significant intermixing of the fresh and contaminated water.

UCG is likely to become more widespread in the future because it has the potential to gasify coal more cheaply and to access unminable coal seams.

THE FUTURE OF COAL

There is considerable opposition to expanded coal use because it is perceived as a dirty fuel that has caused a lot of air pollution in the past. Nevertheless, coal production is increasing anyway and is likely to continue to increase. This is because the largest energy reserves in the world are tied up in coal resources. As the world needs and demands ever increasing energy supplies, coal reserves will be developed and burned in some form. The fact that coal can be converted into liquid and gaseous fuels and petrochemicals and fertilizers makes it a very useful energy resource, and it will be utilized. Moreover, the development of clean coal technologies is likely to dampen opposition to coal use.

China is currently the world's largest producer and consumer of coal, and its production is likely to peak in the next 10 years. As its production peaks, its coal imports will rise rapidly and drive a thriving international trade in coal. As well as being the world's largest producer and consumer of coal, China is also the world's largest importer of coal. Their current coal imports, however, are miniscule compared to their production and consumption.

China is currently importing 200 million tons of coal per year, but it is currently consuming 3.5 billion tons per year. By 2030, it could be importing as much as 5 billion tons per year at a cost of $500 billion/year. The immediate beneficiaries of this trade would be their closest current suppliers, Australia and Indonesia. The largest coal reserves though are in the United States, and it is likely that the United States will also be a significant supplier to the world coal trade, especially China. Given that the United States has a large trade deficit with China, supplying coal to the large potential demand in China makes sound economic sense. Russia also has large proven coal reserves and even larger unproven coal resources and, given its proximity to China, Russia will also be a likely participant in a booming coal trade.

The United States also has the potential to use its large coal resources to decrease its crippling oil import bill. Doubling its coal production to 5 million tons per day (1.825 billion tons per year) and diverting the extra 2.5 million tons per day to make transportation fuels would displace about 4.5 million barrels of imported oil per day.

Carbon Capture and Storage

The vast majority of energy in the world comes from hydrocarbons, which generate carbon dioxide (CO_2) when they are burned. As a consequence, the carbon dioxide levels in the atmosphere have risen dramatically (see Figure 4.5) and the prevailing opinion is that this is leading to global warming. The Kyoto accords are international agreements designed to reduce carbon dioxide levels. These accords call for generators to reduce carbon dioxide emissions to the atmosphere. The principal way to do this is to capture the carbon dioxide before it gets into the atmosphere, transport it to a storage site and sequester it underground. The technologies for doing this will be examined in this chapter.

There is no doubt that the world can generate all of its electricity without pumping any carbon dioxide into the atmosphere. This could even be achieved in a 10-year time frame if the political will was there. To seriously reduce the carbon dioxide emanating from automobiles, there needs to be a major switch to electric vehicles or low emission vehicles. Even if electricity is generated from hydrocarbon fuels, it needs to be done at a point source where the carbon dioxide can be captured and sequestered. While these are not cheap processes, they have to be undertaken to reduce the amount of carbon dioxide in the atmosphere.

Some people doubt that capturing and sequestering can be done in a safe and effective manner, claiming that the technology is unproven. The oil industry, however, has been injecting carbon dioxide into oil reservoirs on a large scale for 50 years. They know how to do it so that the carbon dioxide remains sequestered. As storage is ramped up, the depleted oil and gas fields are the first places to put the CO_2. The next places are in known traps and deep saline aquifers. While these locations represent the largest repositories for CO_2, they are also the least proven technology. Moreover, in order to create enough space in these deep formations and trapped formations, water may have to be removed and used

The Future of Energy
http://dx.doi.org/10.1016/B978-0-12-801027-3.00014-2

or disposed of. The other places where CO_2 can be sequestered are in deep unminable coal beds and deep in the ocean.

CARBON EMISSIONS

In 2010, the world emitted 33.2 billion tons of carbon dioxide into the atmosphere. Of this, 25% was emitted by China and 18.5% was emitted by the United States. These two countries are the largest CO_2 emitters by far, accounting for 43.5% of the world's CO_2 generation. China over-took the United States as the top CO_2 emitter in 2006. Table 14.1 shows the top 15 CO_2 emitting countries in the world.

The vast majority of these emissions come from burning oil, coal and natural gas (see Figure 4.6). Not all of these emissions end up in the atmo-sphere, however, and contribute to the 400 ppm of CO_2 content already in existence. The carbon cycle is a complicated process. Some of the carbon dioxide emitted into the atmosphere is: utilized by plants to grow cell walls in the photosynthesis process; dissolved in the ocean water; removed from the atmosphere by the world's food crops (although the human beings and other animals that eat those crops breathe out more

TABLE 14.1 CO_2 Emissions in 2010 (Mt/yr)

1	China	8332.5
2	United States	6144.9
3	India	1707.5
4	Russia	1700.2
5	Japan	1308.4
6	Germany	828.2
7	South Korea	715.8
8	Canada	605.1
9	Saudi Arabia	562.5
10	Iran	557.7
11	United Kingdom	547.9
12	Brazil	464.0
13	Mexico	447.0
14	Italy	439.4
15	South Africa	437.2

carbon dioxide than is removed by crops); and, removed by the new growth forests. Nevertheless, the rising CO_2 levels in the atmosphere are likely to be due to the rising CO_2 emissions over the past 150 years.

Though the world has been trying to find ways to reduce these emissions, Figure 14.1 shows that CO_2 emissions in the world continue to increase. The United States and Europe have begun to reduce their emissions, but China and other developing countries continue to increase their emissions, spurred by their rapid development. The reader may be confused by an apparent discrepancy in the numbers shown in Figure 14.1 compared to those shown in Figure 4.6. Careful comparison of the two figures reveals that Figure 4.6 shows the emissions of carbon (in million tons per year), whereas Figure 14.1 shows the emissions of carbon dioxide (in million tons per year). Each ton of carbon dioxide contains 0.273 tons of carbon.

While carbon dioxide is not the only greenhouse gas emitted by the industrial world, it is by far the most important. Even though methane, nitrous oxide and chlorofluorocarbons have a stronger greenhouse effect, much smaller volumes of these are emitted each year. Nevertheless, there are programs to reduce these emissions as well.

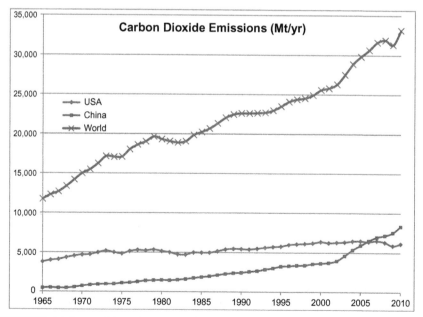

FIGURE 14.1 World Carbon Dioxide Emissions Since 1965. *Data Source: BP Annual Energy Review, 2011.*

PHASE BEHAVIOR OF CARBON DIOXIDE

It is useful to understand a little about the phase behavior of CO_2 because it is not only interesting; it is also relevant to how it is handled for transport and storage. The phase diagram for any pure substance will show the pressures and temperatures where the substance is solid, liquid or gas. There is also a fourth physical state above the critical temperature and critical pressure where the substance is called "supercritical," which is also relevant to its handling.

Figure 14.2 shows the CO_2 phase diagram. Two important points are shown on that diagram: the critical point and the triple point. At the critical point, gas and liquid CO_2 have the same properties (density, viscosity, etc.). It is impossible to distinguish between the liquid and gas phases at this point. At the triple point, the solid, liquid and gas phases are all in equilibrium with one another. Below the triple point, the liquid phase does not exist. It is in this region that the solid CO_2 ice sublimates into a gas phase when it is heated. The triple point of CO_2 occurs at a pressure of 5.11 atmospheres (76.4 pounds-force per square inch absolute or 76.4 psia) and a temperature of $-56.4\,°C\,(-69.6\,°F)$. At sea level, the atmospheric pressure is one atmosphere (14.7 psia). This means that CO_2 cannot be a liquid at atmospheric pressure. It must either be a very cold solid (known as dry ice) or a gas, which mixes with the atmosphere. Solid CO_2, or dry ice, is widely used to keep things cold because it has those two properties

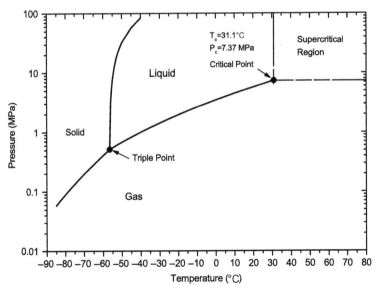

FIGURE 14.2 Schematic Of CO_2 Phase-Diagram. *Source: Akbarabadi and Piri, 2011.*

that make it very useful: it is very cold (less than $-70°F$) and, when it is heated, it does not melt to form a liquid, but rather sublimates into the atmosphere. By contrast, the triple point of water occurs at 0.006 atm (0.09 psia); consequently, at atmospheric pressure, ice melts into liquid water when it is heated.

The critical point for CO_2 is $31.1°C$ ($88°F$) and 7.37 MPa (1,073 psia). A system pressure of more than 7.37 MPa (which is about 73 atmospheres) and a system temperature of more than $31.1°C$ will result in a supercritical condition. In the supercritical region, the properties of the substance are somewhere between a liquid and a compressed gas. More importantly, the CO_2 does not undergo a phase change when the temperature and pressure change. If the temperature is decreased, the CO_2 smoothly transforms into a liquid. If the pressure is decreased, the CO_2 smoothly transforms into a gas. Both of these changes occur without undergoing any sudden or rapid phase change.

Between the critical point and the triple point, the CO_2 exists in either a liquid or gaseous state, depending on the temperature and pressure. When in the liquid state, if the temperature is increased or the pressure is decreased, the liquid will reach the boiling point line and be transformed into a gas through the boiling process. Conversely, from the gaseous state, if the temperature is decreased or the pressure is increased, the gas will reach the same line and be transformed into a liquid through the condensation process.

Technically speaking, CO_2 can be transported through pipelines in the form of a gas, a supercritical fluid or in the sub-cooled liquid state. Gas phase transport is disadvantageous because of the lower density and consequently larger pipe diameter required. It is not used for pipelines of any significant capacity and length. Furthermore, in storage, the CO_2 must be injected into deep wells at its destination. It must also be supplied at the appropriate pressure, which is usually around 3,000 psia (204 atmospheres) or more. To be transported in a pipeline, CO_2 should be compressed to ensure that single phase flow is achieved for the entire length of the pipeline. A transmission pipeline can experience a wide range of ambient temperatures, so maintaining stability of this single phase is important in order to avoid considerations of two-phase flow that could result in pressure surges which damage the pipeline.

Most CO_2 pipelines are operated in the liquid region (high pressures and low temperatures). In very hot climates, however, the liquid flowing in the pipeline can be heated above the critical temperature ($31.1°C$). The liquid thus becomes a supercritical fluid, which oddly enough is easier to handle. Most CO_2 pipelines are buried in the earth a few feet, which keeps the operating temperature below the critical temperature. In that case, care must be taken to ensure the pressure in the pipeline remains higher than the boiling point pressure so that a phase change from liquid to gas does not occur. If this happens, it results in pressure surges that can possibly damage the

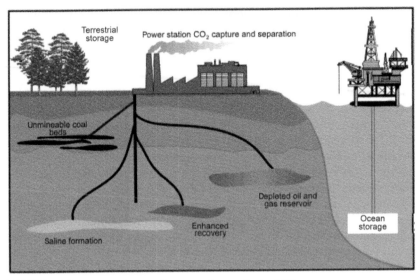

FIGURE 14.3   Carbon Storage Concepts. *Source: Fowler 2007.*

pipeline. On the other hand, once the CO_2 is injected into the reservoirs under the ground, the earth temperature is usually above the critical temperature ($31.1\,^\circ$C ($88\,^\circ$F,)) and the CO_2 becomes a supercritical fluid.

In Figure 14.3 in the storage concepts section, some people have proposed that CO_2 be sequestered at the bottom of the deep ocean where the temperatures can be very cold ($35\,^\circ$F to $40\,^\circ$F) and the pressures are very high [4,450 psia (303 atm)] at a depth of 10,000 feet. At these conditions, CO_2 is a liquid, slightly less dense than seawater; however, CO_2 is soluble in seawater. When sequestered on the ocean floor in this manner, it will gradually dissolve in the ocean rather than rise to the surface. Another method of sequestering CO_2 in the ocean would be to convert it to dry ice. The density of dry ice is heavier than seawater so it could be dumped into the deepest part of the ocean. It would then sink to the ocean floor where the pressure would be above the triple point pressure. The ocean would then transfer heat to the dry ice, melting it into a liquid form which would then begin rising and dissolving in the ocean water. There are certain disadvantages to this method of storage which will be explained later.

CARBON CAPTURE METHODS

Most of the CO_2 that is emitted into the atmosphere comes from burning hydrocarbons in air. Air contains about 80% nitrogen and about 20% oxygen, with many other components of much smaller values. The CO_2

component in the air is currently about 400 ppm, or 0.04%. The flue gas from a coal-burning power plant, however, contains 10-12% CO_2. This can be captured by various methods. The most widely used commercial technology for doing this is to absorb it in Mono-Ethanolamine (MEA) using a contacting column. This process is illustrated in Figure 4.7. After absorbing the CO_2 from the flue gas, the MEA is pumped to a stripping column where it is heated and the CO_2 exits at the top of the column as a mostly pure CO_2 stream that is ready for sale or storage. Using this technology, CO_2 can be stripped out of flue gas from a power plant for about $50 per ton of CO_2 recovered. It costs another $25 per ton to compress the CO_2 and sequester it underground. This increases the cost of coal-fired or gas-fired electricity by about 30%.

Another method of capturing CO_2 is to use solid absorbents, instead of MEA, in the process described above. There are many substances that will absorb and capture CO_2, ranging from activated carbon to poly-ionic liquids. This is the focus of intensive research programs around the world.

A final method of capturing the CO_2 is oxy-combustion, which is illustrated in Figures 13.13 and 13.15. In this process, hydrocarbons are burned in an oxygen environment with recycled CO_2 (instead of nitrogen) as the carrier gas. This results in a flue gas that is mostly pure CO_2, ready for storage.

CARBON STORAGE CONCEPTS

Figure 14.3 shows the concept of how carbon dioxide is sequestered in the earth or in the ocean. The CO_2 is pumped into depleted oil or gas reservoirs, deep saline formations, coal beds or into the ocean itself.

CO_2 is currently used to increase the recovery of oil from oil reservoirs, a process illustrated in Figure 14.4. The CO_2 is used to sweep oil towards the producing wells, mobilizing oil that would otherwise remain in the reservoir, held in place by capillary forces. At the end of the life of the Enhanced Oil Recovery (EOR) project, the CO_2 remains sequestered in the reservoir. CO_2 is more soluble in oil than it is in water. When it is pumped into the oil reservoir, it dissolves in the oil. If the pressure is high enough, it becomes miscible with the oil. This swells up the volume of the oil and makes it more mobile and able to be swept out of the corners of the porous rock and flow towards the producing well.

Though there are many depleted oil and gas fields, there is not enough space in such reservoirs to sequester all the CO_2 that will need to be sequestered. Consequently, coal beds and saline aquifers will also have to be used.Coal beds use a different physical principle to sequester the CO_2. Coal has a chemical attraction for CO_2 and, when it is injected into the coal bed, it adsorbs onto the surface of the coal (releasing methane

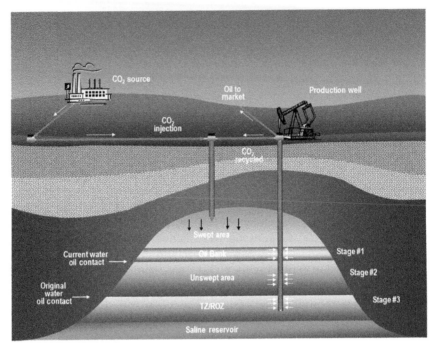

FIGURE 14.4 CO_2 Storage in Enhanced Oil Recovery Operations. *Source: DOE.*

in the process). It therefore does not require much additional space in which to store the CO_2. Large volumes of CO_2 can be stored in coal beds, but it must be unminable coal to ensure that the CO_2 remains sequestered. Chapter 13 discussed the large mass of coal reserves in the world; however, there is much more coal in the earth that is too deep to be mined. This coal can also store a large quantity of CO_2.

When sequestered underground, most of the formations capable of holding CO_2 are deep enough that the formation temperature is greater than the critical temperature of CO_2 (31.1°C). The pressure in the formation is also likely to be greater than the critical pressure (73 atm). This means that most of the CO_2 is sequestered in a dense-phase supercritical state. There are some shallow formations where the temperature would be cooler than the critical temperature. In such formations, however, it is unlikely that the pressure could be maintained high enough to keep the CO_2 in a liquid state. Consequently, porous shallow formations are not suitable for storage.

The largest spaces available for carbon storage are in deep saline aquifers. When CO_2 is injected into these formations, there are a number of different mechanisms that contribute to the storage process. Two of the principal mechanisms are the dissolution of CO_2 in the water and the chemical reaction between the CO_2 and the rock minerals and mineral salts present in the water.

When CO_2 reacts with the mineral salts in the formation, it produces carbonate minerals that frequently precipitate into the formation, trapping the CO_2 in-situ. This can be a relatively slow process, but once completed, it keeps the CO_2 trapped and immobile in the formation. If the CO_2 is not mineralized in this manner, some of the CO_2 will dissolve in the water and the rest will remain as a supercritical CO_2 phase that displaces some of the formation water. In this case, it can be beneficial to produce some of the formation water to the surface to create room for the supercritical CO_2 phase in the formation. The produced water can be: desalinated at the surface and used; disposed of in the ocean; or re-injected back into the same aquifer trapping the CO_2 by capillary action. As long as the supercritical CO_2 saturation is kept low (less than about 25%), the CO_2 remains immobile and trapped and cannot migrate or leak out of the containing formation. Over time, the CO_2 saturation gradually dissolves and mineralizes in the aquifer water.

On the bottom of the deep oceans, high enough pressures and low enough temperatures exist that would be sufficient to maintain the CO_2 in a liquid state. Liquid CO_2, however, is lighter than seawater. It would rise and become dissolved in the ocean water. The problem with this method of storage is that the CO_2 is more soluble in cold water. As the ocean currents carry the CO_2-laden water to warmer regions, the CO_2 could be released back into the atmosphere. Solid CO_2 (dry ice) is heavier than seawater. If it was dropped into the ocean, it would sink to the bottom. It would then rapidly dissolve in the ocean water and suffer the same fate.

Another problem with ocean storage is that increasing the CO_2 levels in the ocean will make the ocean more acidic. When CO_2 dissolves in water, it makes carbonic acid which lowers the pH of the ocean water. This will adversely affect the corals and many of the marine species, killing them off. For these reasons, ocean storage of CO_2 is not a viable option on a large scale or on a long-term basis.

CARBON STORAGE PROJECTS

The Global CCS Institute, based in Australia's capital city of Canberra, has identified 74 large-scale integrated CCS projects around the world. In their annual survey, undertaken in May–August 2011, they published a detailed assessment of project status, including an analysis of project dynamics, challenges and opportunities. They define large-scale and integrated projects as those that involve the capture, transport and storage of CO_2 at a scale of more than 800,000 tons of CO_2 annually for a coal-based power plant and more than 400,000 tons of CO_2 annually for other emission-intensive industrial facilities (including natural gas–based power generation). There are many more projects around the world which are of a smaller scale or only focus on part of the CCS chain. Twenty-five of these projects are located in the United States and 21 are located in Europe. Of these 74 projects, 8 are currently in operation and another 6 are currently being constructed. These 14 projects are shown in Table 14.2. Eight of the projects are located in the United States.

TABLE 14.2   Current Large Scale CCS Projects

Name	Location	Size (Mt/yr)	Type	Date
In Operation				
Shute Creek Gas Processing Facility	United States	7	EOR	1986
Sleipner CO_2 Injection	Norway	1	Deep saline formation	1996
Val Verde Natural Gas Plants	United States	1.31	EOR	1972
Great Plains Synfuels Plant and Weyburn-Midale Project	United States/ Canada	3	EOR with MMV	2000
Enid Fertilizer Plant	United States	0.7	EOR	1982
In Salah CO_2 Storage	Algeria	1	Deep saline formation	2004
Snøhvit CO_2 Injection	Norway	0.7	Deep saline formation	2008
Century Plant	United States	5	EOR	2010
Under Construction				
Lost Cabin Gas Plant	United States	1	EOR	2012
Illinois Industrial Carbon Capture and Storage (ICCS) Project	United States	1	Deep saline formation	2013
Boundary Dam with CCS Demonstration	Canada	1	EOR	2014
Agrium CO_2 Capture with ACTL	Canada	0.6	EOR	2014
Kemper County IGCC Project	United States	3.5	EOR	2014
Gorgon Carbon Dioxide Injection Project	Australia	4	Deep saline formation	2015

THE FUTURE OF CARBON CAPTURE AND STORAGE

Currently, the world is emitting more than 33 billion tons of CO_2 annually. If this were to be sequestered in deep saline aquifers at an average pressure of 3,500 psia and an average temperature of 180°F, the average specific gravity of the supercritical fluid would be 0.65. This correlates to a density of 650 kg/m^3, or 40.5 lb/ft^3. At this density, the 33 billion tons of CO_2 per year would have a volumetric rate of injection into the formation of 875 million barrels of supercritical CO_2 per day. This is 10 times the volume of oil that is being produced per day in the world. At the surface conditions of 2,600 psia and 75°F, the specific gravity of the liquid phase would be 0.9, giving a density of 56 lb/ft^3. At these conditions, the total volume of fluid being injected in liquid form at the surface would be 632 million barrels of liquid CO_2 per day. Consequently, the CO_2 storage business could be larger than today's oil business. Even if only 10% of the CO_2 produced is sequestered underground, it is still a very large business that today is a relatively small business. If the world chooses to invest more time and energy into carbon capturing and storage, the business potential could be huge.

The Kyoto Accords failed to make much headway in reducing the CO_2 emissions. Any such future agreements would have to begin with a bilateral agreement between United States and China, the two largest CO_2 emitters. Europe, Russia and India would also have to be eventual signatories to the agreement. With this critical mass, most of the other countries in the world would likely follow suit.

15

Hydrogen

Many people have proposed hydrogen as a clean energy source because its only by-product is water when it is burned or reacted with oxygen. Rifkin (author of "The Hydrogen Economy") has even gone as far as to say that the end of the hydrocarbon age has been reached and that hydrogen is the energy source of the future, the "forever fuel." This claim, however, contradicts the fact that there are no indigenous hydrogen sources in the world. Hydrogen has to be generated from other energy sources before it can be made available for use. The major methods of generating hydrogen are from natural gas and coal. Consequently, hydrogen is not really an energy source but a method of energy storage. It also suffers from serious deficiencies in this regard. It is difficult to store and transport and the technologies for doing this have not yet been developed. Moreover, even when compressed to 10,000 psia, it has an energy density that is only 15% of that of gasoline. For any given fuel tank volume, a hydrogen-fueled vehicle will only travel 15% of the miles that a gasoline-fueled vehicle will travel with the same size fuel tank.

HYDROGEN TECHNOLOGIES

Hydrogen can be burned in internal combustion engines (ICE) or converted to electricity in fuel cells. A fuel cell is a device that uses the chemical reaction between hydrogen and oxygen to produce electricity. The fuel cell has two electrodes sandwiched around an electrolyte. Hydrogen passes over one electrode and oxygen over the other, generating electricity, water and heat. Hydrogen is fed into the negative electrode, or "anode," of the fuel cell. Oxygen (or air) enters the fuel cell through the positive electrode, or "cathode." Under the action of a catalyst, the hydrogen atom splits into a proton and an electron, which take different paths to the cathode. The proton passes through the electrolyte. The electrons create a separate current that can be utilized before they return to the cathode to be reunited with the hydrogen and oxygen in a molecule of water. The electricity can be used directly to power an electric vehicle or to supply electric current to any other users such as houses or factories. There are several different fuel cell technologies, but the polymer electrolyte membrane (PEM) fuel cell, also called a proton exchange membrane fuel cell, is the technology proposed for use in automobiles. Figure 15.1 shows how a PEM fuel cell works. These cells have been widely used to provide electrical power in the space program.

The principle of the fuel cell was first discovered by a German chemistry professor, Christian Schönbein, in 1838. Based on his work, the first fuel cell was demonstrated by Welshman William Grove in February 1839. The fuel cell he made used similar materials to today's phosphoric-acid fuel cell. It wasn't until 1955 that Thomas Grubb, a chemist working

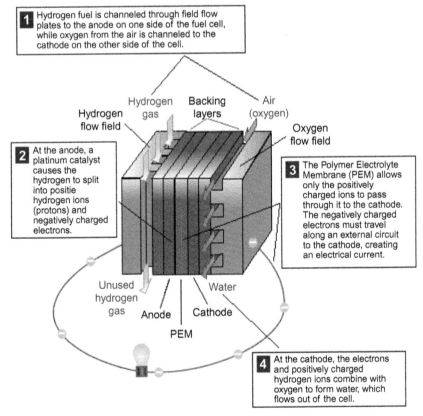

1 Hydrogen fuel is channeled through field flow plates to the anode on one side of the fuel cell, while oxygen from the air is channeled to the cathode on the other side of the cell.

Hydrogen flow field

Hydrogen gas

Backing layers

Air (oxygen)

Oxygen flow field

2 At the anode, a platinum catalyst causes the hydrogen to split into positie hydrogen ions (protons) and negatively charged electrons.

3 The Polymer Electrolyte Membrane (PEM) allows only the positively charged ions to pass through it to the cathode. The negatively charged electrons must travel along an external circuit to the cathode, creating an electrical current.

Unused hydrogen gas

Water

Anode

Cathode

PEM

4 At the cathode, the electrons and positively charged hydrogen ions combine with oxygen to form water, which flows out of the cell.

FIGURE 15.1 Polymer Electrolyte Membrane (PEM) Fuel Cells. *Source: DOE.*

for General Electric Company (GE), further modified the original fuel cell design by using a polystyrene ion-exchange membrane as the electrolyte. Three years later another GE chemist, Leonard Niedrach, devised a way of depositing platinum onto the membrane. This served as the catalyst for the hydrogen oxygen reaction and became known as the "Grubb-Niedrach" fuel cell. GE went on to develop this technology with NASA and McDonnell-Douglas, leading to its use in the first space program entitled "Project Gemini." This was the first commercial use of a fuel cell.

In 1959, British engineer Francis Bacon developed a 5-kW stationary fuel cell and in the same year Harry Ihrig built a 15-kW fuel cell tractor for the Allis-Chalmers Company, which was widely demonstrated in the United States. This system used potassium hydroxide as the electrolyte and compressed hydrogen and oxygen as the reactants. Meanwhile, Bacon and his colleagues demonstrated that their five-kilowatt unit was capable of powering a welding machine.

In the 1960s, Pratt and Whitney adopted Bacon's technology for use in the U.S. space program to supply electricity and drinking water to the space modules. In 1991, the first hydrogen fuel cell automobile was developed by Roger Billings. He also developed the first hydrogen powered internal combustion engine. He did not attempt to commercialize them because he was more focused on his computer storage technology businesses, but to this day he retains a strong interest in hydrogen technologies.

United Technologies Corporation's power subsidiary (UTC Power) was the first company to manufacture and commercialize a large, stationary fuel cell system for use as a co-generation power plant in hospitals, universities and large office buildings. UTC Power continues to market this fuel cell as the PureCell 200, a 200-kW system. In 2009, they added the PureCell 400 which was a 400-kW version. UTC Power is now the sole supplier of fuel cells to NASA for use in space vehicles, having supplied fuel cells for the Apollo missions and the Space Shuttle program. They are currently developing fuel cells for automobiles, buses and cell phone towers. Their PEM fuel cell has been shown to be capable of starting under freezing conditions.

Most fuel cells designed for use in vehicles produce less than 1.16 volts of electricity, clearly not enough to power a vehicle. Multiple cells must be assembled into a fuel cell stack. The potential power generated by this stack depends on the number and size of the individual fuel cells that comprise the stack and the surface area of the PEM. Figure 15.2 shows the Honda fuel cell vehicle with the essential elements (the hydrogen storage tank, the fuel cell stack, the electric motor and the power control unit) cut away.

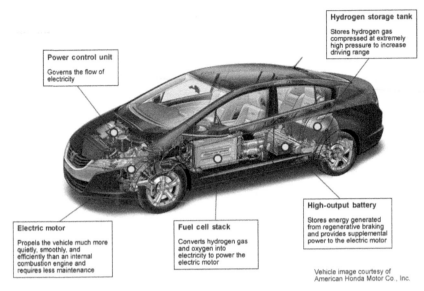

FIGURE 15.2   Elements of a Fuel Cell Vehicle. *Source: Honda.*

THE HYDROGEN ECONOMY

The idea for utilizing hydrogen instead of fossil fuels to power houses and vehicles received a large boost in 2002 from a book by Jeremy Rifkin (2002) called *The Hydrogen Economy*. The jacket of the book claims that:

> Rifkin takes us on an eye-opening journey into the next great commercial era in history. He envisions the dawn of a new economy powered by hydrogen that will fundamentally change the nature of our market, political and social institutions, just as coal and steam power did at the beginning of the industrial age. Rifkin observes that we are fast approaching a critical watershed for the fossil-fuel era, with potentially dire consequences for industrial civilization. While experts had been saying that we had another forty or so years of cheap available crude oil left, some of the world's leading petroleum geologists are now suggesting that global oil production could peak and begin a steep decline much sooner, as early as the end of this decade. Non-OPEC oil producing countries are already nearing their peak production, leaving most of the remaining reserves in the politically unstable Middle East. Increasing tensions between Islam and the West are likely to further threaten our access to affordable oil. In desperation, the U.S. and other nations could turn to dirtier fossil-fuels—coal, tar sand, and heavy oil—which will only worsen global warming and imperil the earth's already beleaguered ecosystems. Looming oil shortages make industrial life vulnerable to massive disruptions and possibly even collapse.
>
> While the fossil-fuel era is entering its sunset years, a new energy regime is being born that has the potential to remake civilization along radical new lines, according to Rifkin. Hydrogen is the most basic and ubiquitous element in the universe. It is the stuff of the stars and sun and, when properly harnessed, it is the "forever fuel." It never runs out and produces no harmful CO_2 emissions. Commercial fuel-cells

powered by hydrogen are just now being introduced into the market for home, office and industrial use. The major automakers have spent more than two billion dollars developing hydrogen cars, buses, and trucks, and the first mass-produced vehicles are expected to be on the road in just a few years.

The hydrogen economy makes possible a vast redistribution of power, with far-reaching consequences for society. Today's centralized, top-down flow of energy, controlled by global oil companies and utilities, will become obsolete. In the new era, says Rifkin, every human being could become the producer as well as the consumer of his or her own energy—a so called "distributed generation." When millions of end-users connect their fuel-cells into local, regional, and national Hydrogen Energy Webs (HEWs), using the same design principles and smart technologies that made possible the World Wide Web, they can begin to share energy—peer-to-peer—creating a new decentralized form of energy use.

Hydrogen has the potential to end the world's reliance on imported oil and help diffuse the dangerous geopolitical game being played out between Muslim militants and Western nations. It will dramatically cut down on carbon dioxide emissions and mitigate the effects of global warming. Because hydrogen is so plentiful and exists everywhere on earth, every human being could be "empowered," making it the first truly democratic energy regime in history." —Jeremy Rifkin (2002), *The Hydrogen Economy*.

These are heady claims, but do they stand up to any scientific scrutiny? While the potential of hydrogen will be examined in this chapter, one answer to that question is contained in a review of Rifkin's book by David Goodstein (Department of Physics, California Institute of Technology) that appeared in the March-April 2003 issue of American Scientist.

"The age of oil is about to end, which will come as a surprise to most people. Depletion of the rest of the fossil fuels may not be far behind. And if we do go on merrily burning up the planet's legacy, the result may be irreversible damage to our climate. This crucially important idea is the starting point of *The Hydrogen Economy*, a new book by Jeremy Rifkin, a former peace advocate who now crusades against biotechnology and various other perceived ills.

Rifkin believes that oil will be replaced by hydrogen fuel cells. Unlike the oil economy, which requires a top-down, capitalist-corporate order, the hydrogen economy will be something like the Internet, he says, with users who are also providers, generating their own hydrogen and sharing any surplus on the Hydrogen Energy Web. After all, unlike oil, hydrogen is everywhere: It's the most common element in the universe, the "forever fuel" that we can never run out of. The revolution brought about by the hydrogen economy will lead to a democratization of society and give a whole new meaning to the word globalization.

But wait a minute. Doesn't Rifkin understand that it takes energy to generate hydrogen from water, or from any other source? Well, yes, he even says so in a couple of places, but he seems to have trouble holding that thought. And when he does come to grips with it, he believes all the energy will come from "renewable resources—photovoltaic, wind, hydroelectric, geothermal, and biomass." There will be no nuclear reactors in a world designed by Jeremy Rifkin. At present, all the renewables on Rifkin's list aside from hydroelectricity collectively generate less than 1 percent of our energy needs.

Is Rifkin's proposed solution physically possible? Well, yes, sort of, but it's extremely implausible that all the power generated today by fossil fuels, about 10 terawatts worldwide, could ever be replaced from those sources. Biomass is a terribly inefficient use of sunlight. There are only a few places on Earth where enough geothermal energy to generate electricity is within drilling distance of the surface. Hydroelectric capacity is already saturated, and wind is an intermittent, low-density (and often ugly) source of power. According to an article by Martin I. Hoffert and colleagues in the November 1, 2002, issue of *Science* (298:981-987), to replace the 10 terawatts with photovoltaics would require an array covering more than 200,000 square kilometers, whereas all the photovoltaic cells shipped between 1982 and 1998 would cover only 3 square kilometers.

Our best hope in the short run is that somebody will start building nuclear power plants in a hurry, before the oil starts to run out. In the longer run, when the fossil fuels are gone or sequestered and the uranium starts to run low, if we haven't yet brought thermonuclear energy under control, heroic measures like huge arrays of photovoltaics on Earth, or somewhat smaller ones in space (where the solar flux is about 8 times the average at the Earth's surface), may be in order. That is not to say we should not do our best to develop renewable energy sources. We certainly should. But they will not replace fossil fuels anytime soon.

Rifkin is certainly right to say that we will soon start running out of oil, that continued burning of fossil fuels is a grave threat to the Earth's climate, and that hydrogen, either in fuel cells or by combustion, is the best bet for the future of transportation. He has correctly identified the biggest problem we have. But this book is not part of the solution." —David Goodstein, Dept of Physics, California Institute of Technology.

Rifkin's book certainly had a broad impact. In his State of the Union Address on January 28, 2003, President Bush embraced hydrogen as the fuel of the future, announcing the Freedom Car and Freedom Fuel Initiative. Then an article entitled *How Hydrogen Can Save America* appeared in the April 2004 issue of *Wired* magazine. The authors proposed a program equal in scope to the Apollo program that put men on the moon claiming:

"The cost of oil dependence has never been so clear. What had long been largely an environmental issue has suddenly become a deadly serious strategic concern. Oil is an indulgence we can no longer afford, not just because it will run out or turn the planet into a sauna, but because it inexorably leads to global conflict. Enough. What we need is a massive, Apollo-scale effort to unlock the potential of hydrogen, a virtually unlimited source of power. The technology is at a tipping point. Terrorism provides political urgency. Consumers are ready for an alternative. From Detroit to Dallas, even the oil establishment is primed for change. We put a man on the moon in a decade; we can achieve energy independence just as fast." —*Wired* magazine, April 2004

President Bush's Hydrogen Initiative of 2003 provoked a response from the public radio show *Car Talk* with Tom and Ray Magliozzi, who are also syndicated nationally on many commercial radio stations. Here was their take on the Bush Administration's program on hydrogen vehicles:

Dear Tom and Ray:
President Bush talked about a "hydrogen car" in his State of the Union address. Is this a realistic possibility during the Bush administration?

Ray: Maybe during the Jenna Bush administration, Jim. The technology itself works, but people "in the know" say it's going to be at least 20 years before hydrogen-powered cars are viable on a large scale—if then.

Tom: The main problems are: (1) the fuel cell "stacks" are still incredibly expensive to build, (2) the range of the cars is insufficient and (3) there's no national infrastructure (like gas stations) to support hydrogen. So it's not going to happen anytime soon.

Ray: So, why is the president talking about hydrogen-powered cars? Well, in my humble opinion, he's creating a distraction.

Tom: I think so, too. You probably know that we now import boatloads of foreign oil every day. And almost everybody agrees that this is not a good thing (except for the countries that sell us the oil). So what do you do about it?

Ray: Well, you can try to find more oil here at home, by drilling in Alaska's forests, for instance. Or you can force people to use less oil. The president knows that both of these options are pretty unpopular. So he's doing what any good politician would do: He's changing the subject.

Tom: Here's another reason why he might want to distract us from thoughts of fuel economy and foreign oil. With no pressure on American car companies to increase gas mileage, the Japanese have taken a significant lead in the most important new propulsion technology in decades: hybrid engines. Hybrid engines use battery power some of the time and gasoline power at other times, and they never have to be plugged in. They're a great way to increase mileage without sacrificing power or convenience. And you're going to see Americans adopting them in big numbers over the next five to 10 years.

Ray: Who makes the best-selling hybrid cars in America? Honda and Toyota. So, instead of urging America to make more fuel-efficient cars and cut down on foreign oil by raising the Corporate Average Fuel Economy (CAFE) standards, or urging U.S. manufacturers to catch up with the Japanese on hybrids—which would make a huge difference right away—the president's talk about hydrogen cars is, essentially, the old "Hey, everybody, look over there!" —*from* **Car Talk *with* Tom *and* Ray *Magliozzi* (2003)**

Another very useful report written in response to Rifkin's book is *The Future of the Hydrogen Economy: Bright or Bleak?* by Bossel et al. (2003). Their response was presented at the 2003 Fuel Cell Seminar, 3-7 November 2003 in Miami Beach, Florida. This report discusses hydrogen sources, hydrogen storage and hydrogen distribution. They state: "*It seems that the recent surge of interest in a hydrogen economy is based more on visions than on hard facts. We are not at all against hydrogen, but we would like the discussion about the synthetic energy carrier to return to facts and physics. Hopefully, this study will help to identify chances and limitations of a hydrogen economy.*"

Hydrogen has had important applications for a century, but it has never been an energy source or energy carrier for the reasons described by Bossel, Eliasson and Taylor. While hydrogen has important uses, it first has to be manufactured before it can be used. Hydrogen fuel does not exist on or in the Earth at all. In 1905, the Germans developed the Haber process to make ammonia from nitrogen and hydrogen. Ammonia is the starting point for making fertilizers and explosives. Hydrogen is also necessary for the refining of crude oil into high octane-low sulfur gasoline and high cetane-low sulfur diesel fuel. Consequently, one of the largest generators and users of hydrogen are petroleum refineries.

Hydrogen can be produced by several methods. To make hydrogen, you need an energy source and a hydrogen source. Today, the cheapest source of hydrogen is natural gas, which is both the energy source and the hydrogen source. The problem is that the resulting hydrogen has only 70% of the chemical energy of the original natural gas. Hydrogen can also be produced from oil, but the energy efficiency is a little worse. If hydrogen is derived from any fossil fuel, carbon dioxide will be created.

The primary ways in which natural gas, mostly methane (CH_4), is converted to hydrogen involve reaction with either steam (steam reforming), oxygen (partial oxidation) or both in sequence (auto-thermal reforming). The overall reactions are:

$$CH_4 + 2H_2O \rightarrow CO_2 + 4H_2$$

$$CH_4 + O_2 \rightarrow CO_2 + 2H_2$$

In practice, gas mixtures containing carbon monoxide (CO) as well as carbon dioxide (CO_2) and unconverted methane (CH_4) are produced and require further processing. The reaction of CO with steam (the water-gas shift reaction) over a catalyst produces additional hydrogen and carbon dioxide (CO_2). After separation, high-purity hydrogen (H_2) is recovered.

The United States has lots of coal, and it can be used to make hydrogen. Coal is a cheap energy source, but not a good hydrogen source. Water must be used as a supplemental hydrogen source because coal has more energy than it has hydrogen. Syngas, a mixture of carbon monoxide and hydrogen, is made by gasifying the coal by reacting the water (as steam) with hot coal particles (see Chapter 13). This method has been used since the nineteenth century, but the energy efficiency is only 35%. The resulting hydrogen has 35% of the energy of the original coal.

Electricity can also be used to make hydrogen from water through electrolysis, an efficient but expensive process. Electrolysis of water has been known for centuries, and it is a common classroom demonstration. The reason why this method turns out to be expensive is once again due to the second law of thermodynamics. The electrical energy has been generated by burning fuels in a heat engine and converting the rotating energy to electrical energy, with an efficiency that is reduced by the second law of thermodynamics. In electrolysis, the electrical energy is then turned back into a hydrogen fuel.

Fossil fuel is the source of most of the energy used to make electricity in the United States. Hydrogen produced by electrolysis may be viewed as a convoluted way to convert fossil fuel to hydrogen. Fossil fuel (coal, natural gas and oil) can be converted to hydrogen directly without making electricity first. The direct methods use less fossil fuel for the same amount of produced hydrogen. Electricity from wind turbines or solar arrays can also make hydrogen through electrolysis, but this is inefficient and expensive.

Hydrogen is also expensive to store because it has a very low volumetric energy density. It is the simplest and lightest element; it is even lighter than helium. At atmospheric pressure, hydrogen is 2.93 times less energy dense than natural gas and 2933 times less energy dense than gasoline. Hydrogen, however, contains 3.1 times more energy than gasoline on a weight basis. This makes it useful for rocket fuel in the space program. To be useful for any transportation purposes, however, it must be made more energy dense. There are three ways to do this: hydrogen can be compressed, liquefied or chemically combined (Tables 15.1 and 15.2).

Hydrogen compressed to a pressure of 10,000 psia occupies 6.05 times more volume than gasoline for the same energy, as shown in Table 15.3. This table also shows the energy density of liquid hydrogen, hydrogen at 10,000 psia, gasoline and ethanol on a volume basis as well as a mass basis. It is necessary to liquefy hydrogen or compress it to a pressure of 10,000 psia if a vehicle is to carry enough hydrogen fuel to be practical. It is very difficult to contain such high pressures safely in a lightweight tank. Catastrophic tank failures release as much energy as an equal weight of dynamite. A 10,000 psia tank made of high strength steel would weigh 100 times more than the hydrogen it contains. A truck or an automobile using a steel tank would be impractical as the tank would weigh nearly as much as the vehicle. High pressure hydrogen tanks made from carbon fiber have been considered as a solution. Carbon fiber is a material used in aircraft and sporting goods. At the present time, carbon fiber tanks are very expensive, too expensive to be practical. The DOE had proposed a performance goal as part of the Freedom Car initiative. Their goal is 4.5% as the ratio of hydrogen to tank weight at 10,000 psia.

A typical 18-wheeled semitruck carries two 90-gallon tanks, providing a driving range up to 750 miles. A typical 4-cylinder sedan has an 18-gallon tank, providing a driving range up to 575 miles. The diesel engine achieves an efficiency of 35% at cruising speeds. The gasoline engine achieves an efficiency of 30% at cruising speed. Both vehicles could be converted to hydrogen operation. ICE could be used, resulting in an efficiency of 35%. Fuel cells could also be used, resulting in an efficiency of 45%. This information is summarized in Table 15.4. The space, weight and expense of steel tanks make them impractical. Any gains in energy efficiency would be offset by the losses incurred in hauling the very heavy tanks. Carbon fiber tanks of this size and performance do not exist; they are only goals. Gasoline, by contrast, requires only a small, low-tech tank.

The laws of thermodynamics govern the amount of energy it takes to compress a gas. The physical properties of hydrogen make it the most difficult of all gases to compress. At 10,000 psia, a perfect, single-stage

TABLE 15.1 Comparison of Properties of Hydrogen with Other Fuels

Properties	Units	Hydrogen	Methane	Propane	Methanol	Ethanol	Gasoline
Chemical Formula		H_2	CH_4	C_3H_8	CH_3OH	C_2H_5OH	C_xH_y (x=4-12)
Molecular Weight		2.02	16.04	44.1	32.04	46.07	100-105
Density (NTP)	kg/m^3	0.0838	0.668	1.87	791	789	751
	lb/ft^3	0.00523	0.0417	0.116	49.4	49.3	46.9
Normal Boiling Point	°C	−253	−162	−42.1	64.5	78.5	27-225
	°F	−423	−259	−43.8	148	173.3	80-437
Vapor Specific Gravity (NTP)	air=1	0.0696	0.555	1.55	N/A	N/A	3.66
Flash Point	°C	<−253	−188	−104	11	13	−43
	°F	<−423	−306	−155	52	55	−45
Flammability Range in Air	vol%	4.0-75.0	5.0-15.0	2.1-10.1	6.7-36.0	4.3-19	1.4-7.6
Auto Ignition Temperature in Air	°C	585	540	490	385	423	230-480
	°F	1085	1003	914	723	793	450-900

(Data Source: DOE)

TABLE 15.2 Energy Equivalencies of Fuels

	Hydrogen	Natural Gas	Crude Oil	Conventional Gasoline	Reformulated Gasoline (RFG)	California RFG	U.S. Conventional Diesel	Low-Sulfur Diesel
	(kg)	(million cubic feet)	(barrel)	(gallon)	(gallon)	(gallon)	(gallon)	(gallon)
Hydrogen, 1 kg =	1	0.000117	0.0211	0.992	1.014	1.011	0.896	0.889
Natural Gas, 1 MMCF =	8538	1	180.5	8468	8653	8628	7653	7591
Crude Oil, 1 barrel =	47.30	0.00554	1	46.91	47.94	47.80	42.40	42.06
Gasoline, 1 gal =	1.008	0.000118	0.0213	1	1.022	1.019	0.904	0.897
RFG, 1 gal =	0.987	0.000116	0.0209	0.979	1	0.997	0.884	0.877
Calif RFG, 1 gal =	0.989	0.000116	0.0209	0.981	1.003	1	0.887	0.880
Diesel, 1 gallon =	1.116	0.000131	0.0236	1.106	1.131	1.127	1	0.992
LS Diesel, 1 gal =	1.125	0.000132	0.0238	1.115	1.140	1.137	1.008	1

(*Data Source: DOE*)

TABLE 15.3 Energy Density of Hydrogen Compared to Other Fuels

Fuels	BTU/lb	BTU/gal
Hydrogen at 10,000 psia	61,127	20,534
Liquid Hydrogen	61,127	38,243
Gasoline	20,007	124,340
Ethanol	12,832	84,530

TABLE 15.4 Hydrogen Fuel Tank Comparisons

Vehicle	Weight of Fuel	Weight of Steel Tank	Weight of Carbon Fiber Tank	Volume of Tank Contents	Volume of Tank
Typical 18-wheel truck (diesel)	1175 lb	200 lb	NA	22.5 ft^3	24.0 ft^3
Typical sedan (gasoline)	108 lb	50 lb	NA	2.25 ft^3	2.5 ft^3
Truck converted to hydrogen	313 lb	31,300 lb	6,960 lb	67.5 ft^3	157 ft^3
Sedan converted to hydrogen fuel cell	17.4 lb	1740 lb	387 lb	4 ft^3	9 ft^3

compressor consumes energy equal to 16% of the chemical energy in the hydrogen (this is the energy that gets instantly released in the event of a tank failure). It might be possible to use a series of multistage compressors with intercoolers to achieve 12%. This is an estimate extrapolated from an actual multistage compressor working at 3,000 psia.

It is technologically challenging to compress hydrogen to 10,000 psia. Higher pressure would not result in much volume reduction. At these pressures, hydrogen acts less like a gas and more like a liquid. The laws of thermodynamics also dictate that energy losses occur when hydrogen is transferred from a storage tank to a vehicle. The design of the transfer lines and the pressure fittings is critical in keeping energy losses low.

Liquid hydrogen also occupies 3.25 times more volume than gasoline for the same energy. Paradoxically, there is more hydrogen in a gallon of gasoline than there is in a gallon of liquid hydrogen. The advantage of hydrogen liquefaction is that a cryogenic hydrogen tank is much lighter. Hydrogen's physical properties mean hydrogen is harder to liquefy than any other gas except helium. There are significant and inevitable energy losses when hydrogen is liquefied.

Liquid hydrogen is colder than any other substance except liquid helium. The advantage of liquid hydrogen is that it can be stored in relatively lightweight tanks. A tank for cryogenic hydrogen is like a thermos bottle, but it must work much better. It consists of a tank within a tank with a vacuum between the two. The inner tank must be supported without conducting heat. This is very difficult to do in a tank designed for a vehicle. Bossel, Eliasson and Taylor estimate that a liquid hydrogen tank designed for automobile use would lose about 5% of its capacity every day through evaporation, in the same manner that LNG evaporates while it is being shipped in tankers. If you are not driving your liquid hydrogen fueled vehicle, its fuel would evaporate in about 20 days. Losses of this magnitude are acceptable for a taxicab fleet, for example, but unacceptable to most people. Hydrogen cannot be vented to the atmosphere because it is an explosion hazard and because it is a greenhouse gas. The vented hydrogen must be burned.

Hydrogen molecules exist in two forms, ortho (the electron spins are antiparallel) and para (the electron spins are parallel). Room temperature hydrogen is a mixture of the two. Liquid hydrogen, however, turns into pure para hydrogen over the course of a few days. The process releases enough heat to turn the liquid hydrogen to gas. Liquid hydrogen can be catalytically converted to all para during the liquefaction process. If this were not done, 30% of the hydrogen would escape in two days even in a perfect cryogenic tank. Conversion to the para form of hydrogen adds to the cost and complexity of liquefaction.

Certain alkali metal hydrides release hydrogen when exposed to water. These metal hydrides hold enough hydrogen to make them useful for transportation. 70% of the energy, however, is lost in the creation of the hydrides, making them unacceptable for widespread use. Certain metals (platinum, zirconium, lanthanum) can be formed into "sponges" for hydrogen. These sponges can only hold 1% of their weight in hydrogen and they are very expensive.

In both cases, the energy storage tank is heavy and expensive not to mention energy inefficient. These are not the qualities needed for vehicles designed for mass transportation. It is possible though for the technology to have a niche application. Long lasting, but expensive, batteries for laptop computers may use these hydrogen sources to power small fuel cells.

Another issue is the distribution of the hydrogen fuel. A 40-ton truck can deliver 26 tons of gasoline to a conventional gasoline filling station. One daily delivery is sufficient for a busy station. A 40-ton truck carrying compressed hydrogen can deliver only 400 kilograms. This is due to the weight of the tank, which is only capable of holding 3,000 psia of pressure. An empty truck will weigh almost as much as a full one. The compressed hydrogen tank must be robust. The energy used to compress the hydrogen to 3,000 psia would be released instantly if a tank ruptured. The fireball would cover a football field. Hydrogen is more energy dense than gasoline (by weight) and hydrogen-powered transportation is more energy efficient. Yet the hydrogen filling station will require 15 deliveries every day, everything else being equal. The energy cost of truck transport becomes unacceptable unless the source of hydrogen is very close to the point of use. A cryogenic truck could carry more hydrogen, but the energy cost to liquefy hydrogen makes this impractical in most cases.

Because hydrogen is a very small and reactive molecule, it is difficult to store and transport. It tends to diffuse through many materials and it makes most metals brittle and prone to failure. According to Bossel, Eliasson and Taylor, if it is transported by pipeline it takes about four times more energy to move hydrogen through the pipeline compared to natural gas. This leads them to make several significant conclusions, among them being:

> Meanwhile, we find that the conversion of natural gas into hydrogen cannot be the solution of the future. Hydrogen produced by reforming natural gas may cost less (in both money and energy) than hydrogen obtained by electrolysis, but for most applications, natural gas is as good as, if not better than hydrogen. For use in road transport, if natural gas were converted to hydrogen, the well-to-wheel efficiency would be reduced and hence, for given final energy demand, the emission of CO_2 would be increased. Moreover, for all stationary applications, the distribution of energy as electricity would be energetically superior to the use of hydrogen as an energy carrier.

After the release of Rifkin's book and the announcement of the hydrogen initiatives by the Bush Administration, the National Research Council in conjunction with the National Academy of Engineering formed a Committee on Alternatives and Strategies for Future Hydrogen Production and Use to study the matter. The report on their findings entitled *The Hydrogen Economy: Opportunities, Costs, Barriers, and R&D Needs* was released in early 2004 (National Research Council and National Academy of Engineering, 2004). The report examined hydrogen end-use technologies, the distribution and storage of hydrogen, estimated costs of hydrogen supply, transition to hydrogen in vehicles, hydrogen production technologies and crosscutting issues (such as safety, research and international partnerships). Some of their conclusions were:

- There is a potential for eliminating almost all CO_2 and criteria pollutants from vehicular emissions; however, there are currently many barriers to be overcome before that potential can be realized.
- There are other strategies for reducing oil imports and CO_2 emissions. The DOE should keep a balanced portfolio of R&D efforts and continue to explore supply-and-demand alternatives that do not depend upon hydrogen. If battery technology improved dramatically, for example, all-electric vehicles might become the preferred alternative.

Furthermore, hybrid electric vehicle technology is commercially available today, and benefits from this technology can therefore be realized immediately. Fossil-fuel-based or biomass-based synthetic fuels could also be used in place of gasoline.

They went on to identify the four principal research needs of a hydrogen economy:

1. *To develop and introduce cost-effective, durable, safe, and environmentally desirable fuel cell systems and hydrogen storage systems.* Current fuel cell lifetimes are much too short and fuel cell costs are at least an order of magnitude too high. An on-board vehicular hydrogen storage system that has an energy density approaching that of gasoline systems has not been developed; thus, the resulting range of vehicles with existing hydrogen storage systems is much too short.

2. *To develop the infrastructure to provide hydrogen for the light-duty-vehicle user.* Hydrogen is currently produced in large quantities at reasonable costs for industrial purposes. The committee's analysis indicates that at a future, mature stage of development, hydrogen (H_2) can be produced and used in fuel cell vehicles at reasonable costs. The challenge, with today's industrial hydrogen as well as tomorrow's hydrogen, is the high cost of distributing H_2 to dispersed locations. This challenge is especially severe during the early years of a transition when demand is even more dispersed. The costs of a mature hydrogen pipeline system would be spread over many users as the cost of the natural gas system is today. The transition is difficult to imagine in detail. It requires many technological innovations related to the development of small-scale production units. Nontechnical factors such as financing, siting, security, environmental impact and the perceived safety of hydrogen pipelines and dispensing systems will also play a significant role. All of these hurdles must be overcome before there can be widespread hydrogen use. An initial stage during which hydrogen is produced at small scale near the small user seems likely. In this case, production costs for small production units must be sharply reduced, which may be possible with expanded research.

3. *To reduce sharply the costs of hydrogen production from renewable energy sources over a time frame of decades.* Tremendous progress has been made in reducing the cost of making electricity from renewable energy sources. Making hydrogen from renewable energy through the intermediate step of making electricity, a premium energy source, requires further breakthroughs in order to be competitive. Basically, these technology pathways for hydrogen production make electricity, which is converted to hydrogen,

which is later converted by a fuel cell back to electricity. These steps add costs and energy losses that are particularly significant when the hydrogen competes as a commodity transportation fuel leading the committee to believe that most current approaches— except possibly that of wind energy—need to be redirected. The committee believes that the required cost reductions can be achieved only by targeted fundamental and exploratory research on hydrogen production by photo-biological, photochemical, and thin-film solar processes.

4. *To capture and store ("sequester") the carbon dioxide by-product of hydrogen production from coal.* Coal is a massive domestic U.S. energy resource that has the potential for producing cost-competitive hydrogen. Coal processing, however, generates large amounts of CO_2. In order to reduce CO_2 emissions from coal processing, massive amounts of CO_2 would have to be captured and safely and reliably sequestered for hundreds of years. Key to the commercialization of a large-scale, coal-based hydrogen production option (and also for natural-gas-based options) is achieving broad public acceptance, along with additional technical development, for CO_2 sequestration.

For a viable hydrogen transportation system to emerge, all four of these challenges would have to be addressed. The committee, however, failed to identify the fundamental flaws of a hydrogen economy detailed by the Bossel, Eliasson and Taylor report. In the years since the report was released, none of their four challenges have been met. Consequently, a hydrogen-based economy has not become viable in terms of these criteria and nor is it likely to become viable in the future.

It is useful to document some of the data from the committee report. Table 15.5 shows the total investment required to manufacture hydrogen from natural gas, as well as the H_2 production costs and overall thermal efficiency.

Figure 15.3 shows the unit investment cost (in 2004 \$/kg/day) as a function of plant size. As expected, the economies of scale demand that plant sizes be large—greater than 1 million kg/day.

Table 15.6. shows the total costs of manufacturing hydrogen from methane and distributing it for sale to consumers in 2004. The current price is \$1.99/kg (noting that 1 kg of H_2 has about the same energy content as one gallon of gasoline). In 2004, the total cost of gasoline for reference was \$1.12/gal. The distributed on-site cost of hydrogen using electrolysis was \$6.58/kg, which is far too high to be useful. Figure 15.4 shows the hydrogen production cost as a function of plant size, and Figure 15.5 shows the hydrogen production cost as a function of both plant size and gas price.

TABLE 15.5 Hydrogen Production Cost Data

| | Plant Size (Kilograms of Hydrogen per Stream Day [SD]) and Case | | | | | |
| | 1,200,000[a] | | 24,000[b] | | 480[c] | |
	Current	Possible Future	Current	Possible Future	Current	Possible Future
Investment (no sequestration), $/kg/SD	411	297	897	713	3847	2001[d]
Investment (with sequestration), $/kg/SD[c]	520	355	1219	961	-	-
Total H_2 cost (no sequestration), $/kg	1.03[e]	0.92[e]	1.38[e]	1.21[e]	3.5[f]	2.33[f]
Total H_2 cost (with sequestration), $/kg[g]	1.22[e]	1.02[e]	1.67[e]	1.46[e]	-	-
CO_2 emissions (no sequestration), kg/kg H_2	9.22	8.75	9.83	9.12	12.1	10.3
CO_2 emissions (with sequestration), kg/kg H_2	1.53	1.30	1.71	1.53	-	-
Overall thermal efficiency (no sequestration), %[h]	72.3[a]	77.9[a]	46.1	53.1	55.5	65.2
Overall thermal efficiency (with sequestration), %[g,h]	61.1	68.2	43.4	49.0	-	-

[a] Includes compression of production hydrogen to pipeline of 75 atm.
[b] Includes compression of H_2 prior to transport.
[c] Includes compression of H_2 to 400 atm for storage/fueling vehicles.
[d] Includes estimated benefits of mass production.
[e] Based on natural gas at $4.50/million Btu.
[f] Based on natural gas at $6.50/million Btu.
[g] Includes capture and compression of CO_2 to 135 atm for pipeline transport to sequestration site.
[h] Based on lower heating values for natural gas and hydrogen: includes hydrogen generation, purification, and compression, and energy imported from offsite as well as distribution and dispensing.
(Source: National Research Council and National Academy of Engineering Committee on Alternatives and Strategies for Future Hydrogen Production and Use, 2004)

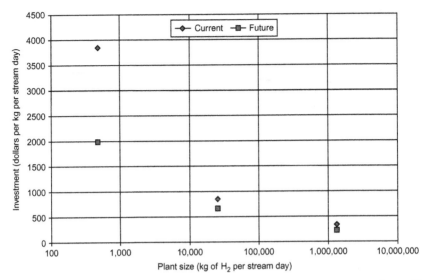

FIGURE 15.3   Investment Cost (2004 US$) as a Function of Plant Size to Manufacture H₂ from Natural Gas. *Source: National Research Council and National Academy of Engineering Committee on Alternatives and Strategies for Future Hydrogen Production and Use, 2004.*

THE FUTURE OF HYDROGEN AND THE HYDROGEN ECONOMY

Currently, hydrogen is being produced in chemical plants and petroleum refineries. It is being utilized to upgrade gasoline and diesel fuels and to make fertilizer and petro-chemicals. Those uses will continue to grow. It has future appeal to many people as a transportation fuel both in fuel cell-powered electric vehicles and for internal combustion engine-powered vehicles. The appeal emanates from the fact that the only product created from burning hydrogen is water. The hydrogen economy proposed by Rifkin, however, will never be realized for all the reasons documented by the Bossel, Eliasson and Taylor report. It is cheaper and better to transport and burn natural gas than convert the natural gas to hydrogen and burn hydrogen. Hydrogen is expensive to manufacture and distribute, and it is difficult to handle and store. If the goal is to make vehicles that have zero emissions, electric vehicles are probably a better alternative. Electric vehicles need better and cheaper batteries, and that seems to be a more controllable problem than all of the problems associated with a hydrogen economy. Hydrogen is also used to propel rockets into space, and it will continue to be used for that application. To eliminate carbon dioxide as a by-product of plane travel, hydrogen may find an additional application in plane travel, but at increased passenger costs.

TABLE 15.6 Total H$_2$ Costs

Case	Production Costs ($/kg)	Distribution Costs ($/kg)	Dispensing Costs ($/kg)	Total Dispensing and Distribution Costs ($/kg)	Total Costs ($/kg)	Total Energy Efficiency (%)
Centralized Production,						
Pipeline Distribution						
Natural gas reformer						
Today	1.03	0.42	0.54	0.96	1.99	72
Future	0.92	0.31	0.39	0.70	1.62	78
Natural gas+CO$_2$ capture						
Today	1.22	0.42	0.54	0.96	2.17	61
Future	1.02	0.31	0.39	0.70	1.72	68
Coal						
Today	0.96	0.42	0.54	0.96	1.91	57
Future	0.71	0.31	0.39	0.70	1.41	66
Coal+CO$_2$ capture						
Today	1.03	0.42	0.54	0.96	1.99	54
Future	0.77	0.31	0.39	0.70	1.45	61

Continued

TABLE 15.6 Total H₂ Costs—cont'd

Case	Production Costs ($/kg)	Distribution Costs ($/kg)	Dispensing Costs ($/kg)	Total Dispensing and Distribution Costs ($/kg)	Total Costs ($/kg)	Total Energy Efficiency (%)
Distributed Onsite Production						
Natural gas reformer						
Today					3.51	56
Future					2.33	65
Electrolysis						
Today					6.58	30
Future					3.93	35
Liquid H₂ Shipment						
Today		1.80	0.62	2.42		
Future		1.10	0.30	1.40		
Gasoline (for reference)	$0.93/gal refined			$0.19/gal	$1.12/gal	Well to tank: 79.5%

Notes: The energy content of 1 kilogram of hydrogen (H2) approximately equals the energy content of 1 gallon of gasoline. Details of the analysis of the committee's estimates in this table are presented in Chapter 5 and Appendix E of this report—see the discussion in this chapter.
(Source: *National Research Council and National Academy of Engineering Committee on Alternatives and Strategies for Future Hydrogen Production and Use, 2004*)

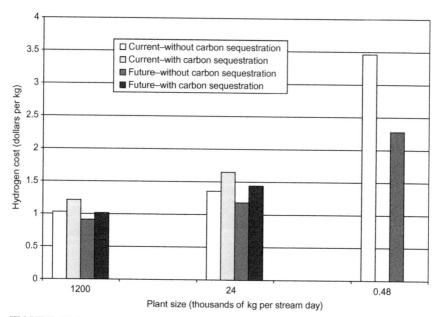

FIGURE 15.4   Hydrogen Production Cost from Natural Gas as a Function of Plant Size. *Source: National Research Council and National Academy of Engineering Committee on Alternatives and Strategies for Future Hydrogen Production and Use, 2004.*

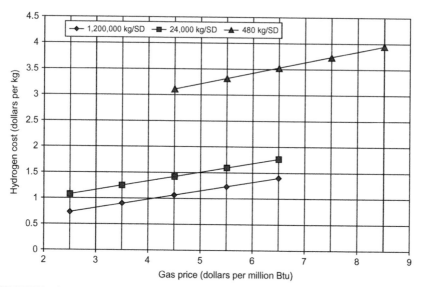

FIGURE 15.5   Hydrogen Production Cost from Natural Gas as a Function of Natural Gas Price and Plant Size. *Source: National Research Council and National Academy of Engineering Committee on Alternatives and Strategies for Future Hydrogen Production and Use, 2004.*

What Is the Future of Energy? An Energy Policy for the United States

In order to maintain a robust economy, each country individually—as well as the entire world collectively—needs access to adequate supplies of energy at a competitive price. As the supplies of cheap oil and gas are used up, many people fear that the world will be plunged into an energy war as countries compete for access to the remaining dwindling energy reserves. Nothing could be further from the truth as there are adequate supplies of energy for all concerned. At the same time, many are grappling with the desire for cleaner energy supplies and are advocating renewable energy sources regardless of the cost. They particularly call for wind and solar energy as the solution to all the energy needs of the world. As was previously discussed in this book, there are major problems with these sources. First, there is the issue of cost. Solar energy is extremely expensive and is unlikely to be ever cost competitive, except in remote and small niche markets. Wind and solar power also suffer from intermittency and reliability of supply, and both have environmental impacts that do not seem to be fully appreciated.

Security of the oil supply is another major issue at the moment. In the past the majority of U.S. oil supplies were imported from other countries: Middle Eastern countries, African countries, South American countries and countries that are prone to political instability. These imports are now trending down rapidly. But Europe and many Asian economies (China, Japan and South Korea) are facing the same situation. When civil war breaks out in Libya or Nigeria, crucial supplies are shut down. If Saudi Arabia was ever plunged into civil war and the ten million barrels per day that it currently produces were removed from the market, the price of oil would rise dramatically. When civil unrest was spreading throughout the Middle East in 2011 from Algeria to Egypt to Libya to

The Future of Energy
http://dx.doi.org/10.1016/B978-0-12-801027-3.00016-6

Syria and Yemen and Bahrain, the big fear was that Saudi Arabia would be caught up in similar civil unrest that would seriously disrupt world oil supplies. There was also a concern about a protracted civil war in Egypt because Egypt controls the Suez Canal, through which a great volume of oil is shipped. Other factors, such as political differences between the importer and exporter, can also have an effect on the security of supply. Consequently, it is important for large energy users (and this means countries with robust economies) to generate energy supplies internally or at least from countries with stable political systems.

The United States is the second largest energy user in the world, and though it is improving it currently does not have sufficient oil production to satisfy its own needs. It currently imports more than 5 million barrels of oil per day, and some of this oil comes from countries that are prone to political instability. In 2011–2014, imported oil averaged about $100 per barrel. That represented an import bill of $200–300 billion per year that must be sent to foreign suppliers.

Is the United States capable of generating all of its energy supplies internally? The answer is yes. It has the largest reserves of coal in the world, as well as vast reserves of natural gas and sufficient uranium supplies to power the largest network of nuclear power plants in the world. Recently, it also identified large reserves of shale oil at onshore locations, as well as conventional oil in the deep water of the Gulf of Mexico. The technology exists to economically convert its coal and gas reserves into transportation fuels (gasoline, diesel, jet fuel, bunker fuel, etc.). It also has the potential to increase its supplies of renewable energy that are economic, such as wind energy, hydroelectric power and geothermal power. If it chose to invest in fast breeder reactor technology, it could also increase its nuclear power resources.

For those who are concerned that many of these energy sources are unacceptable because they generate carbon dioxide, there are solutions to this issue. Carbon dioxide can be captured at the generating source and sequestered underground. The oil industry has been doing this for more than fifty years and the technology to do it is mature. The first place to do this is at the electric generating plants. In fact, it is feasible to generate all the electricity needs of the United States without releasing any carbon dioxide into the atmosphere. This will raise the price of electricity, but it can be done if that is what is required. Of course, it is more difficult to capture carbon dioxide from moving sources such as automobiles and aircrafts, but a move towards more electric vehicles, plug-in hybrids and hydrogen-fueled vehicles could significantly reduce the carbon dioxide from these sources as well.

In this chapter, the future of energy will be discussed, as well as how current technology can harness sufficient energy supplies to power the largest economy in the world. The same technologies are also relevant to Europe, China, India and Australia.

NATURAL GAS

The United States actually peaked in natural gas production in 1973 and hit a local minimum in 1986. Since that time, U.S. gas production has been steadily rising toward a new peak. The growth in gas production started to accelerate further in 2006, when significant shale gas supplies began to come online. Gas producers have now generated so much in the way of natural gas supplies that the local price of natural gas has plummeted and new markets are being sought.

On September 27, 2010, ConocoPhillips Chairman and Chief Executive Officer James Mulva gave an address at Rice University, during which he argued that increasing natural gas production and consumption would help drive U.S. economic recovery and job creation. "Natural gas is an overlooked job-creation machine," Mulva said. "Let's crank it up and step on the accelerator." He was speaking during a two-day conference that discussed the market consequences of carbon management policies worldwide. Mulva is exactly right, and many of the ideas he advocated are directed towards increased natural gas usage. The technology that has unlocked these increased gas supplies in the United States are applicable to many areas of the world and will mean increased gas supplies in many other countries, including Europe (Poland in particular), Russia, China and Australia.

Consequently, Mulva called upon U.S. lawmakers to develop a comprehensive energy policy "...that allows all energy sources to compete on the basis of abundance, cost, efficiency, and environmental merit." He said, "Energy from renewable sources cannot be provided fast enough to instantly replace the energy currently provided by fossil fuels." Further, The oil and gas industry currently supports 9.2 million U.S. jobs," Mulva said, but he also acknowledged that all energy sources will be needed in the future. "So yes, bring on the green jobs. But in doing so, don't destroy the real jobs that we have today in the oil and gas industry." He recommended that state and U.S. lawmakers carefully examine renewable electricity standards, noting that some states require utilities to use particular renewable sources for a large part of their power supply. "Washington is considering a national standard," Mulva said. "Unfortunately, these renewable sources are quite expensive." Mulva also stated that he supports research and development to lower the cost of renewable energy, but he resists energy policy that mandates the use of renewable energy. "Ideally, a balanced policy would create an attractive investment climate for gas," Mulva said. "But oil and gas is a heavily taxed industry through income taxes, lease bonus payments, royalties, and severance taxes. And, the Obama administration has targeted us for new taxes."

According to Mulva, a U.S. energy policy would enable the market to determine the best use of gas.

For example, to reduce greenhouse gas emissions, many in Washington support the use of compressed natural gas in vehicles. But gas could do more good in the industrial and power sectors. We believe we could get larger reductions in the future from using plug-in hybrid-electric and all-electric vehicles. The electricity would come from gas-fired generating plants

In an executive summary of numerous reports from the above-mentioned conference, issued on September 27, 2010, Rice University said that "...business-as-usual, market-related trends might propel the United States toward greater oil and natural gas self-sufficiency over the next 20 years."

Currently, according to a number of studies, the U.S. vehicle fleet involves 250 million oil-fueled vehicles. Recently adopted improvements in car fuel efficiency standards are expected to cut U.S. oil usage by 3 million b/d by 2050. Proposals to replace 30% of the vehicle fleet with electric vehicles by 2050 could cut oil usage by another 2.5 million b/d. Under this electric car scenario, U.S. greenhouse gas emissions would be cut by only 7.4% by 2050 because more electric cars would encourage more coal-fired electricity generation, unless there was also a mandated carbon cap system. If all of the U.S. electricity was generated carbon free (zero carbon emissions), however, the GHG emissions would be cut by a more substantial amount of 50%. Another widely supported change that would significantly reduce carbon emissions is the use of natural gas fueled vehicles. This makes particular economic sense for large fleet vehicles, such as buses and long haul trucks.

Putting a price on carbon dioxide, imposed by either a carbon tax or a cap-and-trade system, could increase electricity costs. In the short term, this could lead to a greater reliance on oil imports if the development of U.S. oil resources were to become less cost competitive. Adjustments to the carbon emissions policy would have to be made to penalize imports, and the carbon tax would mean an additional gasoline tax.

Analysis of the impact on U.S. energy markets of a cap-and-trade system indicates that the price of carbon could be highly unpredictable and dependent on the rules for offset programs or the availability of carbon capture and sequestration technology. All cap-and-trade scenarios ultimately lead to higher U.S. energy prices in general and higher electricity prices in particular; the more binding the policies, the more burdensome the electricity bill for the policy. This would not be a good thing for the country and would be an economic disaster unless the entire world adopted similar policies.

On May 3, 2011 speakers at a panel discussion during the Offshore Technology Conference in Houston made two claims: fossil fuels will remain vital to the global energy supply mix for many years and technology development is crucial to enabling the oil and gas industry to meet future world energy demand.

"Looking for any alternative is the right thing to do, but looking at it to replace hydrocarbons is impractical," said Ali Moshiri, president of Chevron Africa & Latin America Exploration & Production Company.

He predicted that oil and gas will account for half of the world energy supply in 50 years. Currently, it accounts for slightly more than half. Chevron estimates that at least 55 million b/d of additional oil capacity will be needed by 2030 to offset declines and fulfill the minimum projected demand growth.

Zuhair Al-Hussain, Saudi Aramco vice-president of drilling and workovers, also emphasized the role hydrocarbons will play in the future total energy supply mix. He welcomed low-carbon energy sources, saying they need to be developed in "...a manner that is rational and sustainable." Al-Hussain said, "It is my wish that policymakers acknowledge the dominant role of fossil fuels for the coming decades."

Stephen Greenlee, president of ExxonMobil Exploration Company, said technology under development will continue to add new resources. Even with anticipated gains in energy efficiency, Exxon projects that the global energy demand will be 35% higher in 2030 compared to the 2005 world energy demand.

The United States has the capacity and the reserves to double its total natural gas production to about 136 BSCF/day from the current 68 BSCF/day. In order to do that, an additional 34,000 wells would have to be drilled assuming an average initial rate of 2 million SCF/day. In addition to that, about 5,000-10,000 new wells would have to be drilled each year to offset the declines in production from the current wells. The United States has the capacity to drill these wells and get them into production. It also has sufficient natural gas reserves to supply these new wells. The additional 68 BSCF/day of gas by itself would displace about 7 million barrels of imported oil per day from the market. The additional 68 BSCF/day would also bring with it an additional 2 million barrels per day of condensate and other natural gas liquids, which can also replace imported oil. This amounts to the displacement of 9 million barrels per day of imported oil by indigenous energy supplies.

The additional gas production could be used in several ways:

- To generate additional electricity to power-electric vehicles as well as to service increased industrial activity;
- To power vehicles fueled by compressed natural gas;
- To convert the gas to liquid fuels (gasoline, diesel, bunker fuel and jet fuel); and,
- To manufacture plastics and other petrochemicals.

COAL

The United States also has the capacity and the reserves to double its coal production to about 5 million tons/day, from the current 2.5 million tons/day. The majority of this increase would likely come from the

Powder River Basin in Wyoming and Montana. Substantial increases would also be made in the coal basins of Illinois, Kentucky and West Virginia. The additional 2.5 million tons/day of coal would displace 4.5 million barrels of imported oil per day from the market. The additional coal production would also be used in a similar manner as the additional gas production:

- To generate additional electricity to power-electric vehicles as well as increased industrial activity.
- To manufacture plastics and other petrochemicals.
- To convert the coal to liquid fuels (gasoline, diesel, bunker fuel and jet fuel).

Of course, some people will object to this scenario because coal is seen as a dirty fuel that generates all sorts of air pollution. With adequate deployment of clean coal technologies, however, this coal could be converted into electricity and petrochemicals with almost zero additional emissions.

NUCLEAR POWER

Nuclear power could also be doubled using current technology. The plants simply have to be built and could be adequately supplied by both indigenous uranium supplies as well as imports from Australia, Canada and Kazakhstan. Standard nuclear reactors require large amounts of uranium-235, and the supply of uranium-235, while limited, can be increased. In naturally occurring uranium deposits, uranium-235 constitutes about 0.7% of the total uranium. The rest is uranium-238. New supplies of uranium will naturally become available as the price of uranium yellowcake rises. Doubling of the uranium price would at least double the supplies, without greatly increasing the price of nuclear power.

Nuclear power could be greatly increased if the fast breeder reactor technology is commercialized. Fast breeder reactors consume both uranium-235 and uranium-238 and extract much more energy from a given quantity of fuel than current nuclear reactors. At present, this has not been commercialized, but that time is not far off. Concerns do linger about the safety of nuclear power and, in the wake of the nuclear reactor disaster at the Fukushima nuclear power station in Japan (due to damage by the tsunami generated by the huge earthquake on March 11, 2011), public confidence in nuclear power has again been eroded. It is important to recognize that this type of power does have a large and significant future, and work needs to continue on developing this as an energy source. It has the potential for a bright future if these concerns can be overcome. There is a very large and almost unlimited energy source available if and when fast breeder reactors are brought into operation.

HYDROELECTRIC

Hydroelectric power is one of the cheapest and cleanest energy sources available. It is also a very reliable source of energy that can be rapidly turned on and off as the need arises. While it is not without environmental impact, it is probably the most acceptable source of energy that can be harnessed.

To some it might seem that the supply of hydroelectric power cannot be substantially increased because there are very few additional rivers that can be dammed to substantially increase the supply; however, this is not true. According to the U.S. Hydropower Association, the industry could install 60 GW of new capacity by 2025, which is only 15% of the total untapped hydropower resource potential in the United States, meaning that the Hydropower Association has identified 400 GW of additional hydropower potential. At 36% capacity, it would mean that all 400 GW of additional capacity in the United States could generate an additional 1,250 TWh of clean renewable electricity per year, 40% of the U.S. demand.

The Electric Power Research Institute recommends developing about 90 GW of this potential immediately, which would more than double the current hydroelectric power capacity. Developing this hydroelectric potential has a number of added benefits as stated by the U.S. Department of the Interior: "...flood control and river regulation, water storage and delivery (including irrigation), power generation, recreation, and fish and wildlife development."

The current hydroelectric resources that are available need to be protected and maintained. Throughout the world, there are many additional rivers that can be harnessed for hydroelectric power. This could have a significant impact on future world energy supply. Hydroelectric power also has the ability to play a role in the storage of energy from intermittent and unreliable energy sources such as wind-power and solar-power. Pumped storage can be used to store wind-power and solar-power when they are producing at their peak and to release the stored energy when wind and solar energy sources drop too low.

WIND

At present, wind-power provides a tiny fraction (3%) of U.S. energy needs, and it has the potential to be substantially increased to perhaps 10% of the electricity needs of the country. It too is not without environmental impact, but it is perceived as a clean form of energy and is mostly popular for that reason. The biggest issue is that it is intermittent and unreliable, so it has to be backed by more reliable energy sources such as gas, coal and hydroelectric power. If the supply of wind-generated electricity was increased to more than 10% of the electric supply, substantial energy storage would have be combined with the wind power energy source to provide sufficient reliability for the entire electricity resource. This substantially increases the cost of wind power to uneconomic levels. Below 10% of the total electricity supply, the storage can be provided by hydroelectric pumped storage and turning coal- and gas-fired turbines up and down. More costly storage methods would have to be deployed beyond that level. Nevertheless, optimistically speaking, the wind energy generation could be tripled to about 30 TWh/mo. This could be used to displace gas or coal from the power generation, which would then be able to displace about 0.5 million barrels of oil per day of imported oil.

SOLAR

Solar power is currently a niche player and provides a tiny fraction of U.S. energy supplies. This is unlikely to increase substantially because solar power is way too expensive to be competitive. It is unlikely to be cost competitive in the foreseeable future because solar energy is too diffuse and cannot be concentrated cheaply. It is not a matter of pouring money into research to make it cost competitive; the second law of thermodynamics militates against this.

GEOTHERMAL

Geothermal power is also a clean and relatively reliable source of energy where available. It is unlikely to be a bigger player in the U.S. energy supply in the near future because most geothermal resources, which are easily accessible at a completive price and in an environmentally acceptable manner, have already been developed. There are few further supplies available. Some have proposed substantially increasing geothermal energy supplies by harnessing the earth's heat energy, by drilling deep wells into hot spots in the earth and pumping water under pressure to capture this energy. This is known as enhanced geothermal systems (EGS). While this idea has promise, it has yet to be commercialized.

OIL

The oil supply in the United States is a big problem for energy security and balance of payments. Currently, the U.S. imports about 5 million barrels of oil per day, which (at $100/barrel) contributes $180 billion to the balance of payment problems for the country. In order to achieve energy security, this imported oil has to be replaced by indigenous energy supplies. The United States can increase its own oil production substantially, and it is possible that they can add 5 million barrels to the daily oil production. In addition it is also feasible to replace all of this imported oil with indigenous energy.

In the analysis above, it was shown that:

- Doubling the current gas production is feasible and this would displace 9 million barrels of imported oil per day.
- Doubling the coal production would displace 4.5 million barrels of oil per day.
- Developing tight oil and shale oil and deepwater Gulf of Mexico oil could add 4 million barrels of oil per day.

Canada is also capable of increasing its oil exports to the United States by at least 2 million barrels of oil per day, which would come from the oil sands region of Athabasca in Alberta. Currently, Canada exports 2 million barrels of oil per day to the United States and increasing this to 4 million barrels of oil per day is eminently feasible. While this would still be imported oil and it would mean sending $130 billion per year to Canada, the oil would be coming from a more stable trading partner. It is likely that wind energy can also be tripled to about 30 TWh/mo. This could be used to displace gas or coal from the power generation, which would then be able to displace about 0.5 million barrels of oil per day. This amounts to an additional 23 million barrels of oil per day, much more than the 5 million barrels per day that is currently being imported and the 3 million barrels per day that is not currently being imported from Canada. It is quite feasible to do this, and it would benefit the nation both economically and strategically.

U.S. ENERGY POLICY DEVELOPMENT

To implement the strategy discussed in the above section requires the political will to develop an energy policy directed towards achieving it. But what else will be required? The drilling of more oil and gas wells; the construction of coal mines, hydro-electric dams and wind turbines; the infrastructure to handle the transport of the additional oil and gas and coal being produced; and the infrastructure to convert some gas and coal into liquid fuels and/or electricity. The economic benefits would be enormous. There is also a strategic benefit which translates to an economic benefit because the nation would not need to protect energy sources around the world, particularly in unstable regions.

The elements of the energy policy needed to achieve this goal will not be detailed here, but many of the things currently being done are not conducive to a policy of energy independence. Current policies are directed towards developing expensive energy sources that do not have the potential to move the United States towards energy independence. These policies favor wind power and solar power because they are perceived as clean energy sources. This book has shown that these policies are misguided.

Some of the changes that are needed have been alluded to in statements from Conoco's Jim Mulva. He advocates developing all energy sources, including "green" energy sources. No one wants to develop dirty energy sources and environmental and safety standards need to be maintained and further refined. Beyond that, the large energy sources that are available are oil, gas and coal. These can be developed using clean technologies that adhere to environmental standards. There is a great need to develop policies that level the playing field and allow further development of the energy sources that are already in existence.

In a poll conducted in 2010, people were asked, "What is the primary source of energy we will be using in 50 years' time?" The most popular answer was "...something that has not been discovered yet." While no one can predict the future with any certainty, this statement is unlikely to be true. It is also unlikely that solar power and nuclear fusion will be cost competitive even in 50 years' time. While the energy mix might change because of reduced fossil fuel supplies, it is highly likely that the energy sources in use today will still be our primary energy sources in 50 years' time. While people may be using less oil and more nuclear power, wind power and possibly more gas- and coal-derived fuels, the sources of energy will not change that much. There will be no need for wars to secure our energy supplies. The world has adequate energy supplies; they simply need to be harnessed.

At one time everyone needed access to adequate supplies of cheap oil, but the days of cheap oil are gone and are unlikely to return. The only question that remains is: what should be used to replace the cheap oil? The answer should not be expensive imported oil. Instead, it should be indigenous energy supplies that are now cost competitive, such as gas, coal, hydroelectric, wind, geothermal and nuclear power.

A

The Carnot Cycle

The **Carnot cycle**, when acting as a heat engine, consists of the following four steps:

1. **Reversible and isothermal expansion of the gas at the "hot" temperature, T_H (isothermal heat addition).** During this step (A to B in Figure A.1, 1 to 2 in Figure A.2), the expanding gas makes the piston do work on the surroundings. The gas expansion is propelled by the absorption of heat, designated Q_1, from the high temperature reservoir.

2. **A reversible and adiabatic (isentropic) expansion of the gas (isentropic work output).** Remember that adiabatic means that there is no heat transferred. Isentropic means that the entropy of the system remains constant. For this step (B to C in Figure A.1, 2 to 3 in Figure A.2), the piston and cylinder are assumed to be thermally insulated (adiabatic), and, as a result, they neither gain nor lose heat. The gas continues to expand, working on the surroundings. Expansion causes the gas to cool to the 'cold' temperature, T_C.

3. **Reversible isothermal compression of the gas at the "cold" temperature, T_C (isothermal heat rejection).** This is represented by C to D in Figure A.1 and 3 to 4 in Figure A.2. Now the surroundings do work on the gas, causing the quantity, or Q_2, of heat to flow out of the gas to the low temperature reservoir.

4. **Isentropic compression of the gas (isentropic work input).** This is represented by D to A in Figure A.1 and 4 to 1 in Figure A.2. Once again, the piston and cylinder are assumed to be thermally insulated (or adiabatic). During this step, the surroundings do work on the gas, compressing it and causing the temperature to rise to T_H. At this point, the gas is in the same state as at the start of step one.

The antithesis of a heat engine is a refrigerator. A heat engine burns fuel as part of a thermodynamic cycle to create heat that is converted into mechanical energy. A refrigerator sends the cycle in the opposite direction and uses electrical energy to create mechanical energy that then pumps heat from the cold body to the hotter body. The behavior of a Carnot

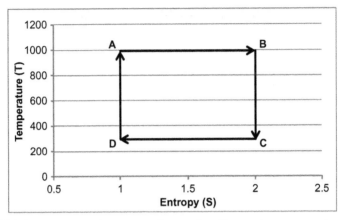

FIGURE A.1   A Carnot cycle acting as a heat engine, illustrated on a temperature-entropy diagram. The cycle takes place between a hot reservoir at temperature T_H and a cold reservoir at temperature T_C. The vertical axis is temperature, and the horizontal axis is entropy.

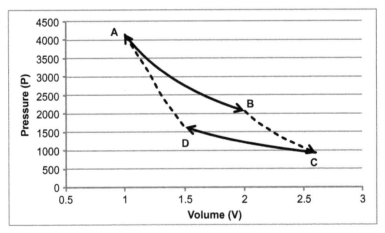

FIGURE A.2   A Carnot cycle acting as a heat engine, illustrated on a pressure-volume diagram to illustrate the work done. The vertical axis is pressure, and the horizontal axis is volume.

engine or refrigerator is best understood by using a temperature-entropy (TS) diagram, in which the thermodynamic state is specified by a point on a graph with entropy (S) as the horizontal axis and temperature (T) as the vertical axis. For a simple system with a fixed amount of mass, any point on the graph will represent a particular state of the system. A thermodynamic process will consist of a curve connecting an initial state (A) and a final state (B). The area under the curve will be the heat.

$$Q = \int_A^B T dS \qquad \text{A.1}$$

This is the amount of thermal energy transferred in the process. Equations A.1 and A.2 will use a little calculus to determine the area under the curves in Figures A.1 and A.2.

If the process moves to greater entropy, the area under the curve will be the amount of heat absorbed by the system in that process. If the process moves towards lesser entropy, it will be the amount of heat removed. For any cyclic process, there will be an upper portion of the cycle and a lower portion. For a clockwise cycle, the area under the upper portion will be the thermal energy absorbed during the cycle, while the area under the lower portion will be the thermal energy removed during the cycle. The area inside the cycle will then be the difference between the two. Given that the internal energy of the system must have returned to its initial value, this difference must be the amount of work done by or on the system over the cycle. Referring to Figure A.1 mathematically, for a reversible process we may write the amount of work done over a cyclic process as:

$$W = \oint P dV = \oint (dQ - dU) = \oint (T dS - dU) \qquad \text{A.2}$$

As dU is an exact differential, its integral over any closed loop is zero, and it follows that the area inside the loop on the T-S diagram is equal to the total work performed if the loop is traversed in a clockwise direction. If the loop is traversed in a counterclockwise direction, it will be equal to the total work done on the system.

An evaluation of the above integral is particularly simple for the Carnot cycle. The amount of energy transferred as work is:

$$W = \oint P dV = (T_H - T_C)(S_B - S_A) \qquad \text{A.3}$$

The total amount of thermal energy transferred between the hot reservoir and the system will be:

$$W = \oint P dV = (T_H - T_C)(S_B - S_A) \qquad \text{A.4}$$

The total amount of thermal energy transferred between the system and the cold reservoir will be:

$$Q_C = T_C(S_B - S_A) \qquad \text{A.5}$$

The efficiency, η, is defined as the work produced divided by the heat input from the hot reservoir:

$$\eta = \frac{W}{Q_H} = 1 - \frac{T_C}{T_H} = \frac{T_H - T_C}{T_H} \qquad \text{A.6}$$

Where,

W is the work done by the system (energy exiting the system as work).

Q_H is the heat put into the system (heat energy entering the system).

T_C is the absolute temperature of the cold reservoir.

T_H is the absolute temperature of the hot reservoir.

S_B is the maximum system entropy.

S_A is the minimum system entropy.

B

Hubbert's Peak Oil Theory from Chapter 5

HUBBERT'S MODEL FOR WORLD OIL PRODUCTION

The equations Hubbert used to model world oil production start with the equation for the cumulative production (Q_t) at any time (t). The production rate (q_t) at any time (t) is the derivative of Q_t with respect to time (t):

$$Q_t = \frac{Q_{max}}{1 + ae^{-bt}} \tag{B.1}$$

Where
Q_t = cumulative production at time (t), usually in barrels.
q_t = production rate at time (t) usually in barrels/day.
Q_{max} = ultimate cumulative production, usually in barrels.
a = growth or decline coefficient, dimensionless.
b = growth or decline exponent where dimensions are the reciprocal of time, such as 1 per day.
The production rate (q_t) at any time (t) is the derivative of the cumulative and so is given by

$$q_t = \frac{Q_{max}abe^{-bt}}{(1 + ae^{-bt})^2} \tag{B.2}$$

Equation B.2 is the equation for the production plots shown in Figures 5.2-5.5. If this model is the correct model for the production data, as long as there is early time data, it can be fit to the model to determine the ultimate recovery (Q_{max}) as well as the parameters a and b. There are several ways to determine the parameters from the data. The method that Hubbert recommended was to combine Equations B.1 and B.2 to give:

$$q_t = bQ_t \left(1 - \frac{Q_t}{Q_{max}} \right) \tag{B.3}$$

Equation B.3 can then be rearranged to give:

$$\frac{q_t}{Q_t} = b - \frac{bQ_t}{Q_{max}} \tag{B.4}$$

Equation B.4 then says that if you were to plot q_t/Q_t versus Q_t, the intercept on the y axis would be the parameter b and the intercept on the x axis would be the ultimate recoverable oil, Q_{max}. In other words, the production rate is divided by the cumulative production at any time (t) and plotted versus the cumulative production. From this plot, the intercept on the x axis is the ultimate recoverable oil. This plot can be done at any time, even before the peak rate is reached. The timing of the peak can be calculated by taking the derivative of Equation B.2 and setting it equal to zero. This leads to:

$$t_{peak} = \frac{1}{b} \ln(a) \tag{B.5}$$

To use Equation B.5, you need to know the parameter a, which was not determined from the plot of Equation B.4. The easiest way to determine the parameter a is to rearrange Equation B.1 into this form:

$$\frac{Q_{max}}{Q_t} - 1 = ae^{-bt} \tag{B.6}$$

Equation B.6 says that if you plot $Q_{max}/Q_t - 1$ versus e^{-bt}, the slope of the plot will be equal to a. To do this second plot, I assume that I have already determined b and Q_{max} from the plot of Equation B.4. Deffeyes showed his plot of Equation B.4 using the world production data up until 2005, which is reproduced here as Figure B.1. His extrapolation to the x axis shows an intercept of 2 trillion barrels, which is his estimate of the ultimate recoverable oil (Q_{max}). Prior to 1983, the data was increasing and decreasing and had not settled down to form a straight line. From 1983 to 2005, the data dots lie on the straight line that Hubbert's theory predicts. Deffeyes was also able to deduce his estimate for the time of the peak oil rate directly from his plot. If the ultimate recovery is projected to be 2 trillion barrels, the peak oil rate will occur when the cumulative production is 1 trillion barrels. This occurred in 2005 and led Deffeyes to predict that oil production would peak in that year.

World oil production, however, did not peak in 2000 or in 2005. Figure B.2 shows annual world oil production from 1965 to 2010, and at that time, the production was still increasing. Figure B.3 shows world

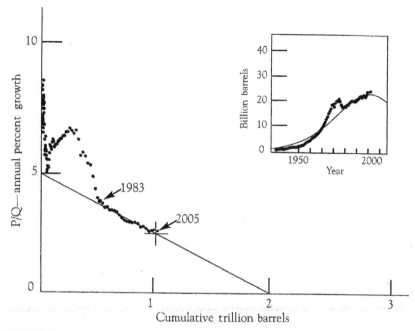

FIGURE B.1   Deffeyes plot of Equation 5.4 for world oil production. *Source: Deffeyes, Beyond Oil.*

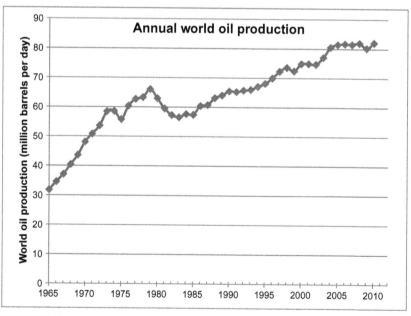

FIGURE B.2   Annual world oil production since 1965. *Data Source: BP Annual Review.*

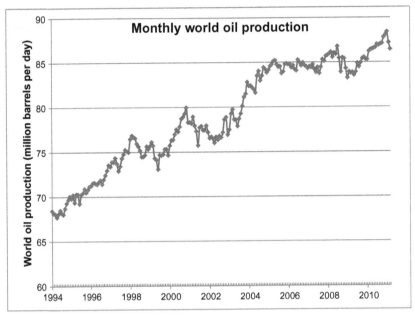

FIGURE B.3   Monthly world oil production Since 1994. *Data Source: EIA.*

oil production on a monthly basis from the beginning of 1994 to early 2011, and though there are occasional declines, the overall trend on both plots is upwards and there is no evidence that a peak has occurred or will occur. The occasional dips in production that have occurred in the past 45 years are due to decreases in demand rather than decreases in supply. This does not, however, stop the pundits from declaring that world oil production has peaked every time the production decreases from one year to the next or even from one month to the next. It should be noted that Figures B.2 and B.3 use data from two different sources. The data for Figure B.2 comes from the *BP Annual Statistical Review*, while the data for Figure B.3 comes from the *U.S. Energy Information Agency (EIA)*. Any discrepancies that are perceived between the two plots are due to the different data sources. What is important is that both plots show a consistent upward trend in world oil production.

Production did show a secondary peak in 1979, and from 1979 to 1983, production did decrease. This was due to a steep increase in the world oil price in the 1970s, which resulted in decreased demand in the early 1980s as people reduced their use of transportation fuels. From 1983 to the present day, there has been a steady increase in oil demand, which has been matched by supply. The world economic crisis that was precipitated in the middle of 2008 tempered demand, which caused a decrease in production in late 2008. From January 2009 to January 2011, there was a persistent increase in production. The uprising in Libya and other countries in the Middle East in early 2011, which is being called the "2011 Arab Spring," has once again caused a new round of price increases and a resulting decrease in demand for oil; however, this too is expected to be temporary.

If the production data shown in Figure B.2 is plotted according to Equation B.4, the result is Figure B.4. Extrapolating this to the x-axis gives a value for Q_{max} of 2.54 trillion barrels, compared with the value of 2 trillion barrels that Deffeyes derived using this same data up until 2005. If the monthly data published by the EIA and shown plotted in Figure B.3 is used to do the same plot, an even more startling revelation emerges. This plot is shown in Figure B.5. This is the monthly data from January 1994 to March 2011 and comes from a very reliable source—the US Energy Information Administration. When this data is extrapolated to the x-axis, the Q_{max} is 3.055 trillion barrels. Of this 3 trillion barrels, 1.3 trillion barrels has already been produced. This leaves about 1.7 trillion barrels still to be found and produced. According to the EIA, the current amount of the listed world oil proved reserves is 1.3 trillion barrels, meaning that there are only 400 billion barrels still to be found. Remember, these numbers are a function of oil price and are likely to be conservative. Given that this data is probably the most reliable available data, this estimate is probably the most reliable estimate using current economics. It would seem

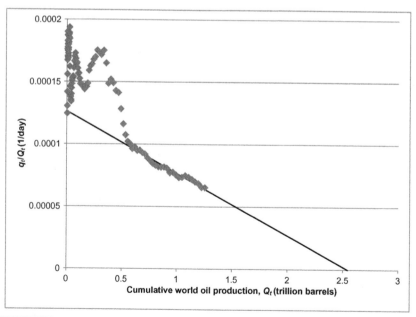

FIGURE B.4 Plot of Equation B.4 for world oil production data shown in Figure B.2. *Data Source: BP Annual Review.*

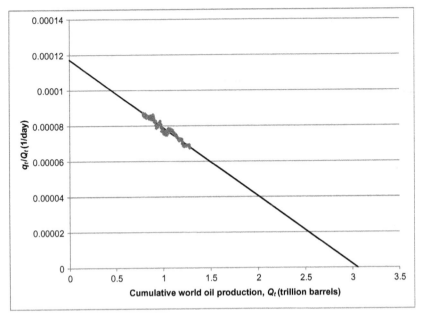

FIGURE B.5   Plot of Equation B.4 for world oil production data shown in Figure B.3. *Data Source: EIA.*

that more oil is appearing all the time and this is due to economics. If and when the price of oil goes up again, and stays up, these numbers will increase again as well.

The value for the exponential coefficient parameter (*a*) is obtained using a plot of Equation B.6. Using this plot and the monthly data from EIA, a value is obtained for *a*:

$$a = 162.66$$

This is a dimensionless parameter. The intercept on the *y*-axis in Figure B.5 gives a value for *b*:

$$b = 0.00011733 \, \text{days}^{-1}$$

Using the above values for *a* and *b*, these numbers can be inserted into Equation B.5 to calculate an updated date for the timing of the world peak oil rate.

$$t_{\text{peak}} = \frac{1}{b} \ln(a) = \frac{1}{0.00011733} \ln(162.66) = 43{,}396 \, \text{days}$$

The 43,396 days is measured from an arbitrary time zero, used in the plots of 1/1/1900. This equation gives a peak oil date of October 23, 2018. Note that I am not saying that the world oil production will peak in October 2018. I am saying that, using the Hubbert model and the latest and most accurate monthly world oil production data, the current projection for the peak oil rate is October 23, 2018. If the price of oil remained constant for the next 7 years and there were no big technological breakthroughs during that time, this may prove to be an approximately accurate forecast. I don't expect that this will be the case and so this date will probably move again.

HUBBERT'S MODEL FOR WORLD COAL PRODUCTION

Hubbert's model can be applied to finite resources, such as coal. According to Hubbert's model, the total cumulative production that can be expected for coal is given by the same Equation B.1:

$$Q_t = \frac{Q_{max}}{1 + ae^{-bt}} \tag{B.1}$$

Where now
Q_t = cumulative coal production at time (t) in tons
q_t = coal production rate at time (t) usually in tons/year
Q_{max} = ultimate cumulative coal production, in tons
a = growth or decline coefficient, dimensionless
b = growth or decline exponent, dimensions are the reciprocal of time such as 1 per year
The production rate (q_t) at any time (t) is the derivative of the cumulative and so is given by

$$q_t = \frac{Q_{max}abe^{-bt}}{(1 + ae^{-bt})^2} \tag{B.2}$$

If this model is the correct model for the production data, as long as there is early time data, the data can be fitted to the model to determine the ultimate recovery (Q_{max}) as well as the parameters a and b. There are several ways to determine the parameters from the data. The first method, which Hubbert recommended, was to rearrange Equations B.1 and B.2 to give:

$$\frac{q_t}{Q_t} = b - \frac{bQ_t}{Q_{max}} \tag{B.3}$$

Using Equation B.3, q_t/Q_t versus Q_t is plotted. Then, the intercept on the y axis is the parameter b and the intercept on the x axis is the ultimate recoverable coal, Q_{max}, the ultimate cumulative coal production. The parameter, q_t/Q_t, represents the growth rate of the cumulative world coal production. This is plotted in Figure B.6, and what it shows is that the growth rate of the cumulative has been constant at about 2.2% since 1931. Moreover, it has not established a trend that can be extrapolated to an ultimate cumulative production, Q_{max}. The fact that this Hubbert

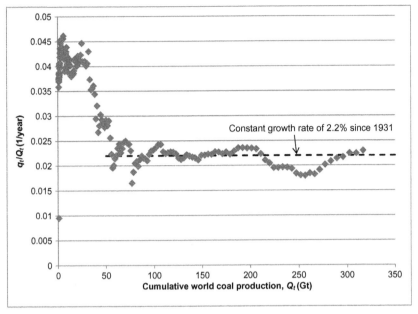

FIGURE B.6   Plot of Equation B.3 for world coal production. *Data Source: David Rutledge, Caltech.*

trend has not been established suggests that perhaps the coal production peak is still many years away. Despite the failure of this method of parameter estimation, it is still possible to fit Rutledge's data to Hubbert's model to determine the Hubbert parameters and make a forecast of future production.

Another method of doing this is to use nonlinear regression techniques. This is not a simple task, but can be accomplished using the "Solver" function in an Excel spreadsheet.

Using the nonlinear regression method, Rutledge's data has been fitted to Hubbert's model, and what is obtained is the forecast shown in Figure 13.3. The red line in this chart is a plot of the Hubbert model, and the blue points are the Rutledge data points. The parameters obtained from nonlinear regression and used to plot Hubbert's model are:

$Q_{max} = 3529$ Gt, $a = 1042.15$, and $b = 0.02203$ per year.

The timing for the peak coal production can be calculated using Equation B.5

$$t_{peak} = \frac{1}{b}\ln(a) = \frac{1}{0.02203}\ln(1042.15) = 315.43 \; years$$

Measured from 1800, this puts the peak of coal production in the year 2115. This estimate has an accuracy limited by the model fit of the data, which is about 15%. Basically, the peak could occur anywhere from 2070 to 2160.

Bibliography

Chapter 1

White, L.A., 1943. Energy and the evolution of culture. Am. Anthropol. 45, 335–356, No. 3, Part 1.

White, L.A., 1959. The Evolution of Culture. McGraw-Hill, New York.

Chapter 3

Carnot, S., 1824. Réflexions Sur La Puissance Motrice Du Feu Et Sur Les Machines Propres À Développer Cette Puissance (Reflections on the Motive Power of Fire and on the Machines Needed to Develop this Power), Self Published.

Clausius, R., 1850. Ueber die bewegende Kraft der Wärme und die Gesetze, welche sich daraus für die Wärmelehre selbst ableiten lassen. Annalen der Physik und Chemie (Poggendorf, Leipzig). 155 (3), 368–394.

Clausius, R., 1854. Ueber eine veränderte Form des zweiten Hauptsatzes der mechanischen Wärmetheoriein. Ann. Phys. Chem. 93 (12), 481–506.

Clausius, R., 1865. Über die Wärmeleitung gasförmiger Körper. Ann. Phys. 125, 353–400.

Thompson (aka Count Rumford), B., 1798. An experimental enquiry concerning the source of heat which is excited by friction. Philos. Trans. R. Soc. 88, 102.

Thomson, W. (Lord Kelvin), 1851. On the dynamical theory of heat, with numerical results deduced from Mr Joule's equivalent of a thermal unit, and M. Regnault's observations on steam. Excerpts [§§1–14 & §§99–100], Transactions of the Royal Society of Edinburgh, March, 1851; and Philosophical Magazine IV. 1852. [from Mathematical and Physical Papers, vol. i, art. XLVIII, pp. 174].

Chapter 5

Deffeyes, K.S., 2001. Hubbert's Peak, The Impending World Oil Shortage. Princeton University Press, Princeton NJ.

Deffeyes, K.S., 2005. Beyond Oil, the View from Hubbert's Peak. Hill and Wang, New York, NY.

Deffeyes, K.S., 2010. When Oil Peaked. Hill and Wang, New York, NY.

Hubbert, M.K., 1956. Nuclear energy and the fossil fuels. In: American Petroleum Institute, Drilling and Production Practices, Proceedings of Spring Meeting.

Hubbert, M.K., 1969. Energy resources. In: National Research Council Committee (Ed.), Resources and Man. W.H. Freeman, San Francisco, CA.

Klare, M., 2008. Rising Powers, Shrinking Planet. Metropolitan Books, New York, NY.

Simmons, M., 2005. Twilight in the Desert. John Wiley and Sons, Hoboken, NJ.

Chapter 13

Patzek, T.W., Croft, G.D., 2010. A Global Coal Production Forecast with Multi-Hubbert Cycle Analysis. Energy. 35, 3109–3122.

Rutledge, D., 2011. Estimating Long-Term World Coal Production with Logit and Probit Transforms. Int. J. Coal Geol. 85, 23–33.

Chapter 15

Bossel, U., Eliasson, B., Taylor, G., 2003. The future of the hydrogen economy: bright or bleak? In: Fuel Cell Seminar, 3–7 November 2003, Miami Beach, FL.

National Research Council and National Academy of Engineering, 2004. The Hydrogen Economy: Opportunities, Costs, Barriers, and R&D Needs Committee on Alternatives and Strategies for Future Hydrogen Production and Use. National Academies Press, Washington, DC.

Rifkin, J., 2002. The Hydrogen Economy. Penguin/Putnam, New York, NY.

Glossary

AC See Alternating Current.

Acceleration The rate of change of velocity with time.

Adiabatic Process A thermodynamic process in which the net heat transfer to or from the working fluid is zero.

Alternating Current Electrical current that alternates directions in a repetitive pattern.

Anaerobic Environment An environment that does not contain any oxygen.

Atomic Vapor Laser Isotope Separation A method by which specially tuned lasers are used to separate isotopes of uranium using selective ionization of hyperfine transitions.

Auto Refrigeration A process in which liquefied natural gas (LNG) is kept at its boiling point so that any heat that is conducted through the tanker walls is countered by the latent heat of vaporization, energy that is lost from the LNG as it vaporizes.

AVLIS See Atomic Vapor Laser Isotope Separation.

Bagasse The fibrous matter that remains after sugarcane or sorghum stalks are crushed to extract their juice. It is currently used as a biofuel and in the manufacture of pulp and paper products and building materials.

Base Gas The volume of gas intended as permanent inventory in a gas storage reservoir needed to maintain adequate pressure and flow rates throughout the winter withdrawal season. It is also called cushion gas.

Betz Law The theory which details the maximum possible energy to be derived from a wind turbine. According to Betz's law, no turbine can capture more than 59.3% of the kinetic energy in wind.

Betz Limit See Betz Law.

Binary Cycle Power Plants A type of geothermal power plant that allows cooler geothermal reservoirs to be used than with dry steam and flash steam plants.

Biomass Biological material from living, or recently living, organisms. As an energy source, it can either be used directly or converted into other energy products such as biofuel.

Breeder Reactor A nuclear reactor that is capable of generating more fissile material than it consumes because its neutron energy is high enough to breed fissile from fertile material like uranium-238 and thorium 232.

Caloric Theory of Heat An obsolete scientific theory that heat consists of a self-repellent fluid called caloric that flows from hotter bodies to colder bodies.

Calorie A unit of energy of French origin that, as originally defined, was the amount of energy needed to heat 1 g of water by 1 °C. Its value can vary depending on the temperature.

Cap and Trade An environmental policy tool that delivers results with a mandatory cap on emissions while providing sources flexibility in how they comply.

Carbon Capture See Carbon Capture and Storage.

Carbon Capture and Storage (CCS) The technology that prevents large quantities of CO_2 from being released into the atmosphere during the use of fossil fuel in power generation and other industries. The process is based on capturing CO_2 from large point sources and storing it in such a way that it does not enter the atmosphere.

Carbon Dioxide Miscible Flooding The process by which carbon dioxide is injected into an oil reservoir in order to increase the output when extracting oil.

CFCs See Chlorofluorocarbons.

Chlorofluorocarbons Organic compounds that contain carbon, chlorine, and fluorine, produced as a volatile derivative of methane and ethane. They have been implicated in the accelerated depletion of the ozone layer.

Clean Coal Technologies The collection of technologies that are being developed to reduce the environmental impact of coal energy generation. Technologies include chemically washing minerals and impurities from the coal; gasification; treating the flue gases with steam to remove sulfur dioxide; carbon capture and storage technologies to capture the carbon dioxide from the flue gas; and, dewatering lower rank coals (brown coals) to improve the calorific value and thus the efficiency of the conversion into electricity.

Coal-Bed Methane Gas A type of natural gas extracted from coal beds.

Coal to Liquids Technologies Processes for converting coal to liquid fuels.

Compression Ignition Engine An internal combustion engine that uses pressure to initiate ignition to burn the fuel which is injected into the combustion chamber. It was developed by Rudolf Diesel in 1893 and is also known as a diesel engine.

Concentrated Solar Power A type of solar power system that uses mirrors or lenses to concentrate a large area of sunlight, or solar thermal energy, onto a small area.

Conduction The transfer of heat between two substances that are in direct contact with one another.

Conservation of Energy The law which states that the total amount of energy in an isolated system will remain constant over time.

Conservation of Mass The law which states that the mass of an isolated system (closed to all matter and energy) will remain constant over time.

Conservation of Momentum The law which states that if no external force acts on a closed system of objects, the momentum of the closed system will remain constant.

Convection The primary method by which heat moves through gases and liquids. It is essentially the up and down movement of gases and liquids caused by heat transfer.

Critical Point The set of conditions under which the properties of a liquid and its vapor become identical.

Critical Pressure The pressure required to liquefy a gas at its critical temperature.

Critical Temperature The temperature at and above which the vapor of a substance cannot be liquefied, no matter how much pressure is applied.

CSAPR The U.S. Environmental Protection Agency's Cross-State Air Pollution Rule that is due to take effect in 2014.

CSP See Concentrated Solar Power.

Cushion gas See Base Gas.

DC See Direct Current.

Dew Point The temperature at which humid air must be cooled, at constant barometric pressure, for water vapor to condense into liquid water. The condensed water is called dew when it forms on a solid surface.

Direct Current The unidirectional flow of electric charge.

Dry Steam Plants Plants that produce energy directly from the steam that emerges at the earth's surface.

EIA See U.S. Energy Information Administration.

Electric Power Research Institute The independent, nonprofit organization that conducts research and development relating to the generation, delivery and use of electricity for the benefit of the public.

Electromagnetic Force The fundamental force that is associated with electric and magnetic fields and is responsible for atomic structure, chemical reactions, the attractive and repulsive forces associated with electrical charge and magnetism and all other electromagnetic phenomena. It is carried by the photon.

Endothermic Reaction Chemical reactions that must absorb energy in order to proceed. Endothermic reactions cannot occur spontaneously. Work must be done in order to get these reactions to occur. When endothermic reactions absorb energy, a temperature drop is measured during the reaction. Endothermic reactions are characterized by positive heat flow (into the reaction) and an increase in enthalpy.

Energy The ability a physical system has to do work on other physical systems. Given that work is defined as a force acting through a distance (a length of space), energy is always equivalent to the ability to exert pulls or pushes against the basic forces of nature, along a path of a certain length.

Enhanced Geothermal Systems A type of system in which heat is extracted by creating a subsurface fracture system to which water can be added through injection wells. Creating an enhanced or engineered geothermal system requires improving the natural permeability of rock.

Enhanced Oil Recovery The techniques that can be used to increase the amount of crude oil that can be extracted from an oil field. It may also be termed improved oil recovery or tertiary recovery.

Enthalpy The measure of the total energy of a thermodynamic system.

Entropy The thermodynamic property that can be used to determine the energy not available for work in a thermodynamic process. It is defined as the heat transferred divided by the temperature at which it is transferred.

EOR See Enhanced Oil Recovery.

EPRI See Electric Power Research Institute.

Exothermic Reaction Chemical reactions that release energy in the form of heat, light, or sound. Exothermic reactions may occur spontaneously and result in higher randomness or entropy of the system. They are denoted by a negative heat flow (heat is lost to the surroundings) and decrease in enthalpy.

External Combustion Engine A type of heat engine where an internal working fluid is heated by combustion in an external source, through the engine wall or a heat exchanger.

Fast Breeder Reactor A nuclear reactor that utilizes fast neutrons to split atoms of uranium or thorium as part of the process. Fast neutrons have an energy level of approximately 1 MeV (100 TJ/kg), resulting from a particular neutron velocity of 14,000 km/s. Like all breeder reactors it is capable of generating more fissile material than it consumes because it generates additional neutrons that are able to breed fissile atoms from fertile material like uranium-238 and thorium-232.

Federal Energy Regulatory Commission The independent agency that regulates the interstate transmission of electricity, natural gas, and oil. The agency also reviews proposals to build LNG terminals and interstate natural gas pipelines as well as licensing hydropower projects.

FERC See Federal Energy Regulatory Commission.

First Law of Thermodynamics The law which dictates the specifics for the movement of heat and work. Energy cannot be created or destroyed; it can only be transformed from one type of energy to another. The total energy output (as that produced by a machine) is equal to the amount of heat supplied. As energy (generally) can neither be created nor destroyed, the sum of mass and energy is always conserved.

Fischer-Tropsch Synthesis A set of chemical reactions that convert a mixture of carbon monoxide and hydrogen into liquid hydrocarbons.

Flash Steam Plant The most common form of geothermal power plant where hot water is pumped under great pressure to the surface. When it reaches the surface, the pressure is reduced and as a result some of the water changes to steam. This produces a "blast" of steam. The cooled water is returned to the reservoir to be heated by geothermal rocks again.

Flue Gas The gas exiting to the atmosphere via a flue, which is a pipe or channel for conveying exhaust gases from a fireplace, oven, furnace, boiler, or steam generator.

Force Any influence that causes an object to undergo a change in speed, a change in direction or a change in shape. In other words, a force is that which can cause an object with mass to change its velocity or which can cause a flexible object to deform.

Fracking See Hydraulic Fracturing Process.

Fracture Stimulation The technology that enables natural gas producers to safely and effectively recover natural gas from hard-to-produce resources trapped in deep shale and other unconventional formations thousands of feet below ground.

F-T synthesis See Fischer-Tropsch Synthesis.

Fuel Cells the device that converts chemical energy from a fuel into electricity through a chemical reaction with oxygen or another oxidizing agent.

Futures Contracts A standardized contract between two parties to exchange a specified asset of standardized quantity and quality for a price agreed today (the futures price or the strike price) with delivery occurring at a specified future date, the delivery date.

Gas Centrifuge The device that performs isotope separations of gas.

Gas Condensate The light, oil-like liquid that drops out of the gas when the pressure drops below the dew point.

GDP See Gross Domestic Product.

Geothermal Energy Thermal energy generated and stored in the earth. Thermal energy is the energy that determines the temperature of matter. Earth's geothermal energy originates from the original formation of the planet and from radioactive decay of minerals.

Geothermal Heat Pump A central heating and/or cooling system that pumps heat to or from the ground.

Global Warming The rising average temperature of the earth's atmosphere and oceans and its projected continuation.

Gravity The natural phenomenon by which physical bodies attract with a force proportional to their mass.

Greenhouse Gases Gases in the atmosphere that absorb and emit radiation within the thermal infrared range. The primary greenhouse gases in the earth's atmosphere are water vapor, carbon dioxide, methane, nitrous oxide, and ozone.

Gross Domestic Product The market value of all final goods and services produced within a country in a given period.

Ground Source Heat Pumps See Geothermal Heat Pumps.

GSHP See Ground Source Heat Pumps.

Heavy Water Water that is chemically the same as regular (light) water, but with the two hydrogen atoms (as in H_2O) replaced with deuterium atoms (hence the symbol D_2O).

Horizontal Drilling A type of drilling in which the borehole is drilled at least 80° from vertical so that it penetrates a productive formation in a manner parallel to the formation.

Hubberts Peak Theory M. King Hubbert's theory that for any given geographical area, the production rate of any finite resource (such as petroleum, coal, or any mineral) tends to follow a bell-shaped curve. It is based on the observation that, because the resource amount in any region is finite, the rate of discovery which initially increases quickly must reach a maximum and decline following a bell-shaped curve.

Hydraulic Fracturing Process A process in which water, chemicals, and proppant material are pumped into the well to release the gas trapped in low permeability formations by creating fractures in the rock and allowing oil or natural gas to flow from the rock into the well.

Hydroelectric Power The production of electrical power through the use of the gravitational force of falling or flowing water. It is the most widely used form of renewable energy. It is also termed Hydroelectricity.

IGCC See Integrated Gasification Combination Cycle.

Integrated Gasification Combination Cycle The technology that turns coal into gas-synthesis gas (syngas). It then removes impurities from the coal gas before it is combusted and attempts to turn any pollutants into re-usable by-products. This results in lower emissions of sulfur dioxide, particulates, and mercury.

Internal Combustion Engine A type of heat engine where the combustion of a fuel (normally a fossil fuel) occurs with an oxidizer (usually air) in a combustion engine.

International Organization for Standardization The organization that is the world's largest developer and publisher of international standards.

In-Situ Leaching The mining process that is used to recover minerals such as copper and uranium through boreholes drilled into a deposit, in-situ.

Isentropic Process A thermodynamic process that takes place from initiation to completion without an increase or decrease in the entropy of the system. In other words, the entropy of the system remains constant.

ISL See In-Situ Leaching.

ISO See International Organization for Standardization.

Kyoto Accord The international treaty whereby countries agree to reduce the amount of greenhouse gases they emit if their neighbors do likewise.

Liquefied Natural Gas Natural gas that has temporarily been converted to liquid form for ease of storage or transport.

LNG See Liquefied Natural Gas.

MEA See Monoethanolamine.

Mine Safety and Health Administration The federal enforcement agency responsible for the health and safety of the nation's miners.

MLIS See Molecular Laser Isotope Separation.

Molecular Laser Isotope Separation The method of isotope separation where specially tuned lasers are used to separate isotopes of uranium using selective ionization of hyperfine transitions of uranium hexafluoride molecules.

Monoethanolamine An organic chemical compound that is both a primary amine and a primary alcohol (due to a hydroxyl group). Like other amines, monoethanolamine acts as a weak base. It is often used to remove carbon dioxide from flue gas.

MSHA See Mine Safety and Health Administration.

NAFTA See North American Free Trade Agreement.

National Oceanic and Atmospheric Administration A scientific agency within the U.S. Department of Commerce that is focused on the conditions of the oceans and the atmosphere.

Newton's Laws of Motion The three laws of motion composed by Sir Isaac Newton. The first law states that an object will remain at rest or in uniform motion in a straight line unless compelled to change its state by the action of an external force. The second law explains how the velocity of an object changes when it is subjected to an external force. The third law states that for every action (force) in nature there is an equal and opposite reaction.

NOAA See National Oceanic and Atmospheric Administration.

North American Free Trade Agreement The agreement signed by the governments of Canada, Mexico, and the United States to create a trilateral trade bloc in North America.

Nuclear Fission Either a nuclear reaction or a radioactive decay process in which the nucleus of an atom splits into smaller parts (lighter nuclei), often producing free neutrons and photons (in the form of gamma rays), and releasing a very large amount of energy.

Nuclear Force The force between two or more nucleons. It is responsible for the binding of protons and neutrons into atomic nuclei. The energy released causes the masses of nuclei to be less than the total mass of the protons and neutrons which form them.

Nuclear Fusion The process by which two or more atomic nuclei join together, or "fuse," to form a single heavier nucleus. This is usually accompanied by the release or absorption of large quantities of energy.

Nuclear Power The use of sustained nuclear fission to generate heat and electricity.

Occupational Safety and Health Administration The main federal agency charged with the enforcement of safety and health legislation.

Octane Rating The standard measure of the performance of a motor or aviation fuel. The higher the octane number, the more compression the fuel can withstand before detonating. In broad terms, fuels with a higher octane rating are used in high-compression engines that generally have higher performance.

OPEC See Organization Of The Petroleum Exporting Countries.

Operating Temperature The temperature at which an electrical or mechanical device operates.

Organization of the Petroleum Exporting Countries The global organization that is dedicated to stability in and shared control of the petroleum markets.

OSHA See Occupational Safety and Health Administration.

Otto Cycle Engine A type of internal combustion engine in which the piston completes four separate strokes—intake, compression, power, and exhaust—during two separate revolutions of the engine's crankshaft and one single thermodynamic cycle.

Oxy-Fuel Combustion The process of burning a fuel using pure oxygen instead of air as the primary oxidant.

Partial Oxidation A process that is used to convert natural gas to syngas.

Peak Oil The point in time when the maximum rate of global petroleum extraction is reached, after which the rate of production enters terminal decline. This concept is based on the observed production rates of individual oil wells, projected reserves and the combined production rate of a field of related oil wells.

Peak Oil Theory The theory proposed by M. King Hubert that any finite resource—oil, gas, coal, or uranium—follows a bell-shaped curve in its production history. At some point, it reaches a peak and begins to decline, and the decline in production will mirror the rise in production on the way up. The peak production rate and the timing of the peak production depend on the total reserves that exist and are to be discovered in the future.

PEM Fuel Cells See Proton Exchange Membrane Fuel Cell.

Permeability The measure of the ease with which a fluid can move through a porous rock.

Photoelectric effect The principle which states that when a photon of light strikes a solar cell, its energy is transferred to an electron in the semiconductor material in the top layer. If enough energy is absorbed by the electron, it can escape from its normal position in the atom.

Photovoltaic The method of generating electrical power by converting solar radiation into direct current electricity using semiconductors that exhibit the photovoltaic effect.

Photovoltaic effect The creation of voltage or electric current in a material upon exposure to light.

Porosity The measure of how much rock is an open space or a void. This space can be between grains or within cracks or cavities of the rock.

Produced Water The term used in the oil and gas industry to describe the water that is produced along with the oil and gas.

Proppants Particles of a particular size that are mixed with fracturing fluid to hold fractures open after a hydraulic fracturing treatment. The most common proppant is sand coated with resin, but other proppants can be aluminum bauxite or special ceramics.

Proton Exchange Membrane Fuel Cell A type of fuel cell that is being developed for transport applications as well as for stationary fuel cell applications and portable fuel cell applications.

Pumped Storage The wind power storage method where the excess energy generated while the wind is blowing is used to pump water from a low-lying reservoir uphill to a higher reservoir. When the wind is not blowing, the water is allowed to flow from the higher reservoir to the lower reservoir, driving a water-powered turbine to generate electricity.

Radiation The process by which electromagnetic waves travel through space. When electromagnetic waves come into contact with an object, heat is transferred to that object.

Rankin Cycle The external combustion steam engine cycle that converts heat from burning fuels such as coal and wood into work. In this case the working fluid is water.

Reserve Life The number of years it would take a country to produce their reserves if they keep producing at the current rates. To obtain this number, you divide their reserves (in barrels) by their producing rate (in barrels/year) to get their reserve life in years.

Ring of Fire An area where large numbers of earthquakes and volcanic eruptions occur in the basin of the Pacific Ocean.

SAGD See Steam-Assisted Gravity Drainage.

Second Law of Thermodynamics The law that puts a limit on how efficient the energy conversion processes can be. It states that it is impossible to convert heat from a heat source totally into work without rejecting some of that heat to a heat sink. There are many alternative statements of the second law of thermodynamics.

Sequestration See Carbon Capture and Sequestration.

Shale The fine-grained, clastic sedimentary rock composed of mud that is a mix of flakes of clay minerals and tiny fragments (silt-sized particles) of other minerals, especially quartz and calcite. The ratio of clay to other minerals is variable.

Shale Gas Natural gas that is produced from shale.

Slick Water See Slick Water Fracturing.

Slick Water Fracturing A type of fracturing in which chemicals are added to water to lower the viscosity and increase the fluid flow.

Solar Cell A solid-state electrical device that converts the energy of light directly into electricity by the photovoltaic effect.

Solar Collectors Devices that are used to collect heat by absorbing sunlight.

Solar Energy The radiant light and heat from the sun.

Solar Power The conversion of sunlight into electricity, either directly using photovoltaics (PV) or indirectly using concentrated solar power (CSP).

Sour Gas Natural gas or any other gas that contains significant amounts of hydrogen sulfide.

Spot Pricing The price that is quoted for immediate payment and delivery.

Steam-Assisted Gravity Drainage The enhanced oil recovery technology that is used to produce heavy crude oil and bitumen. It is an advanced form of steam stimulation in which a pair of horizontal wells is drilled into the oil reservoir, one a few meters above the other. Low pressure steam is continuously injected into the upper wellbore to heat the oil and reduce its viscosity, causing the heated oil to drain into the lower wellbore, where it is pumped out.

Steam Reforming A process that is used to convert natural gas to syngas. It is carried out in a fired heater with catalyst-filled tubes that produce a syngas with at least 5:1 hydrogen to carbon monoxide ratio. To adjust the ratio, hydrogen can be removed by a membrane or pressure swing adsorption system. This process is widely used if the surplus hydrogen is used in a petroleum refinery or for the manufacture of ammonia in an adjoining plant.

Supercritical Fluid Any substance at a temperature and pressure above its critical point, where distinct liquid and gas phases do not exist.

Sweet Gas Natural gas or other gas that does not contain significant amounts of hydrogen sulfide.

Thermodynamic Cycle A series of thermodynamic processes transferring heat and work, while varying pressure, temperature, and other state variables, eventually returning a system to its initial state. In the process of going through this cycle, the system may perform work on its surroundings, thereby acting as a heat engine.

Tight Gas Unconventional natural gas which is difficult to access because of the nature of the rock and sand surrounding the deposit.

Tight Oil Petroleum that consists of light crude oil contained in petroleum-bearing formations of relatively low porosity and permeability (shales). It uses the same horizontal well and hydraulic fracturing technology used in the production of shale gas. It should not be confused with oil shale as it differs by the thermal maturity of the oil and the method of extraction.

Towler Principle The principle which states that it not possible to extract energy from the environment without having an impact on the environment.

Trans-Esterfication Reaction The reaction of a triglyceride (fat/oil) with an alcohol to form esters and glycerol.

Triple Point The temperature and pressure of a substance at which the three phases (gas, liquid, and solid) of that substance coexist in thermodynamic equilibrium.

UCG See Underground Coal Gasification.

Unconventional Oil and Gas Petroleum that is produced or extracted using techniques other than the conventional (oil-well) method. It usually involves the fracture stimulation of very low permeability reservoirs.

Underground Coal Gasification The in-situ gasification process carried out in non-mined coal seams using injection of oxidants and bringing the product gas to surface through production wells drilled from the surface.

Ultrasonic Frequency Generator Any sounds above the frequencies of audible sound and nominally includes anything over 20,000 Hz.

U.S. Department of Energy The government agency whose goal is to advance energy technology and promote related innovation.

U.S. DOE See U.S. Department of Energy.

U.S. Energy Information Administration The government agency that collects, analyzes, and disseminates independent and impartial energy information to promote sound policymaking, efficient markets, and public understanding of energy and its interaction with the economy and the environment.

U.S. Geological Survey The science organization that provides impartial information on the health of our ecosystems and environment, the natural hazards that threaten us, the natural resources we rely on, the impacts of climate and land-use change, and the core science systems that help us provide timely, relevant, and useable information.

USGS See U.S. Geological Survey.

Velocity The speed in a given direction.

Volumetrics The method of determining the amount of oil that was originally in place in a formation by calculating the volume that it occupies.

Wind Power The conversion of wind energy into a useful form of energy, such as using wind turbines to make electricity, windmills for mechanical power, wind pumps for water pumping or drainage or sails to propel ships.

Wind Turbine A device that converts kinetic energy from the wind into mechanical energy. If the mechanical energy is used to produce electricity, the device may be called a wind generator, or wind charger. If the mechanical energy is used to drive machinery, such as for grinding grain or pumping water, the device is called a windmill, or wind pump.

Work The amount of work done by, or energy transferred, by a force acting through a distance.

Working Gas The volume of gas in a storage gas reservoir above the level of base gas that is available to be withdrawn in the winter. Working gas capacity is equal to the total gas storage capacity minus the base gas.

Index

Note: Page numbers followed by *f* indicate figures and *t* indicate tables.

Printed and bound by CPI Group (UK) Ltd, Croydon, CR0 4YY

03/10/2024

01040426-0002